LONDON MATHEMATICAL SOCIETY LECTURE NOTE SERIES

Managing Editor: Professor N.J. Hitchin, Mathematical Institute,
University of Oxford, 24–29 St Giles, Oxford OX1 3LB, United Kingdom

The titles below are available from booksellers, or, in case of difficulty, from Cambridge University Press at www.cambridge.org.

Transcendental Aspects of Algebraic Cycles

Proceedings of the Grenoble Summer School, 2001

Edited by

S. MÜLLER-STACH
Johannes Gutenberg Universität Mainz

and

C. PETERS
Université de Grenoble I

CAMBRIDGE
UNIVERSITY PRESS

University Printing House, Cambridge CB2 8BS, United Kingdom

One Liberty Plaza, 20th Floor, New York, NY 10006, USA

477 Williamstown Road, Port Melbourne, VIC 3207, Australia

314-321, 3rd Floor, Plot 3, Splendor Forum, Jasola District Centre, New Delhi - 110025, India

103 Penang Road, #05-06/07, Visioncrest Commercial, Singapore 238467

Cambridge University Press is part of the University of Cambridge.

It furthers the University's mission by disseminating knowledge in the pursuit of education, learning and research at the highest international levels of excellence.

www.cambridge.org
Information on this title: www.cambridge.org/9780521545471

First published 2004

A catalogue record for this publication is available from the British Library

Library of Congress Cataloging in Publication data
Transcendental aspects of algebraic cycles: proceedings of the Grenoble Summer School, 2001/edited by S. Müller-Stach and C. Peters.
p.   cm. – (London Mathematical Society lecture note series: 313)
Includes bibliographical references and index.
ISBN 0 521 54547 1
1. Algebraic cycles – Congresses.   I. Müller-Stach, Stefan, 1962–
II. Peters, C. (Chris)   III. Series.
QA564.T73 2004
516.3′5 – dc22   2003062526

ISBN 978-0-521-54547-1 Paperback

Contents

Participants

Name	University	Email address
Archava, S.	Purdue	archava@math.purdue.edu
Asakura, M.	Kyushu	asakura@math.kyushu-u.ac.jp
Aubry, E.	Institut Fourier	Erwann.Aubry@ujf-grenoble.fr
Bauer, T.	Bayreuth	Thomas.Bauer@uni-bayreuth.de
Bayle, L.	Angers	Lionel.Bayle@univ-angers.fr
Bender, A.	Padova	bender@dpmms.cam.ac.uk
Ben Kilani, A.	La Marsa	—
Bertin, J.	Institut Fourier	Jose.Bertin@ujf-grenoble.fr
Bonavero, L.	Institut Fourier	Laurent.Bonavero@ujf-grenoble.fr
Boudaoud, F.	Oran	f_boudaoud@univ-oran.dz
Brion, M.	Institut Fourier	Michel.Brion@ujf-grenoble.fr
Buczynski, J.	Warsaw	jb176466@zodiac.mimuw.edu.pl
Colin-de-Verdière, Y.	Institut Fourier	Yves.Colin-de-verdiere@ujf-grenoble.fr
Collino, A.	Torino	alberto.collino@unito.it
Colombo, E.	Milano	Elisabetta.Colombo@mat.unimi.it
Dahari, A. S.	Bar Ilan	dahari@macs.biu.ac.il
Dan, N.	IMAR, Bucharest	atac@dnt.ro
Del Angel, P.	CIMAT, Guanajuato	luis@cimat.mx
Eckl, T.	Bayreuth	Thomas.Eckl@uni-bayreuth.de
Elbaz-Vincent, P.	Montpellier II	pev@math.univ-montp2.fr

Elizondo, J.	UNAM,	
	Mexico	javier@math.unam.mx
Erdil, A.	Chicago	erdil@math.uchicago.edu
Ferreira dos Santos, P.	IST Lisboa	pedfs@math.ist.utl.pt
Fichou, G.	Angers	fichou@tonton.univ-angers.fr
Funar, L.	Institut Fourier	Louis.Funar@ujf-grenoble.fr
Galluzzi, F.	Torino	galluzzi@dm.unito.it
Gonzalez-Sprinberg, G.	Institut Fourier	Gerard.Gonzalez-Sprinberg @ujf-grenoble.fr
Hanamura, M.	Kyushu	hanamura@math.kyushu-u.ac.jp
Hazrat, R.	Bielefeld	rhazrat@mathematik.uni-bielefeld.de
Iyer, J.	Essen	bms051@sp2.power.uni-essen.de
Joshua, R.	Ohio	joshua@math.ohio-state.edu
Jun, B.	Essen	mat9f0@uni-essen.de
Katsura, T.	Tokyo	tkatsura@ms.u-tokyo.ac.jp
Kerr, M.	Princeton	matkerr@princeton.edu
Klingler, B.	ETH	klingler@math.ethz.ch
Kosarew, S.	Institut Fourier	Siegmund.Kosarew@ujf-grenoble.fr
Laterveer, R.	IRMA, Strasbourg	laterv@math.u-strasbg.fr
Lewis, J.	Alberta	lewisjd@gpu.srv.ualberta.ca
Lima-Filho, P.	Texas A&M	plfilho@math.tamu.edu
Lombardo, G.	Torino	lombardo@dm.unito.it
Madonna, C.	Roma II	madonna@mat.uniroma2.it
Maican, M.	Notre Dame	Maican.1@nd.edu
Manivel, L.	Institut Fourier	Laurent.Manivel@ujf-grenoble.fr
Marini, G.	Roma II	marini@axp.mat.uniroma2.it
Mavlyutov, A.	MPI Bonn	anvar@mpim-bonn.mpg.de
Milman, P.	Toronto	milman@math.toronto.edu
Mouroukos, E.	Michigan	emourouk@math.lsa.umich.edu
Müller-Stach, S.	Essen	mueller-stach@uni-essen.de
Münstermann, B.	Göttingen	mail@BjoernMuenstermann.de
Murre, J.	Leiden	murre@math.leidenuniv.nl
Nagel, J.	Lille I	nagel@agat.univ-lille1.fr
Otwinowska, A.	ETH	ania@clipper.ens.fr

Paksoy, V. E.	Brandeis	emrah@brandeis.edu
Paluch, M.	Lisboa IST	mike@math.ist.utl.pt
Park, S.-S.	Erlangen-Nürnberg	s2park@yahoo.com
Penacchio, O.	Toulouse III	penacchi@picard.ups-tlse.fr
Peters, C.	Institut Fourier	Chris.Peters@ujf-grenoble.fr
Piontkowski, J.	Düsseldorf	piontkow@uni-duesseldorf.de
Puydt, J.	Institut Fourier	Julien.Puydt@ujf-grenoble.fr
Reznikov, A.	Durham	alexander.reznikov@durham.ac.uk
Rosenschon, A.	Duke	axr@math.duke.edu
Saito, S.	Nagano	sshuji@msb.biglobe.ne.jp
Schwarzhaupt, A.	Essen	Alexander.Schwarzhaupt@uni-essen.de
Skiadas, C.	Chicago	skiadas@math.uchicago.edu
Stalij, M.	Warsaw	mstalij@mimuw.edu.pl
Térouanne, S.	Institut Fourier	Sophie.Terouanne@ujf-grenoble.fr
Van Geemen, L.	Pavia	geemen@dragon.ian.pv.cnr.it
Vasilyev, S.	Chicago	vasilyev@math.uchicago.edu
Veliche, R.	Purdue	rveliche@math.purdue.edu
Vu, N. S.	Institut Fourier	San.Vu-Ngoc@ujf-grenoble.fr

Preface

General facts

The Institut Fourier has been organizing summer schools in mathematics for several years. They are intended for researchers as well as graduate students from all countries. One of the main aims is to transmit fundamental knowledge in the field of mathematics and to promote the exchange of ideas between researchers. The subject of each summer school is closely linked with the research themes of the Institut Fourier's teams. During these schools, several foreign and French lecturers are invited. The graduate students will get an efficient and useful complementary training and have an opportunity to get in touch with the most current research in the world. The participants will discover the Grenoble department and may pursue their research in collaboration with Grenoble researchers.

Each summer school takes place at the Institut Fourier, lasts three weeks and hosts about 70 participants (upon application, a selection will be made). Twenty to 25 hours of lectures or seminars are given each week on different themes.

The subject of the Summer School 2001 was 'Transcendental Aspects of Algebraic Cycles'.

Organizers: Chris Peters (Institut Fourier, Grenoble) and Stefan Müller-Stach (University Essen, Germany).

About these proceedings

Introductory material

The first week of the Summer School was devoted to explaining some basic material in order to make up for the different backgrounds and levels of the participants.

Firstly Stefan Müller-Stach explained Griffiths' theory of the period map starting from elliptic curves, moving up to higher genus curves and then to the general situation of a family of projective varieties. This was preceded by a users guide to the de Rham, singular and simplicial (co)homology presented by Chris Peters.

We decided not to publish our notes from these two lecture series in the current Proceedings since they form the beginning of a book by Jim Carlson and the editors of these proceedings (see [1]). For this there are also alternative sources: the book [2] by Clemens treats periods of elliptic curves, the first chapters of the book [4] by Griffiths and Harris can serve as an initiation to cohomological methods in algebraic geometry and finally, for period maps the readers can consult the original papers [5, 6]

Secondly, Javier Elizondo presented a series of lectures on algebraic cycles, their equivalence relations, and Chow varieties. He started out by explaining the various ways of assigning intersection multiplicities to the components of a set-theoretic intersection. He then went on to treat the most important equivalence relations among cycles and showed how this leads to an intersection ring in the case of non-singular quasi-projective varieties. All this foundation material can be found in [3] and is therefore not published in these Proceedings. Instead of equivalence classes of cycles, one can also consider cycles of a fixed dimension and fixed degree (in some projective embedding). This is the set underlying a Chow variety, as next explained by Elizondo in his contribution published here. Chow varieties become the underlying moduli spaces for cycles in Lawson homology. One can combine the topological Euler characteristics of the components of the Chow varieties in a formal series, the Euler–Chow function of the variety. This series has been computed in a number of cases which are explained in detail in Elizondo's notes.

Finally, Siegmund Kosarew and Chris Peters gave an introduction to Lawson homology for projective varieties. Classically one introduces flexibility in the study of cycles by introducing suitable equivalence relations. In Lawson homology this flexibility is created by looking at the homotopy type of cycle spaces, i.e. the group of cycles equipped with the Chow topology coming from the topology on the Chow varieties. The contribution of Kosarew and Peters starts out by recalling the basics from homotopy theory. It is followed by one of the definitions of Lawson homology. Since functoriality can only be understood in the framework of continuous algebraic maps, this notion is explained next. This is followed by a discussion of the basic ingredient in the theory, the Chow topology. The article ends with various other useful definitions of Lawson homology which are shown to be equivalent. This is based on some basic but sophisticated simplicial constructions which the authors only sketch but for which references are given.

Lawson (co)homology

Paulo Lima-Filho gave a series of lectures treating various foundational aspects of Lawson homology and cohomology for algebraic varieties, not just for projective varieties.

An axiomatic set-up of Chow topology is at the basis of Lima-Filho's treatment which he presents in the first lecture. He also treats the cycle class maps which compare Lawson homology to ordinary singular homology and the s-maps which link Lawson homology groups mapping to the same homology group under the class map.

One cannot directly put a mixed Hodge structure on Lawson homology, but one has to do this through colimits as explained in the second lecture.

Although there is no good intersection theory for cycles, there is one at the level of homotopy groups, i.e. the Lawson homology groups. This, as well as morphic cohomology, the cohomological counterpart of Lawson homology, is explained in the final lecture.

Motives and motivic cohomology

In Jacob Murre's article, pure motives in the sense of Grothendieck are discussed, i.e. motives arising from smooth, projective varieties. Here, first the categories of pure motives are defined using correspondences and a suitable equivalence relation. This construction does not depend on any conjectures. However, the properties of these categories depend very much on the so-called standard conjectures. This is discussed in detail in Murre's lectures. Further topics are: Manin's identity principle, Jannsen's theorem on motives modulo numerical equivalence and conjectural filtrations on Chow groups arising from Chow–Künneth decompositions of the diagonal. The article contains many illustrating examples, such as the motive of curves, surfaces, Albanese and Picard motives and Chow–Künneth decompositions of elliptic modular varieties.

The article by Philippe Elbaz-Vincent consists of a concise approach to motivic cohomology via higher Chow groups. Both concepts are known to be the same by a theorem of Voevodsky. Historically, higher Chow groups were the first candidate for motivic cohomology. These were defined by Bloch around 1985. Elbaz-Vincent first summarizes the basic functorial properties of higher Chow groups including localization and the Mayer–Vietoris property. Then he discusses in detail higher Chow groups of smooth local rings and the

Gersten conjecture, which leads to a local-to-global principle for motivic cohomology. Also the Bloch–Lichtenbaum–Levine–Friedlander–Suslin spectral sequence from higher Chow groups to algebraic K-theory is presented. At the end of this paper many examples and applications of the previous results are explained.

Hodge theoretic invariants of cycles

In James Lewis' 'Three lectures on the Hodge conjecture', the Hodge conjecture, in its classical form as well as in the general (Grothendieck's amended) form are presented. The topological cycle class map is revisited and the classical Hodge conjecture is first discussed together with several examples, including uniruled fourfolds after Conte and Murre. Lefschetz' original approach using normal functions is also treated, culminating in Zucker's theorem on the cubic fourfold. However, owing to failure of Jacobi inversion this approach is very limited. In his second lecture, Lewis explains the geometric approach using cylinder maps. This proves the Hodge conjecture for many types of hypersurfaces like the quintic fourfold. Finally, he discusses the approach going back to Colliot-Thélène, Bloch and Srinivas where a decomposition of the diagonal in $X \times X$ is used. All explanations are covered by many examples and additional related material.

Jan Nagel presents the foundations of some infinitesimal Hodge theoretic methods going back to Griffiths. These can be applied to study algebraic cycles. He first explains the concept of normal functions by introducing all the complex differential geometry around families of algebraic cycles and their Abel–Jacobi classes living in families of complex tori. The fundamental concept here is the infinitesimal invariant of a normal function, first defined by Griffiths and later refined by Green and Voisin. He proceeds to give a proof of the original theorem of Griffiths about what is now called the Griffiths' groups. In Section 7.3, the theorem of Green and Voisin on the Abel–Jacobi map for hypersurfaces of large degrees is discussed and proved. This uses Koszul theoretic and monodromy methods. Finally, an effective version of Nori's connectivity theorem and a sketch of its proof is presented together with a nice set of applications, including higher Chow groups of certain complete intersections of large degree.

The article 'Beilinson's Hodge and Tate conjectures' by Shuji Saito is devoted to an application of Hodge theoretic invariants of non-compact varieties to the study of algebraic cycles. He reviews his results with Asakura on the computation of Gauss–Manin complexes on families of open complete intersections via a generalization of the Dwork–Griffiths methods of residues. This leads to a proof of the Beilinson Hodge and Tate conjectures in several cases and has applications to the injectivity of cycle class maps for higher Chow groups which are also discussed. The paper finishes with an application to a Noether–Lefschetz problem for K_2 of open surfaces. This shows once again the strength of Hodge theoretic methods applied to interesting mathematical problems.

Schedule of the summer school

First week: June 18–June 22, 2001

Lecturers: J. Elizondo (JE), S. Kosarew (SK), S. Müller-Stach (SM), J. Nagel (JN), C. Peters (CP)

Day	Hour	Speaker	Title
Monday June 18	08:00–09:00		*Welcome*
	09:00–10:30	CP	Cohomology of compact Kähler manifolds (1)
	10:30–11:00		*Coffee break*
	11:00–12:30	JE	Algebraic cycles, moving lemma, intersection theory
	14:00–15:00	CP	Cohomology of compact Kähler manifolds (2)
	15:00–16:00	SM	Periods of elliptic curves (1)
Tuesday June 19	08:30–10:00	CP	Holomorphic invariants (1)
	10:00–10:30		*Coffee break*
	10:30–12:00	SM	Periods of elliptic curves (2)
	13:30–14:30	CP	Holomorphic invariants (2)
	14:30–15:30	JE	Equivalence relations of cycles
Wednesday June 20	08:30–10:00	SM	Period maps (1)
	10:00–10:30		*Coffee break*
	10:30–12:00	JE	Chow varieties (1)
	13:30–14:30	SM	Period maps (2)
	14:30–15:30	CP	Cycle spaces
Thursday June 21	08:30–10:00	CP	Lawson homology: introduction (1)
	10:00–10:30		*Coffee break*
	10:30–12:00	SM	Infinitesimal period maps
	13:30–14:30	JN	Normal functions
	14:30–15:30	SK	Lawson homology: introduction (2)
Friday June 22	08:30–10:00	SK	Lawson homology: introduction (3)
	10:00–10:30		*Coffee break*
	10:30–12:00	JN	Infinitesimal invariants of normal functions
	13:30–14:30	JN	The theorem of Green–Voisin
	14:30–15:30	JE	Chow varieties (2)

Second week: June 25–June 29, 2001
Lecturers: A. Collino (AC), P. Elbaz-Vincent (PE), J. Elizondo (JE), J. Lewis (JL), P. Lima-Filho (PL), J. Murre (JM), J. Nagel (JN)

Day	Hour	Speaker	Title
Monday June 25	08:30–10:00	JN	The effective Nori theorem (1)
	10:00–10:30		*Coffee break*
	10:30–12:00	PE	Higher Chow theory (1)
	13:30–14:30	AC	Abel–Jacobi maps (1)
	14:30–15:30	JM	Motives (1)
Tuesday June 26	08:30–10:00	JM	Motives (2)
	10:00–10:30		*Coffee break*
	10:30–12:00	AC	Abel–Jacobi maps (2)
	13:30–14:30	PE	Higher Chow theory (2)
	14:30–15:30	JN	The effective Nori theorem (2)
Wednesday June 27	08:30–10:00	JM	Motives (3)
	10:00–10:30		*Coffee break*
	10:30–12:00	JL	The Hodge conjecture (1)
	13:30–14:30	AC	Abel–Jacobi maps (3)
	14:30–15:30	PL	Lawson homology (1)
Thursday June 28	08:30–10:00	JN	The effective Nori theorem (3)
	10:00–10:30		*Coffee break*
	10:30–12:00	PL	Lawson homology (2)
	13:30–14:30	PE	Higher Chow theory (3)
	14:30–15:30	JL	The Hodge conjecture (2)
Friday June 29	08:30–10:00	JL	The Hodge conjecture (3)
	10:00–10:30		*Coffee break*
	10:30–12:00	PL	Lawson homology (3)
	13:30–14:30	PL	Lawson homology (4)
	14:30–15:30	JE	Euler–Chow series

Third week: July 2–July 6, 2001
Lecturers: E. Mouroukos (EM), A. Reznikov (AR), S. Saito (SS)

Day	Hour	Speaker	Title
Monday July 2	08:30–10:00	SS	Applications of Hodge theory to algebraic cycles (1)
	10:30–12:00	EM	Motives (4)
Tuesday July 3	10:30–12:00	EM	Motives (5)
Wednesday July 4	08:30–10:00	AR	Bloch's conjecture (1)
	10:30–12:00	SS	Applications of Hodge theory to algebraic cycles (2)
Thursday July 5	08:30–10:00	AR	Bloch's conjecture (2)
	10:30–12:00	EM	Motives (6)
Friday July 6	08:30–10:00	AR	Work of Bloch–Esnault, Esnault–Srinivas
	10:30–12:00	SS	Applications of Hodge theory to algebraic cycles (3)

Invited speakers: S. Archava (SA), M. Asakura (MA), E. Colombo (EC), P. Del Angel (PD), R. Joshua (RJ), M. Kerr (MK), R. Laterveer (RL), A. Otwinowska(AO), O. Penacchio (OP), P. F. dos Santos (PdS)

Day	Hour	Speaker	Title
Monday July 2	13:30–14:30	MA	On the regulator image of K_2
	14:30–15:30	RL	Relative version of Mumford's theorem on 0-cycles
Tuesday July 3	08:30–10:00	RJ	Applications of the strong Künneth decomposition of the diagonal
	13:30–14:30	EC	Mixed Hodge structures on the fundamental group of a hyperelliptic curve and algebraic cycles
	14:30–15:30	SA	Arithmetic Hodge structures on homotopy groups and intersection theory of algebraic cycles
Wednesday July 4	13:30–14:30	PD	K-theory and the Painlevé equation
	14:30–15:30	PdS	Lawson homology for real varieties
Thursday July 5	13:30–14:30	JM	Chow–Künneth decomposition for elliptic modular varieties
	14:30–15:30	AO	Hodge locus of a family of hypersurfaces
Friday July 6	13:30–14:30	MK	Milnor regulators and higher Abel–Jacobi maps
	14:30–15:30	OP	The R-split level of variations of mixed Hodge structures

References

[1] Carlson J., Müller-Stach, S. and Peters, C. *Period Maps and Period Domains*, Cambridge Studies in Advanced Mathematics, Cambridge University Press, Oxford, 2003.

[2] Clemens, H. *A Scrapbook of Complex Curve Theory*, Plenum, New York, 1980.

[3] Fulton, W. *Intersection Theory*, Springer Ergebnisse der Mathematik, Springer Verlag, Heidelberg, 1984.

[4] Griffiths, P. and Harris, J. *Principles of Algebraic Geometry*, John Wiley, New York, 1978.

[5] Griffiths, P. Periods of integrals on algebraic manifolds I, II, *Am. J. Math.* **90** (1968) 568–626, 805–865.

[6] Griffiths, P. Periods of rational integrals on algebraic manifolds III, *Publ. Math. Inst. Hautes Etudes. Sci.* **38** (1970) 125–180.

Acknowledgements

The following institions and persons have financed this school: the University 'Joseph-Fourier' (University of Grenoble I), the Town of Grenoble, the General Board of the Department Isère, J.-P. Demailly's grant for scientific exchange with Germany and, finally, the Department of Mathematics of the University of Grenoble (both the Institut Fourier and the CNRS-unit UMR 5582).

Our thanks go to all of the above and in addition to the secretarial staff of the Institut Fourier, in particular to Myriam Charles. For the excellent atmosphere during the school we also thank the participants; by engaging in various activities they contributed in an essential way to the success of the school.

One of the requirements from the granting institution is to have lecture notes available for the students in advance when they arrive at the school. These were the basis for all the articles in this volume. Several speakers handed in hand-written notes which were typed by A. Combes and A. Kahl at Essen. We express our thanks for their help.

Part I

Introductory material

1

Chow varieties, the Euler–Chow series and the total coordinate ring

E. Javier Elizondo

Instituto de Matemáticas, Ciudad Universitaria, UNAM, México DF 04510, Mexico

Introduction

Chow varieties play an important role in the geometry and topology of algebraic varieties. However, their geometry and topology are not well understood. It is important to mention the work of Blaine Lawson, Eric Friedlander, Paulo Lima-Filho and others on the homotopy and topology of the space of cycles with fixed dimension. Chapter 3 by Lima-Filho is relevant in this aspect.

In this article we would like to present other aspects of the geometry and topology of Chow varieties.

In Section 1.1 we introduce Chow varieties and give some important examples. In the last part of this section we mention the case of zero cycles, and we state a theorem where it is shown that they are isomorphic to a certain symmetric product.

In Section 1.2 we study the Euler–Chow series. These are a class of invariants for projective varieties arising from the Euler characteristic of their Chow varieties. It is a series that in a way generalizes the Hilbert series and also appears in many different problems in algebraic geometry. It is also worth mentioning that it belongs, for the correct dimension, as an element in the quantum cohomology of the variety. We do not know what role it plays here, but it shows that a lot more has to be understood before we can get a clear picture of the role of this series in geometry.

In this section we start with simple and interesting examples, then we go through some formalism in order to understand the product of two Euler–Chow series of different varieties. Then some properties of the series are shown and the series is computed for some important examples, like toric varieties, abelian varieties, projective closure of line bundles, and other cases. In Section 1.2.4

Transcendental Aspects of Algebraic Cycles ed. S. Müller-Stach and C. Peters.
© Cambridge University Press 2004.

we relate the Euler–Chow series for the Grassmannian varieties with the Chow quotients, another interesting fact that perhaps shows how the series is full of information about the geometry of the variety.

In Section 1.3 we state some open problems and introduce the total coordinate ring of a variety, stating some theorems. This ring is associated to the Euler–Chow series of a projective variety and only for divisors. The ring itself is very interesting; it is related to different classical problems in algebraic geometry, canonical rings and Mori theory, an old classical problem by Zariski, and some others.

1.1 Chow varieties

In this section we will sketch the construction of Chow varieties and give some examples. There are two main references, the first of which is the book of Shafarevich [Sha77]. It is important to note that the last edition which consists of two volumes does not have the construction of the Chow varieties. The second main reference where most of the examples and material can be found is the book of Gelfand, Kapranov and Zelevinsky [GKZ94]. The reader is encouraged to consult the latter for details and proofs of some of the theorems.

We should also mention the book of Kollár [Kol96]; it has an excellent exposition of Hilbert schemes and Chow varieties in characteristic p. Although this is very important, it is not possible to cover it in these notes. Throughout this section we work over an algebraic closed field.

1.1.1 Chow forms of irreducible varieties

Let $X \subset \mathbb{P}^{n-1}$ be an irreducible subvariety of dimension $k - 1$ and degree d. Let $\mathcal{Z}(X)$ be the set of all $(n - k - 1)$-dimensional projective subspaces $L \in \mathbb{P}^{n-1}$ that intersect X. This is a subvariety in the Grassmannian $G(n - k, n)$ parametrizing all the $(n - k - 1)$-dimensional projective subspaces in $\mathbb{P}^{n-1}$. Then we have the following theorem.

Theorem 1.1.1.1 *The subvariety* $\mathcal{Z}(X)$ *is an irreducible hypersurface of degree* d *in* $G(n - k, n)$.

We shall call $\mathcal{Z}(X)$ the *associated hypersurface* of X. Let $\mathfrak{B} = \oplus \mathfrak{B}_m$ be the coordinate ring of the Grassmannian $G(n - k, n)$. It can be proven that $\mathcal{Z}(X)$ is defined by the vanishing of some element $R_X \in \mathfrak{B}_d$. This element is called the *Chow form* of X. The coefficients of this form are the *Chow coordinates* of X. It is important to notice that X can be recovered from its Chow coordinates. Consider these examples.

Example 1.1.1.2

1. Let X be a curve in $\mathbb{P}^3$. Its associate hypersurface is the variety of all lines which intersect X.
2. Let X be a hypersurface in $\mathbb{P}^{n-1}$. The Grassmannian $G(n - k, n)$ coincides with $\mathbb{P}^{n-1}$ and the associated hypersurface $Z(X)$ coincides with X.
3. Let X be a point p, then $G(n - k, n)$ is the dual projective space $(\mathbb{P}^{n-1})^*$ and $Z(X)$ is the hyperplane dual to p.
4. Let X be a linear projective subspace. The variety $Z(X)$ is known as the *Schubert divisor* in $G(n - k, n)$. We can assume that the linear equations of X are given by $x_1 = 0, \ldots, x_{n-k} = 0$ where $x_1, \ldots, x_n$ are coordinate functions. Then the associate variety is given by the vanishing of the Plücker coordinates $p_1, \ldots, p_{n-k}$. For explicit formulas of the Plücker coordinates, see [Ful98].

Now, we have the following theorem that tells us that we can recover X from $Z(X)$.

Theorem 1.1.1.3 *A $(k - 1)$-dimensional irreducible subvariety $X \subset \mathbb{P}^{n-1}$ is determined uniquely by its associated hypersurface $Z(X)$. More precisely, a point $p \in \mathbb{P}^{n-1}$ lies in X if and only if any $(n - k - 1)$-dimensional plane containing p belongs to $Z(X)$.*

1.1.2 Definition of Chow variety

In this section we construct for any irreducible $(k - 1)$-dimensional subvariety $X \subset \mathbb{P}^{n-1}$, its Chow form R_X. This is a polynomial $R_X(f_1, \ldots, f_k)$ in coefficients of k indeterminate linear forms on $\mathbb{C}^n$ which vanishes whenever the projective subspace $\{f_1 = \cdots = f_k = 0\}$ of $\mathbb{P}^{n-1}$ intersects X. It also satisfies the following homogeneity property, if $g = (g_{ij})$ is a matrix in $GL(k)$, we have that

$$R_X(g_{11}f_1 + \cdots + g_{1k}f_k, \ldots, g_{k1}f_1 + \cdots + g_{kk}f_k)$$
$$= \det(g)^d \, R_X(f_1, \ldots, f_k)$$

where $d = \deg X$. The space of polynomials with this property is denoted by $\mathfrak{F}_d$.

Let $X = \sum m_i X_i$ be a $(k - 1)$-dimensional effective algebraic cycle in $\mathbb{P}^{n-1}$ of degree d. We define the Chow form of X as

$$R_X = \prod R_{X_i}^{m_i} \in \mathfrak{F}_d.$$

The coordinates of the vector R_X are called *Chow coordinates* of X. Let us denote by $\mathcal{C}_{k-1,d}\left(\mathbb{P}^{n-1}\right)$ the space of all the effective $(k-1)$-cycles in $\mathbb{P}^{n-1}$ of degree d. The main result is a theorem due to Chow and van der Waerden.

Theorem 1.1.2.1 *The map* $X \longmapsto R_X$ *defines an embedding of* $\mathcal{C}_{k-1,d}\left(\mathbb{P}^{n-1}\right)$ *into the projective space* $\mathbb{P}^{\mathfrak{F}_d}$ *as a closed algebraic variety.*

The variety $\mathcal{C}_{k-1,d}\left(\mathbb{P}^{n-1}\right)$ with the algebraic structure defined by the above embedding is called the *Chow embedding*. For a proof of this theorem see [GKZ94, p. 126].

1.1.3 Examples of Chow varieties

Example 1.1.3.1

1. The Chow variety $\mathcal{C}_{k-1,1}\left(\mathbb{P}^{n-1}\right)$ is the Grassmannian $G(k,n)$ and its Chow embedding coincides with the Plücker embedding.
2. Consider the Chow variety $\mathcal{C}_{n-2,d}\left(\mathbb{P}^{n-1}\right)$, parametrizing cycles of degree d and codimension 1 in $\mathbb{P}^{n-1}$, that is hypersurfaces. We saw in Example 1.1.1.2 that the Chow form of an irreducible hypersurface is just its equation which is an irreducible homogeneous polynomial of degree d in n variables. Algebraic cycles of codimension 1 corresond to all non-zero homogeneous polynomials, irreducible or not, of degree d. Therefore, the Chow variety $\mathcal{C}_{n-2,d}\left(\mathbb{P}^{n-1}\right)$ is the projective space of such polynomials, i.e.

$$\mathcal{C}_{n-2,d}\left(\mathbb{P}^{n-1}\right) = \mathbb{P}^{N-1} \quad \text{where} \quad N = \binom{n+d-1}{d}.$$

Example 1.1.3.2 Consider $\mathcal{C}_{1,2}\left(\mathbb{P}^3\right)$. Thus we are considering curves of degree 2 in $\mathbb{P}^3$. There are two cases, either an irreducible curve or two lines. A curve of degree 2 must be a plane quadric, by Bézout. This implies that $\mathcal{C}_{1,2}\left(\mathbb{P}^3\right)$ has two irreducible components C and D corresponding to planar quadrics and pairs of lines. $C \cap D$ consists of pairs of coplanar lines. Now, dim $D = 8$ since one line in $\mathbb{P}^3$ depends on four parameters. What is interesting, and rare, in Chow varieties is that dim $C = 8$. This is easy to see since a plane needs three parameters, and a quadric in a plane needs five parameters.

Example 1.1.3.3 This example is $\mathcal{C}_{1,3}\left(\mathbb{P}^3\right)$ parametrizing 1-dimensional cycles in $\mathbb{P}^3$ of degree 3. The possible curves for those cycles are:

1. an irreducible curve of degree 3; here we have two possible choices (see [Har75b, IV.6]), either a twisted curve or a plane cubic;
2. a line and a planar quadric;
3. three lines.

A twisted cubic is a curve which can be modified by a projective transformation of $\mathbb{P}^3$ to the standard Veronese curve given in homogeneous coordinates by

$$\left\{ \left(x_0^3 : x_0^2 x_1 : x_0 x_1^2 : x_1^3 \right) \,\middle|\, (x_0 : x_1) \in \mathbb{P}^1 \right\}. \tag{1}$$

If we denote by C_1, C_2, C_3, C_4 the subvarieties of $\mathcal{C}_{1,3}\left(\mathbb{P}^3\right)$ parametrizing twisted curves, plane cubics, a line and a planar quadric, and three lines, then we have that

$$\mathcal{C}_{1,3}\left(\mathbb{P}^3\right) = C_1 \cup C_2 \cup C_3 \cup C_4.$$

The dimension of C_4 is 12, since one line depends on four parameters. The dimension of C_3 is also 12; check the last example. The dimension of C_1 is also 12. To see this we observe that all twisted cubics are images of one particular twisted cubic, see equation (1), under projective transformations. The stabilizer of the curve in (1) is the group $PGL(2)$ of projective transformations of $\mathbb{P}^1$ embedded into $PGL(4)$ via the map

$$GL(2) = GL(\mathbb{C}^2) \hookrightarrow GL(4) = GL(S^3\mathbb{C}^2)$$

by the correspondence

$$g \longmapsto S^3 g$$

where $S^3\mathbb{C}^2$ is the space of all homogeneous polynomials of degree d in two variables. Hence, $C_1 = PGL(4)/PGL(2)$, and its dimension is equal to $15 - 3 = 12$.

For C_2, the dimension is given by the number of parameters defining a plane, which is three, plus the dimension of the space of cubics in a given plane, which is nine, therefore, $\dim C_2 = 12$.

It is tempting to conjecture that all components of the variety $\mathcal{C}_{1,d}\left(\mathbb{P}^3\right)$ have dimension $4d$. However, there is this example.

Example 1.1.3.4 Consider the Chow variety $\mathcal{C}_{1,4}\left(\mathbb{P}^3\right)$ of 1-dimensional cycles in $\mathbb{P}^3$ of degree 4. This variety has many components corresponding to the various possibilities that occur for a cycle of degree 4:

1. an irreducible curve of degree 4;
2. a cubic and a line;
3. two quadric curves;
4. a quadric curve and two lines;
5. four lines.

From the previous examples it is reasonably clear that all of the components in cases 2–5 above have dimension 16. Thus, we have to concentrate on the first component, let us call it C. This variety also has reducible components, irreducible curves of degree 4 that can be of three different types (see [Har75b, IV.6]), namely:

1. a planar quartic;
2. a rational curve of degree 4;
3. a spatial elliptic curve of degree 4.

The last one is the intersection of two quadric surfaces. Let C_1, C_2, C_3 be the components corresponding to 1, 2, 3. C_2 and C_3 have dimension 16. However, the number of parameters defining a plane is three, plus the dimension of the space of quartics in a given plane is 14. Therefore $\dim C_1 = 17$.

We mention in passing that Eisenbud and Harris have computed the dimension of the Chow variety of curves [EH92]. A student of Harris, Pablo Azcue, computed in his Ph.D. thesis the dimension of Chow varieties in higher dimensions. In both cases there are small numbers of Chow varieties (of low degree) that cannot be considered by their computations.

1.1.4 Zero cycles

A positive 0-cycle of degree d is just an unordered collection $\{x_1, \ldots, x_d\}$ of d points (not necessarily distinct) in $\mathbb{P}^{n-1}$. Thus, as a set $\mathcal{C}_{0,d}(\mathbb{P}^n)$ is identified with $\mathrm{Sym}^d(\mathbb{P}^{n-1})$, the d-fold symmetric product of $\mathbb{P}^{n-1}$. So we start with a comparison of the dimensions of $\mathcal{C}_{0,d}(\mathbb{P}^{n-1})$ and $\mathrm{Sym}^d(\mathbb{P}^{n-1})$.

Suppose that our projective space $\mathbb{P}^{n-1}$ is $P(V)$ where V is an n-dimensional vector space. The Chow form of a point $x \in P(V)$ is the linear function l_x on V^* given by the scalar product with x:

$$l_x(\xi) = (x, \xi).$$

If $X = \sum m_i x_i$ is a positive 0-cycle in $P(V)$ then, by our convention, the Chow form R_X is the polynomial $\xi \mapsto \prod l_{x_i}^{m_i}(\xi)$. We arrive at the following.

Proposition 1.1.4.1 *The Chow variety $\mathcal{C}_{0,d}(\mathbb{P}^{n-1})$ of positive 0-cycles in $\mathbb{P}^{n-1}$ of degree d is the projectivization of the space of homogeneous polynomials of degree d in n variables which are products of linear forms.*

The set Y of decomposable (into linear factors) polynomials of degree d was already used several times in the course of proving the Chow–van der Waerden theorem. Note that this set has, as its 'odd' analogue, the set of polyvectors from

$\bigwedge^d \mathbb{C}^n$ which are decomposable into wedge products of d vectors. The projectivization of the set of decomposable polyvectors is, as we have seen in Section 1.1.3, nothing more than the Grassmannian $G(d, n)$ in its Plücker embedding. So the variety of 0-cycles is the 'even' analogue of the Grassmannian.

Recall now the definitions of symmetric products. Let X be a quasi-projective algebraic variety. The symmetric product $\text{Sym}^d(X)$ is the quotient of the Cartesian product X^d by the action of the symmetric group S_d permuting the factors. A more precise definition is as follows.

Suppose first that X is an affine variety and R is its coordinate ring. So $R^{\otimes d} = R \otimes \cdots \otimes R$ is the coordinate ring of X^d. The coordinate ring of $\text{Sym}^d(X)$ is, by definition, the subring of S_d-invariants in $R^{\otimes d}$. In other words, this is the ring of regular functions $f(\mathbf{x}_1, \ldots, \mathbf{x}_d)$ of d variables $\mathbf{x}_i \in X$ which are symmetric, i.e. invariant under any permutation of the $\mathbf{x}_i$.

If X is an arbitrary, not necessarily affine, quasi-projective variety then the symmetric product $\text{Sym}^d(X)$ is defined by gluing affine varieties $\text{Sym}^d(U)$ for various affine open subsets $U \subset X$.

It follows from these definitions that we have a regular morphism of algebraic varieties

$$\gamma\colon \text{Sym}^d(\mathbb{P}^{n-1}) \to \mathcal{C}_{0,d}\left(\mathbb{P}^{n-1}\right) \qquad \{\mathbf{x}_1, \ldots, \mathbf{x}_d\} \mapsto \sum x_i, \quad (2)$$

which is set theoretically a bijection. Note that this does not automatically imply that γ is an isomorphism of algebraic varieties: the morphism from the affine line A^1 to the cubic $y^2 = x^3$, given by $x(t) = t^2$, $y(t) = t^3$, is bijective but not an isomorphism since the cubic is singular at $(0, 0)$. So the following fact requires a proof.

Theorem 1.1.4.2 *The morphism* $\gamma\colon \text{Sym}^d(\mathbb{P}^{n-1}) \to \mathcal{C}_{0,d}\left(\mathbb{P}^{n-1}\right)$ *is an isomorphism of algebraic varieties (over the field of complex numbers).*

It is important to note that over a field of finite characteristic the statement is no longer true [Nee91].

Finally, we would like to state two more results.

Theorem 1.1.4.3 *The symmetric product* $\text{Sym}^d(\mathbb{P}^1) = \mathcal{C}_{0,d}\left(\mathbb{P}^1\right)$ *is isomorphic to* $\mathbb{P}^d$.

Theorem 1.1.4.4 *For any d and n, the variety* $\text{Sym}^d(\mathbb{P}^{n-1}) = \mathcal{C}_{0,d}\left(\mathbb{P}^{n-1}\right)$ *is rational, i.e. it is birationally isomorphic to the projective space* $\mathbb{P}^{d(n-1)}$.

1.2 The Euler–Chow series of Chow varieties

1.2.1 General definitions

The use of topological invariants on moduli spaces has played a vital role in various branches of mathematics and mathematical physics in the last two decades. A quick sampling under this vast umbrella includes works in gauge theory, the theory of instantons, various moduli spaces of vector bundles, moduli spaces of curves and their compactifications, Chow varieties and Hilbert schemes.

In this section we study a class of invariants for projective varieties arising from the Euler characteristics of their Chow varieties. We will see that quite nice and elegant behaviour which can often be codified in simple generating functions comes from these Euler characteristics.

Basic examples

In this subsection we follow [ELF98]. As for motivation, we start with some particular cases which are well studied in the literature. Let X be a connected projective variety and let $SP(X)$ denote the disjoint union $\coprod_{d \geq 0} SP_d(X)$ of all symmetric products of X, with the disjoint union topology, where $SP_0(X)$ is a single point. One can define a function $E_0(X) : \mathbb{Z}_+ = \pi_0(SP(X)) \to \mathbb{Z}$ which sends d to the Euler characteristic $\chi(SP_d(X))$ of the d-fold symmetric product of X. This is what we call the *0th Euler–Chow function of* X. The same information can be codified as a formal power series $E_0(X) = \sum_{d \geq 0} \chi(SP_d(X))t^d$, and a result of Macdonald [Mac62] shows that $E_0(X)$ is given by the rational function $E_0(X) = (1/(1 - t))^{\chi(X)}$.

Another familiar instance arises in the case of divisors. Given an n-dimensional projective variety X, let $\mathrm{Div}_+(X)$ denote the space of effective divisors on X and assume that $\mathrm{Pic}^0(X) = \{0\}$. Consider the function $E : \mathrm{Pic}(X) \to \mathbb{Z}$ which sends $L \in \mathrm{Pic}(X)$ to $\dim H^0(X, \mathcal{O}(L))$. Observe that:

1. given $L \in \mathrm{Pic}(X)$, then $E(L) \neq 0$ if and only if $L = \mathcal{O}(D)$ for some effective divisor D;
2. under the given hypothesis, algebraic and linear equivalence coincide, and two effective divisors D and D' are algebraically equivalent if and only if they are in the same linear system.

The last observation implies that $\mathrm{Div}_+(X)$ can be written as $\mathrm{Div}_+(X) = \coprod_{\alpha \in \mathcal{A}_{n-1}^{\geq}(X)} \mathrm{Div}_+(X)_\alpha$, where $\mathcal{A}_{n-1}^{\geq}(X)$ is the monoid of algebraic equivalence classes of effective divisors (cf. [Ful98, Section 12]), and $\mathrm{Div}_+(X)_\alpha$ is the linear system associated to $\alpha \in \mathcal{A}_{n-1}^{\geq}(X)$. The first observation shows that the only data relevant to E is given by $\mathcal{A}_{n-1}^{\geq}(X) \subset \mathrm{Pic}(X)$. Therefore, we might as well restrict E and define the $(n-1)$-*st Euler–Chow function of* X as the function $E_{n-1}(X) : \mathcal{A}_{n-1}^{\geq}(X) \to \mathbb{Z}_+$ which sends $\alpha \in \mathcal{A}_{n-1}^{\geq}(X)$ to the Euler characteristic $\chi(\mathrm{Div}_+(X)_\alpha) = \dim H^0(X, \mathcal{O}(L_\alpha))$, where L_α is the line bundle associated to α.

Example 1.2.1.1 An even more restrictive case arises when $\mathrm{Pic}(X) \cong \mathbb{Z}$, and $\mathcal{A}_{n-1}^{\geq}(X) \cong \mathbb{Z}_+$ is generated by the class of a very ample line bundle L. Then the $(n-1)$-st Euler–Chow function $E_{n-1}(X) = \sum_{d \geq 0} \dim H^0(X; \mathcal{O}(L^{\otimes n}))t^n$ is just the *Hilbert function* associated to the projective embedding of X induced by L. This is once again a rational function.

Preliminary definitions

Let us start with an abelian monoid M, whose multiplication we denote by $*_M : M \times M \longrightarrow M$. When no confusion is likely to arise we use an additive notation $+ : M \times M \longrightarrow M$ with no subscripts attached. We say that M has *finite multiplication* if $*_M$ has finite fibers. Typical examples are the freely generated monoids, such as the non-negative integers $\mathbb{Z}_+$ under addition.

Definition 1.2.1.2 Given a monoid with finite multiplication M, and a commutative ring S, denote by S^M the set of all functions from M to S. If f and f' are elements in S^M, let $f + f' \in S^M$ be defined by pointwise addition, i.e. $(f + f')(m) = f(m) + f'(m)$. Define the product $f * f' \in S^M$ as the 'convolution'

$$(f * f')(m) = \sum_{a *_M b = m} f(a)f'(b).$$

It is easy to see that S^M then becomes a commutative ring with unity, under these operations.

Remark 1.2.1.3 The ring S^M can be identified with the completion $S[\![M]\!]$ of the monoid algebra $S[M]$ at its augmentation ideal. Therefore, the elements of S^M can be written as a formal power series $f = \sum_{m \in M} s_m \cdot t^m$, on variables t^m and coefficients in S. In this form the multiplication is given by the relation $t^m t^{m'} = t^{m+m'}$ for elements $m, m' \in M$.

Definition 1.2.1.4 Given a monoid morphism $\Psi : M \longrightarrow N$, $f \in S^M$ and $g \in S^N$, define $\Psi^\sharp g \in S^M$ and $\Psi_\sharp f \in S^N$ by

$$(\Psi^\sharp g)(m) = g(\Psi(m))$$

and

$$(\Psi_\sharp f)(n) = \sum_{m \in \Psi^{-1}(n)} f(m)$$

if Ψ has finite fibers.

Proposition 1.2.1.5 *Let M and N be monoids with finite multiplication, and let $\Psi : M \longrightarrow N$ be a monoid morphism. Then*

1. *The pull-back map $\Psi^{\sharp} : S^{N} \longrightarrow S^{M}$ is an S-module homomorphism.*
2. *If Ψ has finite fibers then the push-forward map $\Psi_{\sharp} : S^{M} \longrightarrow S^{N}$ is a morphism of S-algebras.*
3. *Any ring homomorphism $\Psi : S \longrightarrow S'$ induces a ring homomorphism $\Psi_{*} : S^{M} \longrightarrow S'^{M}$.*

The last operation we need to introduce is the following *exterior product*.

Definition 1.2.1.6 Given monoids M and N, and a commutative ring S, one can define a map $\odot : S^{M} \otimes_{S} S^{N} \rightarrow S^{M \times N}$. This map sends $f \otimes g$ to the function $f \odot g \in S^{M \times N}$ which assigns to (m, n) the element $f(m)g(n) \in S$.

Proposition 1.2.1.7 *The operation $\odot$ is bilinear and associative. In other words, the following diagram commutes:*

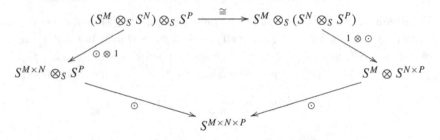

Definition of the Euler–Chow series

Let X be a projective algebraic variety over $\mathbb{C}$, and let p be a non-negative integer such that $0 \leq p \leq \dim X$. The *Chow monoid* $\mathcal{C}_{p}(X)$ of effective p-cycles on X is the free monoid generated by the irreducible p-dimensional subvarieties of X. It is well known that $\mathcal{C}_{p}(X)$ can be written as a countable disjoint union of projective algebraic varieties $\mathcal{C}_{p,\alpha}(X)$, the so-called *Chow varieties*. We summarize, in the following statements, a few basic properties of the Chow monoids and varieties which are found in Hoyt [Hoy66], Friedlander [Fri91], and Friedlander and Mazur [FM91]. For a recent survey and extensive bibliography on the subject, we refer the reader to Lawson [Law95].

Properties 1.2.1.8 Let X be a projective variety, and fix $0 \leq p \leq \dim X$.

1. The disjoint union topology on $\mathcal{C}_{p}(X)$ induced by the classical topology on the Chow varieties is independent of the projective embedding of X, cf. [Hoy66].

2. The restriction of the monoid addition to products of Chow varieties is an algebraic map [Fri91].
3. An algebraic map $f : X \to Y$ between projective varieties (hence a proper map) induces a natural monoid morphism $f_* : \mathcal{C}_p(X) \to \mathcal{C}_p(Y)$ which is an algebraic continuous map when restricted to a Chow variety, cf. [Fri91]. This is the *proper push-forward map*.
4. A flat map $f : X \to Y$ of relative dimension k induces a natural monoid morphism $f^* : \mathcal{C}_p(Y) \to \mathcal{C}_{p+k}(X)$ which is an algebraic continuous map when restricted to a Chow variety, cf. [Fri91]. This is the *flat pull-back map*.

Remark 1.2.1.9 Since we work over $\mathbb{C}$, we can define an *algebraic continuous map* as a continuous map $f : X \to Y$ between varieties which induces an algebraic map $f^\nu : X^\nu \to Y^\nu$ between their weak normalizations, cf. [FM91]. Alternatively, one could define *Chow varieties* as the weak normalization of those we consider here, as in Kollár [Kol96]. This does not alter their topology, but transforms Chow varieties into a functor in the category of projective varieties and algebraic morphisms. Either approach can be used in this paper, without altering the results.

Definition 1.2.1.10

1. We denote by $\Pi_p(X)$, the monoid $\pi_0(\mathcal{C}_p(X))$ of path components of $\mathcal{C}_p(X)$. This is the monoid of 'effective algebraic equivalence classes' of effective p-cycles on X. We use the notation $a \sim_{\mathrm{alg+}} b$ to express that two effective cycles a, b are effectively algebraically equivalent.
2. The group of all algebraic p-cycles on X modulo algebraic equivalence is denoted $\mathcal{A}_p(X)$, and the submonoid of $\mathcal{A}_p(X)$ generated by the classes of cycles with non-negative coefficients is denoted by $\mathcal{A}_p^{\geq}(X)$, cf. [Ful98, Section 12]. We use the notation $a \sim_{\mathrm{alg}} b$ to express that two cycles a, b are algebraically equivalent.
3. Let $c : \mathcal{C}_p(X) \longrightarrow H_{2p}(X, \mathbb{Z})$ be the cycle map into singular homology, cf. [Ful98, Section 19]. The image of c is denoted by $M_p(X)$.

The following result explains the relation between the monoids above.

Proposition 1.2.1.11

1. *The Grothendieck group associated to the monoid $\Pi_p(X)$ is $\mathcal{A}_p(X)$. In particular, there is a natural monoid morphism $\iota_p : \Pi_p(X) \to \mathcal{A}_p(X)$ which satisfies the universal property that any monoid morphism $f : \Pi_p(X) \to G$, from $\Pi_p(X)$ into a group G, factors through $\mathcal{A}_p(X)$.*
2. *The image of ι_p is $\mathcal{A}_p^{\geq}(X)$, and the image of $\mathcal{A}_p^{\geq}(X)$ under the cycle map is $M_p(X)$.*

3. *The monoid surjection* $\bar{\iota}_p : \Pi_p(X) \to \mathcal{A}_p^{\geq}(X)$ *induced by* ι_p *is an isomorphism if and only if* $\Pi_p(X)$ *has cancellation law.*
4. *Both* $\bar{\iota}_p : \Pi_p(X) \to \mathcal{A}_p^{\geq}(X)$ *and* $c_p : \mathcal{A}_p^{\geq}(X) \to M_p(X)$ *are finite monoid morphisms.*

Proof The first assertion is proven in [Fri91], and follows from standard arguments, e.g. in Samuel [Sam71]. The second assertion follows from the definitions and the universal property just described. The third assertion follows from the elementary fact that the universal map from an abelian monoid into its group completion is injective if and only if the monoid has cancellation law. To prove the last assertion, consider a projective embedding of X. The cycles supported in X with a fixed degree d in the ambient projective space form a projective variety, which is then a finite union of components of $\mathcal{C}_p(X)$. The assertion now follows easily from these observations.   □

The following result is found in Elizondo [Eli94].

Proposition 1.2.1.12 *Given a complex projective algebraic variety* X *and* $0 \leq p \leq \dim X$, *the monoids* $\mathcal{C}_p(X)$, $\Pi_p(X)$, $\mathcal{A}_p^{\geq}(X)$ *and* $M_p(X)$ *all have finite multiplication.*

Proof One has surjective monoid morphisms:

$$\mathcal{C}_p(X) \xrightarrow{\pi_p} \Pi_p(X) \xrightarrow{\iota_p} \mathcal{A}_p^{\geq}(X) \xrightarrow{c_p} M_p(X), \tag{3}$$

so that when a projective embedding $j : X \hookrightarrow \mathbb{P}^n$ is chosen one obtains a commutative diagram

$$
\begin{array}{ccccccc}
\mathcal{C}_p(X) & \longrightarrow & \Pi_p(X) & \longrightarrow & \mathcal{A}_p^{\geq}(X) & \longrightarrow & M_p(X) \\
\downarrow{\scriptstyle j_*} & & \downarrow{\scriptstyle j_*} & & \downarrow{\scriptstyle j_*} & & \downarrow{\scriptstyle j_*} \\
\mathcal{C}_p(\mathbb{P}^n) & \longrightarrow & \Pi_p(\mathbb{P}^n) & \xrightarrow{\cong} & \mathcal{A}_p^{\geq}(\mathbb{P}^n) & \xrightarrow{\cong} & M_p(\mathbb{P}^n) \\
& & \| & & \| & & \| \\
& & \mathbb{Z}_+ & & \mathbb{Z}_+ & & \mathbb{Z}_+
\end{array}
$$

where the leftmost vertical arrow is a closed inclusion. Recall that $\mathcal{C}_p(\mathbb{P}^n) = \coprod_{d \in \mathbb{Z}_+} \mathcal{C}_{p,d}(\mathbb{P}^n)$ where $\mathcal{C}_{p,d}(\mathbb{P}^n)$ is a projective connected algebraic (Chow) variety. Furthermore, $j(\mathcal{C}_p(X)) \cap \mathcal{C}_{p,d}(\mathbb{P}^n)$ is a subvariety for all d. It follows that $\Pi_p(X)$, $\mathcal{A}_p^{\geq}(X)$ and $M_p(X)$ are all discrete, and that $\mathcal{C}_p(X)$ has finite multiplication, since it is free. Now, it is not difficult to prove that $\Pi_p(X)$, $\mathcal{A}_p^{\geq}(X)$ and $M_p(X)$ also have finite multiplication.   □

Definition 1.2.1.13 The (algebraic) *pth Euler–Chow function* of X is the function

$$E_p(X) : \Pi_p(X) \longrightarrow \mathbb{Z}$$
$$\alpha \longmapsto \chi(\mathcal{C}_{p,\alpha}(X)), \tag{4}$$

which sends $\alpha \in \Pi_p(X)$ to the topological Euler characteristic of $\mathcal{C}_{p,\alpha}(X)$ (in the classical topology).

Following Remark 1.2.1.3 we associate a variable t^α to each $\alpha \in \Pi_p(X)$ and express the pth Euler–Chow function as a formal power series

$$E_p(X) = \sum_{\alpha \in \Pi_p(X)} \chi(\mathcal{C}_{p,\alpha}(X)) \, t^\alpha. \tag{5}$$

Example 1.2.1.14

1. If X is a connected variety, then $\mathcal{C}_0(X) = \coprod_{d \in \mathbb{Z}_+} SP_d(X)$, where $SP_d(X)$ is the d-fold symmetric product of X. Therefore, the 0th Euler–Chow function is given by

$$E_0(X) = \sum_{d>0} \chi(SP_d(X)) \, t^d = \left(\frac{1}{1-t} \right)^{\chi(X)},$$

according to MacDonald's formula [Mac62].

2. For $X = \mathbb{P}^n$, one has $\Pi_p(\mathbb{P}^n) \cong \mathbb{Z}_+$, with the isomorphism given by the degree of the cycles. In this case, the pth Euler–Chow function was computed in [LY87]:

$$E_p(\mathbb{P}^n) = \sum_{d \geq 0} \chi(\mathcal{C}_{p,d}(\mathbb{P}^n)) \, t^d = \left(\frac{1}{1-t} \right)^{\binom{n+1}{p+1}}.$$

3. Suppose that X is an n-dimensional variety such that $\mathrm{Pic}(X) \cong \mathbb{Z}$, generated by a very ample line bundle L. Then, we have seen in Example 1.2.1.1 that $\Pi_{n-1}(X) \cong \mathbb{Z}_+$ and that $E_{n-1}(X)$ is precisely the Hilbert function for the projective embedding of X induced by L.

1.2.2 The toric case

In this section we survey the results for toric varieties. We will see that the Euler–Chow series is a rational function and where we can get explicit computations. We follow the article [Eli94].

Throughout this section X is a projective algebraic variety, on which an algebraic torus $T := (\mathbb{C}^*)^n$ acts linearly having only a finite number of invariant irreducible subvarieties of dimension p. In particular, we will see that the result is true for any projective toric variety. Let us denote by H the homology group $H_{2p}(X, \mathbb{Z})$. We will write $\mathcal{C}_\lambda(X)$ instead of $\mathcal{C}_{p,\lambda}(X)$ if p is fixed.

The action of T on X induces an action on the Chow variety $\mathcal{C}_{p,d}(X)$. Let λ be an element in H and denote by $\mathcal{C}_\lambda^T$ the fixed point set of $\mathcal{C}_\lambda(X)$ under the action of T. Then its Euler characteristic $\chi(\mathcal{C}_\lambda^T)$ is equal to the number of invariant subvarieties of X with homology class λ. We have

$$E_p(\lambda) \,=\, \chi\left(\mathcal{C}_\lambda\right) \,=\, \chi\left(\mathcal{C}_\lambda^T\right) \tag{6}$$

where the first equality is just the definition of E_p and the second equality is proved in [LY87]. The following theorem tells us that E_p is rational.

Theorem 1.2.2.1 *Let E_p be the Euler series of X. Denote by $V_1, \dots, V_N$ the p-dimensional invariant irreducible subvarieties of X. Let $e_{[V_i]} \in \mathbb{Z}[C]$ be the characteristic function of the subset $\{[V_i]\}$ of C. Then*

$$E_p = \prod_{1 \le i \le N} \left(\frac{1}{1 - e_{[V_i]}} \right). \tag{7}$$

Proof For each V_i define f_i in $\mathbb{Z}[[C]]$ by

$$f_i(\lambda) = \begin{cases} 1 & \text{if } \lambda = n \cdot [V_i], n \ge 0 \\ 0 & \text{otherwise.} \end{cases} \tag{8}$$

It is easy to see from equations (6) and (8) that E_p can be written as

$$E_p \,=\, \prod_{1 \le i \le N} f_i. \tag{9}$$

The theorem follows because of the equality

$$1 = \left(1 - e_{[V_i]}\right) \cdot f_i. \qquad \square$$

The next lemma tells us that the result is true for any projective toric variety.

Lemma 1.2.2.2 *Let X be a projective (perhaps singular) toric variety. Then any irreducible subvariety V of X which is invariant under the torus action is the closure of an orbit. Therefore, any invariant cycle has the form*

$$c \,=\, \sum n_i \, \overline{\mathcal{O}}_i$$

where each n_i is a non-negative integer and each $\overline{\mathcal{O}}_i$ is the closure of the orbit $\mathcal{O}_i$.

Before passing to the proof, let us recall that a toric variety is constructed from a *lattice* N (which is isomorphic to $\mathbb{Z}^n$ for some n), and a *fan* Δ, which is a collection of 'strongly convex rational polyhedral cones' σ in the real vector space $N_{\mathbb{R}} = N \otimes_{\mathbb{Z}} \mathbb{R}$, satisfying conditions analogous to those for a simplicial complex: every face of a cone in Δ is also a cone in Δ, and the intersection of two cones in Δ is a face of each. Now, a *strongly convex rational polyhedral cone* σ in $N_{\mathbb{R}}$ is a cone with apex at the origin, generated by a finite number of vectors and such that there is not a line contained in it.

Proof of Lemma 1.2.2.2 The fan Δ associated to X is finite because X is compact. Hence there is a finite number of cones, therefore a finite number of orbits. Let V be an invariant irreducible subvariety of X. We can express V as the closure of the union of orbits. Since there is a finite number of them we must have that

$$V = \overline{\mathcal{O}}_1 \cup \overline{\mathcal{O}}_2 \cup \cdots \cup \overline{\mathcal{O}}_N$$

where $\overline{\mathcal{O}}_i$ is the closure of an orbit. Finally, since V is irreducible, there must be i_0 such that $V = \overline{\mathcal{O}}_{i_0}$. □

Smooth toric varieties

In this subsection we give an equivariant version of the Euler series and find a relation between the equivariant and non-equivariant Euler series.

The use of equivariant cohomology allows us to analyse the Euler series from a geometrical point of view. This approach might help us to understand other cases. Before we continue, let us recall the definition of equivariant cohomology.

Throughout this subsection, unless otherwise stated, X is a smooth projective toric variety of dimension n, and we use cohomology instead of homology by applying Poincaré duality.

Let G be a Lie group which acts on a manifold X. Since the quotient is not well defined we try to find the right definition in order to compute the cohomology of 'X/G'. The classifying space for a principal G-bundle is by definition a space BG together with a universal principal G-bundle EG over it. By universal we mean that isomorphism classes of principal G-fiber bundles over X correspond naturally to homotopy classes of maps from X to BG. The correspondence is given by pulling-back the universal bundle EG. The cohomology ring $H^*(BG)$ is by definition the *ring of characteristic classes* of G.

In general, if T is a maximal torus in G, we have

$$H^*(BG) = H^*(BT)^W$$

where $W = N_G(T)/T$ is the Weyl group of (G, T) and the right-hand side denotes the subring of invariants.

We are ready to describe the homotopy quotient mentioned earlier. This is given by the *Borel construction* X_G obtained after exchanging the fiber G of the universal bundle EG with X, i.e.

$$X_G = EG \times_G X = (EG \times X)/G$$

where G acts by $g \cdot g(e, x) = (eg^{-1}, gx)$.

Definition 1.2.2.3 The *equivariant cohomology* $H_G^*(X)$ is the cohomology of the Borel construction X_G.

Example 1.2.2.4

1. If G acts freely on X, then $H_G^*(X) = H^*(X/G)$.
2. If T is a maximal torus in G and $W = N_G(T)/T$ is the Weyl group, then

$$H_G(X) = H_T^*(X)^W$$

   where the right-hand side is the subring of W-invariants.
3. If X has a single orbit, then

$$H_G^*(X) \cong H^*(BH)$$

   where H is the stabilizer of any point.
4. If K is a maximal compact subgroup of G, then

$$H_G^*(X) = H_K^*(X).$$

Let H and H_T be the cohomology group $H^{2(n-p)}(X, \mathbb{Z})$ and the equivariant cohomology group $H_T^{2(n-p)}(X, \mathbb{Z})$ of X, respectively. Denote by Δ the fan associated with X and by $\overline{O}_1 \ldots \overline{O}_N$ the p-codimensional orbit closures. Let $\mathcal{C}_\lambda^T$ and $\mathcal{C}_\lambda$ be the spaces of all p-dimensional effective invariant cycles and p-dimensional effective cycles on X with cohomology class λ.

Recall [LY87] that

$$\chi\left(\mathcal{C}_\lambda^T\right) = \chi\left(\mathcal{C}_\lambda\right). \tag{10}$$

The next lemma is crucial for the following results.

Lemma 1.2.2.5 *Let λ be an element in H. Then $\mathcal{C}_\lambda^T$ is a finite set.*

Proof By Lemma 1.2.2.2 we know that any invariant effective cycle c in $\mathcal{C}_\lambda^T$ has the form $c = \sum_{i=1}^N \beta_i \overline{O}_i$ with $\beta_i \in \mathbb{N}$. Hence, we obtain that $\mathcal{C}_\lambda^T$ has a countable number of elements. We know that $\mathcal{C}_\lambda$ is a projective algebraic variety, and since $\mathcal{C}_\lambda^T$ is Zariski closed in $\mathcal{C}_\lambda$, we have that $\mathcal{C}_\lambda^T$ is a finite set. $\square$

Our next step is to define the equivariant Euler series for X. Let $\overline{\mathcal{O}}$ be an irreducible invariant cycle in a smooth toric variety (Lemma 1.2.2.2). Since $\overline{\mathcal{O}} \subset X$ is smooth, we have an equivariant Thom–Gysin sequence

$$\cdots \longrightarrow H_T^{i-2\text{cod}\,\overline{\mathcal{O}}}(\overline{\mathcal{O}}) \longrightarrow H_T^i(X) \longrightarrow H_T^i(X - \overline{\mathcal{O}}) \longrightarrow \cdots$$

and we define $[\overline{\mathcal{O}}]_T$ as the image of 1 under

$$H_T^0(\overline{\mathcal{O}}) \longrightarrow H_T^{2\text{cod}\,\overline{\mathcal{O}}}(X).$$

Let $\{D_1, \ldots, D_K\}$ be the set of T-invariant divisors on X. To each D_i we associate the variable t_i in the polynomial ring $\mathbb{Z}[t_1, \ldots, t_K]$. Let $\mathcal{J}$ be the ideal generated by the (square free) monomials $\{t_{i_1} \cdots t_{i_l} \mid D_{i_1} + \cdots + D_{i_l} \notin \Delta\}$. It is proved in [BDP90] that

$$\mathbb{Z}[t_1, \ldots, t_K]/\mathcal{J} \cong H_T^*(X, \mathbb{Z}). \tag{11}$$

The arguments given there also prove the following.

Proposition 1.2.2.6 *For any T-orbit $\mathcal{O}$ in a smooth projective toric variety X, one has*

$$[\overline{\mathcal{O}}]_T = \prod_{\overline{\mathcal{O}} \subset D_i} [D_i]_T.$$

Furthermore if $\mathcal{O}$ and $\mathcal{O}'$ are distinct orbits, then

$$[\overline{\mathcal{O}}]_T \neq [\overline{\mathcal{O}'}]_T.$$

It is natural to define the cohomology class for any effective invariant cycle $V = \sum m_i \overline{\mathcal{O}}_i$ as $[V]_T = \sum_i m_i [\overline{\mathcal{O}}_i]_T$ where $\mathcal{O}_i \neq \mathcal{O}_j$ if $i \neq j$, and $m_i > 0$. In a form similar to the definition C, $\mathbb{Z}[[C]]$ and $\mathbb{Z}[C]$, we denote by C_T the monoid of equivariant cohomology classes of invariant effective cycles of dimension p, by $\mathbb{Z}[[C_T]]$ the set of functions on C_T, and by $\mathbb{Z}[C_T]$ the set of functions with finite support on C_T. Since $C_T \simeq \mathbb{N}^N$ where N is the number of orbits of dimension p, we obtain that

$$+ : C_T \times C_T \longrightarrow C_T$$

has finite fibers. Observe that if $\pi : H_T \to H$ denotes the standard surjection, we obtain from Lemma 1.2.2.5 that

$$\pi : C_T \longrightarrow C$$

is onto with finite fibers. We arrive at the following definition.

Definition 1.2.2.7 Let X be a smooth projective toric variety and let $H_T(X)$ be the equivariant cohomology of X. Let us denote by $\mathcal{C}_\xi^T$ the space of all invariant effective cycles on X whose equivariant cohomology class is ξ. The *equivariant Euler series* of X is the element

$$E_p^T = \sum_\xi \chi\left(\mathcal{C}_\xi^T\right) \xi \in \mathbb{Z}[[C_T]]$$

where the sum is over C_T.

Let us define the ring homomorphism

$$J : \mathbb{Z}[[C_T]] \longrightarrow \mathbb{Z}[[C]]$$

by

$$J(\xi) = \sum_\lambda \left(\sum_{\pi(\beta)=\lambda} a_\beta \right) \lambda$$

where $\xi = \sum_\beta a_\beta \beta$. This is well defined since π has finite fibers.

Theorem 1.2.2.8 *Let X be a smooth projective toric variety. Denote by E_p, E_p^T and J the Euler series, the equivariant Euler series and the ring homomorphism defined above. Then $J(E_p^T) = E_p$. Furthermore,*

$$E_p^T = \prod_{1 \le i \le N} \left(\frac{1}{1 - e_{[\overline{\mathcal{O}}_i]_T}} \right)$$

and therefore

$$E_p = \prod_{1 \le i \le N} \left(\frac{1}{1 - e_{[\overline{\mathcal{O}}_i]}} \right).$$

Proof We define for each orbit $\mathcal{O}_i$ an element $f_i^T \in \mathbb{Z}[[C_T]]$ by

$$f_i^T(\xi) = \begin{cases} 1 & \text{if } \xi = n \cdot [\overline{\mathcal{O}}_i]_T, n \ge 0 \\ 0 & \text{otherwise,} \end{cases} \tag{12}$$

and denote by e_ξ the characteristic function of $\{\xi\}$. It follows from both Definition 1.2.2.7 and equation (12) that

$$E_p^T = \prod_{1 \le i \le N} f_i^T$$

and

$$\left(1 - e_{[\overline{\mathcal{O}}_i]_T}\right) \cdot f_i^T = 1.$$

Therefore the equivariant Euler series is rational and

$$E_p^T = \prod_{1 \leq i \leq N} \left(\frac{1}{1 - e_{[\overline{O}_i]_T}} \right). \tag{13}$$

For each V_i we defined (see equation (8)) a function f_i in $\mathbb{Z}[[C]]$ by

$$f_i(\lambda) = \begin{cases} 1 & \text{if } \lambda = n \cdot [V_i], n \geq 0 \\ 0 & \text{otherwise.} \end{cases}$$

And we know from Theorem 1.2.2.1 that

$$E_p = \prod_{i=1}^{N} f_i \qquad \text{with} \qquad f_i \cdot \left(1 - e_{[V_i]}\right) = 1.$$

Now, the result follows since $\pi([\overline{O}_i]_T) = [\overline{O}_i]$ and J is a ring homomorphism satisfying

$$J\left(f_i^T\right) = f_i \qquad \text{and} \qquad J(e_\sigma) = e_{\pi(\sigma)}. \qquad \square$$

Example I: the projective space $\mathbb{P}^n$

Let $X = \mathbb{P}^n$ be the complex projective space of dimension n. Let $\{e_1, \ldots, e_n\}$ be the standard basis for $\mathbb{R}^n$. Consider $A = \{e_1, \ldots, e_{n+1}\}$ a set of generators of the fan Δ where $e_{n+1} = -\sum_{i=1}^{n} e_i$. We have the following equality

$$H^*(X, \mathbb{Z}) \cong \mathbb{Z}[t_1, \ldots, t_{n+1}]/I$$

where I is the ideal generated by

(i) $$t_1 \cdots t_{n+1}$$

(ii) $$\sum_{j=1}^{n+1} e_i^*(e_j) t_j \qquad \text{for} \qquad i = 1, \ldots, n,$$

where $e_i^* \in (\mathbb{R}^n)^*$ is the element dual to e_i.

However (ii) says that $t_i \sim t_j$ for all i and j. Therefore

$$H^*(X, \mathbb{Z}) = \mathbb{Z}[t]/t^{n+1}.$$

Consequently, any two cones of dimension $n - p$ represent the same element in cohomology, and Theorem 1.2.2.8 implies that

$$\prod_{i=1}^{\binom{n+1}{n-p}} \left(\frac{1}{1-t} \right) = \left(\frac{1}{1-t} \right)^{\binom{n+1}{n-p}} = \left(\frac{1}{1-t} \right)^{\binom{n+1}{p+1}} = E_p.$$

Example II: $\mathbb{P}^n \times \mathbb{P}^m$

Denote by $X(\Delta)$ the toric variety associated to the fan Δ, and recall that $X(\Delta \times \Delta') \cong X(\Delta) \times X(\Delta')$. Using the same notation as in Example I, we have that a set of generators of $\Delta \times \Delta'$ is given by

$$\{e_1, \ldots, e_n, e_{n+1}, \ldots, e_{n+m}, e_{n+m+1}, e_{n+m+2}\}$$

where $e_{n+m+1} = -\sum_{i=1}^{n} e_i$, $e_{n+m+2} = -\sum_{i=n+1}^{n+m} e_i$ and $\{e_1, \ldots, e_{n+m}\}$ is a basis for $\mathbb{R}^{n+m}$. Then

$$H^*(X, \mathbb{Z}) = \mathbb{Z}[t_1, \ldots, t_{n+m+2}] / I$$

where I is the ideal generated by

(i)
$$\left\{ t_1 \cdots t_n t_{n+m+1}, \quad t_{n+1} \cdots t_{n+m} t_{n+m+2}, \quad \prod_{i=1}^{n+m+2} t_i \right\}$$

(ii)
$$\sum_{j=1}^{n+m+2} e_i^*(e_j) t_j \quad i = 1, \ldots, n+m.$$

From (ii) we obtain,

$$t_i \sim t_{n+m+1} \qquad \text{if } 1 \le i \le n$$
$$t_j \sim t_{n+m+2} \qquad \text{if } n+1 \le j \le n+m.$$

The number of cones of dimension $(n+m) - p$ is equal to $\sum_{k+l=p} \binom{n+1}{n-k} \binom{m+1}{m-l} = \sum_{k+l=p} \binom{n+1}{k+1} \binom{m+1}{l+1}$. Denote by $t_{k,l} = t_{n+m+1}^k t_{n+m+2}^l$. Then

$$\prod_{k+l=p} \left(\frac{1}{1 - t_{k,l}} \right)^{\binom{n+1}{k+1} \binom{m+1}{l+1}} = E_p.$$

Example III: blow-up of $\mathbb{P}^n$ at a point

The fan $\tilde{\Delta}$ associated to the blow-up $\tilde{\mathbb{P}}^n$ of the projective space at the fixed point given by the cone $\mathbb{R}^+ e_2 + \cdots + \mathbb{R}^+ e_{n+1}$ is generated by $\{e_1, \ldots, e_{n+1}, e_{n+2}\}$ where $e_{n+2} = -e_1$. Denote by D_i the 1-dimensional cone $\mathbb{R}^+ e_i$ and by s_i its class in cohomology where

$$H^*(X, \mathbb{Z}) = \mathbb{Z}[s_1, \ldots, s_{n+2}]/I$$

and I is the ideal generated by

(i) $\{s_{i_1} \cdots s_{i_k} \mid D_{i_1} + \cdots + D_{i_k} \text{ is not in } \tilde{\Delta}\}$

(ii) $$\sum_{j=1}^{n+2} e_i^*(e_j) s_j \qquad i = 1, \ldots, n.$$

However, (ii) is equivalent to

$$s_2 \sim \cdots \sim s_3 \sim s_{n+1} \qquad \text{and} \qquad s_1 \sim s_{n+1} + s_{n+2}.$$

Note that a p-dimensional cone cannot contain both D_{n+2} and D_1. The reason is that D_{n+2} is generated by $-e_1$ and D_1 by e_1, but by definition a cone does not contain a subspace of dimension greater than 0. We would like to find a basis for $H^*(\tilde{\mathbb{P}}^n)$ and write any monomial of degree p in terms of it. Consider the monomial $s_{i_1} \cdots s_{i_p}$. There are three possible situations:

1. s_{i_j} is different from both s_{n+2} and s_{n+1}. In this situation we have from (ii) that $s_{i_1} \cdots s_{i_p} = s_{n+1}^p$.
2. s_{n+2} is equal to s_{i_j} for some $j = 1, \ldots, p$. Then from (ii) we obtain that $s_{i_1} \cdots s_{i_p} = s_{n+1}^{p-1} s_{n+2}$.
3. s_1 is equal to s_{i_j} for some $j = 1, \ldots, p$. Then from (ii) we obtain $s_{i_1} \cdots s_{i_p} = (s_{n+1} + s_{n+2}) s_{n+1}^{p+1} = s_{n+1}^p + s_{n+2} s_{n+1}^{p-1}$ which is the sum of 1 and 2.

We conclude that s_{n+1}^p and $s_{n+2} s_{n+1}^{p-1}$ form a basis for H^{2p} if $p < n$. If $p = n$ then $s_{n+1}^p = 0$ and the only generator is $s_{n+2} s_{n+1}^{p-1}$. Let us call s_{n+1} by t_1 and $s_{n+2} s_{n+1}^{p-1}$ by t_2. The Euler series for $\tilde{\mathbb{P}}^n$ is:

$$E_{n-p} = \left(\frac{1}{1 - t_1}\right)^{\binom{n}{p}} \left(\frac{1}{1 - t_1 t_2}\right)^{\binom{n}{p-1}} \left(\frac{1}{1 - t_2}\right)^{\binom{n}{p-1}} \qquad \text{if } p < n$$

$$E_0 = \left(\frac{1}{1 - t_2}\right)^{\binom{n+2}{n}}.$$

Example IV: Hirzebruch surfaces

A set of generators for the fan Δ that represents the Hirzebruch surface $X(\Delta)$ is given by $\{e_1, \ldots, e_4\}$ with $\{e_1, e_2\}$ the standard basis for $\mathbb{R}^2$, and $e_3 = -e_1 + a e_2$, and $e_4 = -e_2$. With the same notation as in the last examples, we have

$$H^*(X(\Delta)) = \mathbb{Z}[t_1, \ldots, t_4]/I$$

where I is generated by

(i) $\{t_1 t_3,\ t_2 t_4\}$

(ii) $\{t_1 - t_3,\ t_2 + a t_3 - t_4\}.$

From (ii) we have the following conditions for the t_i in $H^*(X)$

$$t_1 \sim t_3 \qquad \text{and} \qquad t_2 \sim (t_4 - a t_3). \tag{14}$$

A basis for $H^*(X)$ is given by $\{\{0\}, t_3, t_4, t_4 t_1\}$ (see [Dan78], [Ful93]). The Euler series for each dimension is as follows.

Dimension 0: there are four orbits (four cones of dimension 2), and all of them are equivalent in homology. From Theorem 1.2.2.8 we obtain

$$E_0 = \left(\frac{1}{1-t} \right)^4.$$

Dimension 1: again, there are four orbits (four cones of dimension 1), and the relation among them, in homology, is given by (14). From Theorem 1.2.2.8 we obtain

$$E_1 = \left(\frac{1}{1-t_3} \right)^2 \left(\frac{1}{1-t_4} \right) \left(\frac{1}{1-t_3^{-a} t_4} \right).$$

Dimension 2: the only orbit is the torus itself so

$$E_2 = \left(\frac{1}{1-t} \right).$$

1.2.3 Projective formulas

In this section we will follow [ELF98]. In particular the proofs for the main theorems can be found there.

In this section we exhibit a formula for the Euler–Chow function of certain projective bundles over a variety W, and compute several examples. The basic set-up is the following. Consider two algebraic vector bundles $E_1 \to W$ and $E_2 \to W$ over a complex projective variety W. The various maps involved in our discussion are displayed in the commutative diagram below:

$$
\begin{array}{ccccc}
\mathbb{P}(E_1) & \overset{i_1}{\hookrightarrow} & \mathbb{P}(E_1 \oplus E_2) & \overset{i_2}{\hookleftarrow} & \mathbb{P}(E_2) \\
 & {}_{p_1}\searrow & \downarrow q & \swarrow_{p_2} & \\
 & & W & &
\end{array}
\tag{15}
$$

where p_1, p_2 and q are projections from the indicated projective bundles, and i_1, i_2 are the canonical inclusions.

We will introduce a monoid monomorphism $t_p : \mathcal{C}_{p-1}\big(\mathbb{P}(E_1)\times_W\mathbb{P}(E_2)\big)$ $\longrightarrow \mathcal{C}_p\big(\mathbb{P}(E_1 \oplus E_2)\big)$ in Definition 1.2.3.3 below, which is a closed inclusion. In this way we are equipped with three morphisms of *monoids with finite multiplication*:

$$i_{1p} : \Pi_p\big(\mathbb{P}(E_1)\big) \longrightarrow \Pi_p\big(\mathbb{P}(E_1 \oplus E_2)\big) \tag{16}$$

induced by i_1,

$$i_{2p} : \Pi_p\big(E_2\big) \longrightarrow \Pi_p\big(\mathbb{P}(E_1 \oplus E_2)\big) \tag{17}$$

induced by i_2, and

$$\varphi_p : \Pi_{p-1}\big(\mathbb{P}(E_1)\times_W\mathbb{P}(E_2)\big) \longrightarrow \Pi_p\big(\mathbb{P}(E_1 \oplus E_2)\big) \tag{18}$$

induced by t_p.

These three maps induce a morphism (with finite fibers)

$$\Psi_p : \Pi_{p-1}\big(\mathbb{P}(E_1)\times_W\mathbb{P}(E_2)\big) \times \Pi_p\big(\mathbb{P}(E_1)\big) \times \Pi_p\big(\mathbb{P}(E_2)\big)$$
$$\longrightarrow \Pi_p\big(\mathbb{P}(E_1 \oplus E_2)\big), \tag{19}$$

by sending (a, b, c) to $\Psi_p(a, b, c) = \varphi_p(a) + i_{1p}(b) + i_{2p}(c)$.

The main result in this section is the following.

Theorem 1.2.3.1 *Let E_1 and E_2 be algebraic vector bundles over a connected projective variety W, of ranks e_1 and e_2, respectively, and let $0 \leq p \leq e_1 + e_2 - 1$. Then the pth Euler–Chow function of $\mathbb{P}(E_1 \oplus E_2)$ is given by*

$$E_p\big(\mathbb{P}(E_1 \oplus E_2)\big)$$
$$= \Psi_{p\sharp}\big(E_{p-1}\big(\mathbb{P}(E_1)\times_W\mathbb{P}(E_2)\big) \odot E_p\big(\mathbb{P}(E_1)\big) \odot E_p\big(\mathbb{P}(E_2)\big)\big). \tag{20}$$

Remark 1.2.3.2 In the case $\dim X < p$, the Chow monoid $\mathcal{C}_p(X)$ consists of the zero element only, which is the cycle with empty support. Therefore $\Pi_p(X) = \{0\}$ and $E_p(X) \equiv 1 \in \mathbb{Z}^{\{0\}} \equiv \mathbb{Z}$.

Before providing examples, we must define the map t_p appearing in the formulas above. Let L_1 and L_2 denote the tautological line bundles $\mathcal{O}_{E_1}(-1)$ and $\mathcal{O}_{E_2}(-1)$ over $\mathbb{P}(E_1)$ and $\mathbb{P}(E_2)$, respectively, and let π_1 and π_2 denote the respective projections from $\mathbb{P}(E_1)\times_W\mathbb{P}(E_2)$ onto $\mathbb{P}(E_1)$ and $\mathbb{P}(E_2)$. The $\mathbb{P}^1$-bundle $\pi : \mathbb{P}\big(\pi_1^*(L_1) \oplus \pi_2^*(L_2)\big) \to \mathbb{P}(E_1)\times_W\mathbb{P}(E_2)$ is precisely the blow-up of $\mathbb{P}(E_1 \oplus E_2)$ along $\mathbb{P}(E_1) \amalg \mathbb{P}(E_2)$, which we denote by Q, for short; see Lascou and Scott [LS75] for details. Let $b : Q \to \mathbb{P}(E_1 \oplus E_2)$ denote the blow-up map.

Since π is a flat map of relative dimension 1, and b is a proper map, one has two algebraic continuous homomorphisms (cf. Properties 1.2.1.8), given by the flat pull-back

$$\pi^* : \mathcal{C}_{p-1}\big(\mathbb{P}(E_1)\times_W\mathbb{P}(E_2)\big) \to \mathcal{C}_p(Q) \tag{21}$$

and the proper push-forward

$$b_* : \mathcal{C}_p(Q) \to \mathcal{C}_p\big(\mathbb{P}(E_1 \oplus E_2)\big). \tag{22}$$

Definition 1.2.3.3 The map $t_p : \mathcal{C}_{p-1}\big(\mathbb{P}(E_1)\times_W\mathbb{P}(E_2)\big) \to \mathcal{C}_p\big(\mathbb{P}(E_1 \oplus E_2)\big)$ is defined as the composition $t_p = b_* \circ \pi^*$.

The following lemma will be used as a reference in a few places throughout the paper.

Lemma 1.2.3.4 *Let* $X = \bigsqcup_{i\in\mathbb{N}} X_i$ *and* $Y = \bigsqcup_{j\in\mathbb{N}} Y_j$ *be spaces which are a countable disjoint union of connected projective varieties, and let* $f : X \to Y$ *be a continuous map such that the restriction* $f_{|X_i}$ *is an algebraic continuous map from* X_i *into some* Y_j. *If* f *is a bijection, then it is a homeomorphism in the classical topology.*

Proof Given $j \in \mathbb{N}$, one can write $f^{-1}(Y_j)$ as a disjoint union $\bigsqcup_{k\in\Lambda} X_{i_k}$, for a certain collection of connected components of X. On the other hand, since f is an algebraic continuous map, then $f(X_{i_k})$ is a Zariski closed subset of Y_j, and since f is one-to-one, then Y_j is written as a union of a (at most countable) family of disjoint closed algebraic subsets. This contradicts the connectedness of Y_j, unless there is only one $j(i)$ such that $f^{-1}(Y_j) = X_{j(i)}$, in which case f sends $X_{j(i)}$ homeomorphically onto Y_j. $\square$

Projective closure of line bundles

Here we consider the case where $E_1 = E$ is a line bundle over a projective variety W, and $E_2 = \mathbf{1}_W$ is the trivial line bundle. In this case,

$$\mathbb{P}(E_1) = \mathbb{P}(E_2) = \mathbb{P}(E_1) \times_W \mathbb{P}(E_2) = W,$$

and the inclusions $i_k : \mathbb{P}(E_k) \hookrightarrow \mathbb{P}(E_1 \oplus E_2)$, $k = 1, 2$ (cf. diagram (15)), become sections $i : W \to \mathbb{P}(E \oplus \mathbf{1})$ and $j : W \to \mathbb{P}(E \oplus \mathbf{1})$ of the projective bundle $\mathbb{P}(E \oplus \mathbf{1})$ over W.

If $\xi = c_1\big(\mathcal{O}_{E\oplus 1}(1)\big)$ denotes the first Chern class of the canonical bundle over $\mathbb{P}(E \oplus \mathbf{1})$, then the map

$$T : \mathcal{A}_{p-1}(W) \oplus \mathcal{A}_p(W) \to \mathcal{A}_p\big(\mathbb{P}(E \oplus \mathbf{1})\big)$$

$$(\alpha, \beta) \longmapsto q^*\alpha + \xi \cap q^*\beta \tag{23}$$

is an isomorphism. Note that, since $\xi \cap q^*\beta = i_*\beta$, the isomorphism above restricts to an injection

$$T^\geq : \mathcal{A}^\geq_{p-1}(W) \oplus \mathcal{A}^\geq_p(W) \to \mathcal{A}^\geq_p(\mathbb{P}(E \oplus 1)). \tag{24}$$

Lemma 1.2.3.5 *Let E be a line bundle over W which is generated by its global sections. Then:*

(i) *The injection $T^\geq$ is an isomorphism.*
(ii) *If $\Pi_*(W)$ are monoids with cancellation law for every $*$, then so are $\Pi_*(\mathbb{P}(E \oplus 1))$. Equivalently, if the natural surjections $\Pi_p(W) \to \mathcal{A}^\geq_p(W)$ are isomorphisms for all p, then so are the surjections $\Pi_p(\mathbb{P}(E \oplus 1)) \to \mathcal{A}^\geq_p(\mathbb{P}(E \oplus 1))$.*

We have the following useful corollary.

Corollary 1.2.3.6 *Under the same hypothesis, the homomorphism*

$$\Psi_p : \Pi_{p-1}(W) \oplus \Pi_p(W) \oplus \Pi_p(W)$$
$$\to \Pi_p(\mathbb{P}(E \oplus 1)) \cong \Pi_{p-1}(W) \oplus \Pi_p(W)$$

sends (α, β, γ) to $(\alpha + \beta, \xi \cap \gamma + \alpha)$.

Now we can pass to some examples.

Example 1.2.3.7 Let $W = \mathbb{P}^n$ and $E = \mathcal{O}_{\mathbb{P}^n}(d)$, with $d \geq 0$. We will compute $E_p(\mathbb{P}(\mathcal{O}_{\mathbb{P}^n}(d) \oplus 1))$ for $1 \leq p \leq n$. First observe that W and E satisfy the hypothesis of Lemma 1.2.3.5, and that

$$\Pi_{p-1}(\mathbb{P}(E)) \cong \Pi_p(\mathbb{P}^n) = \mathbb{Z}_+ \cdot [\mathbb{P}^{p-1}] \cong \mathbb{Z}_+ \qquad \text{and}$$
$$\Pi_p(\mathbb{P}(E)) \cong \Pi_p(\mathbb{P}^n) \cong \mathbb{Z}_+ \cdot [\mathbb{P}^p] \cong \mathbb{Z}_+,$$

where $[\mathbb{P}^p]$ denotes the class of a p-dimensional linear subspace of $\mathbb{P}^n$. By definition, the map Ψ_p fits into a commutative diagram

$$
\begin{array}{ccc}
\Pi_{p-1}(\mathbb{P}(\mathcal{O}_{\mathbb{P}^n}(d))) \times \Pi_p(\mathbb{P}(\mathcal{O}_{\mathbb{P}^n}(d))) \times \Pi_p(\mathbb{P}^n) & \xrightarrow{\Psi_p} & \Pi_p(\mathbb{P}(\mathcal{O}_{\mathbb{P}^n}(d) \oplus 1)) \\
\| & & \cong \downarrow T^\geq \\
\Pi_{p-1}(\mathbb{P}^n) \times \Pi_p(\mathbb{P}^n) \times \Pi_p(\mathbb{P}^n) & \xrightarrow{\Psi_p} & \Pi_{p-1}(\mathbb{P}^n) \times \Pi_p(\mathbb{P}^n)
\end{array}
$$

and sends $\Psi_p : (a[\mathbb{P}^{p-1}], b[\mathbb{P}^p], c[\mathbb{P}^p]) \longmapsto (a[\mathbb{P}^{p-1}] + c \cdot c_1(\mathcal{O}_{\mathbb{P}^n}(d)) \cap [\mathbb{P}^p], (b+c)[\mathbb{P}^p])$. Since $c_1(\mathcal{O}_{\mathbb{P}^n}(d)) \cap [\mathbb{P}^p] = d[\mathbb{P}^{p-1}]$ we then identify Ψ_p with $(a, b, c) \mapsto (a + c \cdot d, b + c)$.

One can associate variables t_0, t_1 to the generators of $\Pi_p\big(\mathbb{P}(\mathcal{O}_{\mathbb{P}^n}(d) \oplus \mathbf{1})\big) \cong \mathbb{Z}_+ \oplus \mathbb{Z}_+$, and identify $(\alpha_0, \alpha_1) \in \Pi_p\big(\mathbb{P}(\mathcal{O}_{\mathbb{P}^n}(d) \oplus \mathbf{1})\big)$ with $t_0^{\alpha_0} t_1^{\alpha_1}$. It follows from Theorem 1.2.3.1 and Remark 1.2.1.3 that $E_p\big(\mathbb{P}(\mathcal{O}_{\mathbb{P}^n}(d) \oplus \mathbf{1})\big)$ can be written as

$$E_p(\mathbb{P}(\mathcal{O}_{\mathbb{P}^n}(d) \oplus \mathbf{1}))$$

$$= \sum_{\alpha_0, \alpha_1} E_p\big(\mathbb{P}(\mathcal{O}_{\mathbb{P}^n}(d) \oplus \mathbf{1})\big)(\alpha_0, \alpha_1) t_0^{\alpha_0} t_1^{\alpha_1}$$

$$= \sum_{\alpha_0, \alpha_1} \Psi_{p\#}\big(E_{p-1}\big(\mathbb{P}(\mathcal{O}_{\mathbb{P}^n}(d))\big)\big) \odot E_p\big(\mathbb{P}(\mathcal{O}_{\mathbb{P}^n}(d))\big) \odot E_p(\mathbb{P}^n)\,(\alpha_0, \alpha_1) t_0^{\alpha_0} t_1^{\alpha_1}$$

$$= \sum_{\alpha_0, \alpha_1} \left(\sum_{(a,b,c) \in \Psi_p^{-1}(\alpha_0,\alpha_1)} E_{p-1}(\mathbb{P}^n)(a) \cdot E_p(\mathbb{P}^n)(b) \cdot E_p(\mathbb{P}^n)(c) \right) t_0^{\alpha_0} t_1^{\alpha_1}$$

$$= \sum_{\alpha_0, \alpha_1} \left(\sum_{\substack{a+cd=\alpha_0 \\ b+c=\alpha_1}} E_{p-1}(\mathbb{P}^n)(a) \cdot E_p(\mathbb{P}^n)(b) \cdot E_p(\mathbb{P}^n)(c) \right) t_0^{a+cd} t_1^{b+c}$$

$$= \left(\sum_{a\geq 0} E_{p-1}(\mathbb{P}^n)(a) \cdot t_0^a \right)\left(\sum_{b\geq 0} E_p(\mathbb{P}^n)(b) \cdot t_1^b \right)\left(\sum_{c\geq 0} E_p(\mathbb{P}^n)(c) \cdot (t_0^d t_1)^c \right).$$

In Lawson and Yau [LY87] it was shown that

$$\sum_{k\geq 0} E_p(\mathbb{P}^n)(k)\, t^k = \left(\frac{1}{1-t} \right)^{\binom{n+1}{p+1}}$$

and hence $E_p\big(\mathbb{P}(\mathcal{O}_{\mathbb{P}^n}(d) \oplus \mathbf{1})\big)$ is then written as:

$$E_p\big(\mathbb{P}(\mathcal{O}_{\mathbb{P}^n}(d) \oplus \mathbf{1})\big) = \left(\frac{1}{1-t_0} \right)^{\binom{n+1}{p}} \left(\frac{1}{1-t_1} \right)^{\binom{n+1}{p+1}} \left(\frac{1}{1-t_0^d t_1} \right)^{\binom{n+1}{p+1}}.$$

$$(25)$$

Subexamples 1.2.3.8

1. When $d = 0$ one has $\mathbb{P}(\mathcal{O}_{\mathbb{P}^n}(d) \oplus \mathbf{1}) = \mathbb{P}^n \times \mathbb{P}^1$ and our computations recover the formula

$$E_p(\mathbb{P}^n \times \mathbb{P}^1) = \left(\frac{1}{1-t_0} \right)^{\binom{n+1}{p}} \left(\frac{1}{1-t_1} \right)^{2\binom{n+1}{p+1}}$$

obtained in [LY87]. Furthermore, our general formula in Theorem 1.2.3.1 gives an inductive process to compute $E_p(\mathbb{P}^n \times \mathbb{P}^m)$ in general.

2. When $n = 1$, the $\mathbb{P}^1$-bundle $\mathbb{P}(\mathbb{O}_{\mathbb{P}^n}(d) \oplus 1)$ over $\mathbb{P}^1$ is the Hirzebruch surface $\mathbb{F}_d$, and the formula

$$E_1(\mathbb{F}_d) = \left(\frac{1}{1 - t_0}\right)^2 \left(\frac{1}{1 - t_1}\right)\left(\frac{1}{1 - t_0^d t_1}\right)$$

recovers that obtained in Elizondo [Eli94].

3. When $d = 1$, the $\mathbb{P}^1$-bundle $\mathbb{P}(\mathbb{O}_{\mathbb{P}^n}(1) \oplus 1)$ over $\mathbb{P}^n$ is just the blow-up $\widetilde{\mathbb{P}}^{n+1}$ of $\mathbb{P}^{n+1}$ at a point, and the expression

$$E_p(\widetilde{\mathbb{P}}^{n+1}) = \left(\frac{1}{1 - t_0}\right)^{\binom{n+1}{p}} \left(\frac{1}{1 - t_1}\right)^{\binom{n+1}{p+1}} \left(\frac{1}{1 - t_0 t_1}\right)^{\binom{n+1}{p+1}}$$

recovers the formula obtained in [Eli94].

1.2.4 Chow quotients and Euler–Chow series

In this section we consider a projective algebraic variety X equipped with an algebraic action of $\mathbb{C}^*$. In general, an algebraic action of a torus $\mathbb{T} = (\mathbb{C}^*)^k$ provides stratifications on X, and here we introduce the following, much in the spirit of Kapranov [Kap93].

Definition 1.2.4.1 Let $\mathbb{T} = (\mathbb{C}^*)^k$ act algebraically on X and let $\alpha \in \Pi_p(X)$ be a fixed element with $0 \leq p \leq k$. We say that a p-dimensional orbit $\mathbb{T} \cdot x$ is of *type* α if its closure $\overline{\mathbb{T} \cdot x}$ lies in the component $\mathcal{C}_{p,\alpha}(X)$, when viewed as a p-dimensional effective cycle. One can stratify X according to the orbit type of its elements, introducing the *Chow stratification* of X.

Assume that $\mathbb{T}$ has orbits of maximal dimension k. It is easy to see that when X is irreducible then there is a unique maximal open stratum X^o consisting of points whose orbits have maximal dimension k and are of the same type α_0, for some $\alpha_0 \in \Pi_k(X)$. In particular, there is an embedding $X^o/\mathbb{T} \hookrightarrow \mathcal{C}_{k,\alpha_0}(X)$ and following [GKZ94] we introduce the following notion.

Definition 1.2.4.2 The *Chow quotient* $X/\!\!/\mathbb{T}$ is the weak normalization of the closure of $X^o/\mathbb{T}$ in $\mathcal{C}_{k,\alpha_o}(X)$.

Remark 1.2.4.3 When $X^o/\mathbb{T}$ is proper, the Chow quotient gives a closed embedding of $X^o/\mathbb{T}$ into the appropriate Chow variety.

We combine the use of Chow quotients with the notion of *trace maps* introduced in Friedlander and Lawson [FL92, Section 7.1], which is described as follows. A Chow variety $\mathcal{C}_{k,\alpha}(X)$, being projective, has its own Chow monoid

$\mathcal{C}_p\big(\mathcal{C}_{k,\alpha}(X)\big)$. The trace map is a continuous monoid morphism

$$tr : \mathcal{C}_p\big(\mathcal{C}_{k,\alpha}(X)\big) \to \mathcal{C}_{p+k}(X) \tag{26}$$

that, roughly speaking, associates to an irreducible p-dimensional subvariety $W \subset \mathcal{C}_{k,\alpha}(X)$ its 'total fundamental cycle', which is a $(p+k)$-cycle in X.

Combining the two constructions above, one associates to such a $\mathbb{T}$-action on an irreducible variety X, a trace map

$$t_p : \mathcal{C}_p\big(X /\!\!/ \mathbb{T}\big) \to \mathcal{C}_{p+k}(X). \tag{27}$$

This map, in turn, induces a monoid morphism

$$\varphi_p : \Pi_p\big(X /\!\!/ \mathbb{T}\big) \to \Pi_{p+k}(X), \tag{28}$$

in the level of π_0. In other words, $t_p(\mathcal{C}_{p,\alpha}(X /\!\!/ \mathbb{T})) \subset \mathcal{C}_{p+k,\varphi_p(\alpha)}(X)$. Explicit examples of such monoid morphisms are given in Proposition 1.2.4.11 and Corollary 1.2.4.12 below.

Proposition 1.2.4.4 *Let* $\mathbb{T} \cong (\mathbb{C}^*)^k$ *act on an irreducible projective variety* X *with generic orbits of maximal dimension k. If* $\dim(X /\!\!/ \mathbb{T} - X^o/\mathbb{T}) < p \le \dim X - k$ *then the trace map* $t_p : \mathcal{C}_p\big(X /\!\!/ \mathbb{T}\big) \to \mathcal{C}_{p+k}(X)$ *is injective.*

Proof Let $W \subset X /\!\!/ \mathbb{T}$ be an irreducible variety of dimension p. The hypothesis on dimensions implies that $W^o := W \cap X^o/\mathbb{T}$ is an open dense subvariety of W. In order to define the trace $t_p(W)$ one considers the cycle $Z_W \subset W \times X$, defined as the 'closure of the cycle'

$$Z_W^o = \{(w, c_w) \mid w \in W^o \text{ and } c_w = \text{ cycle whose Chow point is } w\},$$

cf. [FL92, Section 7.1]. By definition, $t_p(W) = pr_{2*}(Z_W)$, where pr_2 is the projection onto the second factor of $W \times X$.

Note that, since the fibers of the projection $Z_W^o \to W^o$ are irreducible and W^o is irreducible, then Z_W is irreducible of dimension $p + k \le \dim X$. Furthermore, if $(a, b), (a', b) \in Z_W^o$ then $a = a'$, for if $a, a' \in W^o \subset X^o/\mathbb{T}$ then the orbits in X represented by a and a' have a common point b, hence $a = a'$. It follows that pr_2 maps Z_W birationally onto its image, and thus $t_p(W) := pr_{2*}(Z_W)$ is an irreducible cycle with multiplicity 1.

Suppose that $t_p(W) = t_p(W')$, where both W and W' are irreducible. An element $w \in W^o$ corresponds to an orbit $\mathbb{T} \cdot x \subset pr_2(Z_W) = pr_2(Z_{W'})$, with $x \in X^o$. Therefore $w \in W'^o$, and hence $W = W'$. Since Chow monoids are freely generated by the irreducible subvarieties, one concludes that the trace map t_p is injective. $\square$

We now consider the situation where X is an irreducible projective variety, on which $\mathbb{C}^*$ acts in such a fashion that the fixed point set $X^{\mathbb{C}^*}$ has only two connected components X_1 and X_2. Let $i_1 : X_1 \hookrightarrow X$ and $i_2 : X_2 \hookrightarrow X$ denote the inclusion maps, and let $t_{p-1} : \mathcal{C}_{p-1}(X /\!/ \mathbb{C}^*) \to \mathcal{C}_p(X)$ be the trace morphism (28). These maps induce a monoid morphism

$$\Psi_p : \Pi_{p-1}(X /\!/ \mathbb{C}^*) \times \Pi_p(X_1) \times \Pi_p(X_2) \to \Pi_p(X)$$
$$(\alpha, \beta, \gamma) \longmapsto \varphi_{p-1}(\alpha) + i_{1*}\beta + i_{2*}\gamma, \quad (29)$$

which yields our next result.

Theorem 1.2.4.5 *Let X be an smooth projective variety on which $\mathbb{C}^*$ acts algebraically. If $X^{\mathbb{C}^*}$ is the union of two connected components X_1 and X_2, then for each $0 \le p \le \dim X$ one has*

$$E_p(X) = \Psi_{p\sharp}(E_{p-1}(X /\!/ \mathbb{C}^*) \odot E_p(X_1) \odot E_p(X_2)). \quad (30)$$

Remark 1.2.4.6 One can prove directly that the Chow quotient $\mathbb{P}(E_1 \oplus E_2) /\!/ \mathbb{C}^*$ is isomorphic to $\mathbb{P}(E_1) \times_W \mathbb{P}(E_2)$, whenever E_1 and E_2 are algebraic vector bundles over a variety W. In this case, Theorem 1.2.3.1 becomes a consequence of the result above whenever W is smooth.

Next, we describe some examples of Chow quotients, trace maps and resulting computations of Euler–Chow series.

Examples

The linear action of $\mathbb{C}^*$ on the last coordinate of $\mathbb{C}^{n+1}$ induces an algebraic $\mathbb{C}^*$ action on the Grassmannian $G(d, n)$ of d-planes in $\mathbb{P}^n$. More generally, one obtains an algebraic action on all partial flag varieties $F(d_1, \ldots, d_r; \mathbb{P}^n)$ of nested linear subspaces $D_1 \subset \cdots \subset D_r$ of $\mathbb{P}^n$, satisfying $\dim D_i = d_i$.

We first describe the orbit structure and the Chow quotient of both $G(d, n)$ and $F(0, 1; \mathbb{P}^n)$, under this $\mathbb{C}^*$ action.

Grassmannian case

We adopt the convention that $G(d, n) = \emptyset$, whenever $d < 0$, and that $\chi(\emptyset) = 0$. Fix $p_\infty = [0 : \cdots : 0 : 1] \in \mathbb{P}^n$, and let $L \in G(d, n)$ be fixed under the above action. Then the corresponding d-dimensional subspace $L \subset \mathbb{P}^n$ can be of two types:

1. either L is contained in $\mathbb{P}^{n-1} = \{[x_0 : \cdots : x_n] \in \mathbb{P}^n \mid x_n = 0\}$,
2. or L has the form $p_\infty \# \ell$, where ℓ is a $(d-1)$-dimensional subspace of $\mathbb{P}^{n-1}$, and $p_\infty \# \ell$ denotes the ruled join of p_∞ and ℓ in $\mathbb{P}^n$.

In other words, the fixed point set $G(d, n)^{\mathbb{C}^*}$ has two connected components which are naturally isomorphic to $G(d, n-1)$ and $G(d-1, n-1)$, and whose inclusions in $G(d, n)$ are denoted by $i : G(d, n-1) \hookrightarrow G(d, n)$ and $j : G(d-1, n-1) \hookrightarrow G(d, n)$, respectively. In particular, this $\mathbb{C}^*$ action on $G(d, n)$ satisfies the hypothesis of Theorem 1.2.4.5. Furthermore, all points in the generic locus $G(d, n)^o = G(d, n) - G(d, n)^{\mathbb{C}^*}$ have the same orbit type.

Proposition 1.2.4.7 *The Chow quotient $G(d, n) /\!\!/ \mathbb{C}^*$ is naturally isomorphic to the flag variety $F(d-1, d; \mathbb{P}^{n-1})$.*

Proof Let $\pi : \mathbb{P}^n - \{p_\infty\} \to \mathbb{P}^{n-1}$ denote the projection onto the hyperplane $\mathbb{P}^{n-1}$. Standard arguments show that

$$q : G(d, n)^o \to F(d-1, d; \mathbb{P}^{n-1})$$
$$L \mapsto (L \cap \mathbb{P}^{n-1}, \pi(L)) \tag{31}$$

is a regular, surjective map. This map descends to the quotient $G(d, n)^o / \mathbb{C}^*$ and produces a closed inclusion $G(d, n) /\!\!/ \mathbb{C}^* \equiv G(d, n)^o / \mathbb{C}^* \hookrightarrow F(d-1, d; \mathbb{P}^{n-1})$; cf. Proposition 1.2.4.4 and Remark 1.2.4.3. This is then easily seen to be an isomorphism. $\qquad\qquad\square$

Corollary 1.2.4.8 *The pth Euler–Chow series of the Grassmannian $G(d, n)$ is given by*

$$E_p\big(G(d, n)\big) = \Psi_{p\sharp}\big(E_{p-1}\big(F(d-1, d; \mathbb{P}^{n-1})\big) \odot E_p\big(G(d, n-1)\big)$$
$$\odot E_p\big(G(d-1, n-1)\big)\big),$$

where Ψ_p is given by (29).

We proceed to describe Ψ_p explicitly in this example. We follow closely the projective notation for the Schubert cycles in Grassmanians and flag varieties, as described in Fulton [Ful98, Section 14.7]. However, since we are dealing with various projective spaces, we add an upperscript in the notation to denote the dimension of the ambient projective space.

First, fix a complete flag $L_0 \subset L_1 \subset \cdots \subset L_n = \mathbb{P}^n$ of linear subspaces, and associate to a sequence of length d $\mathbf{a} : 0 \leq a_0 < a_1 < \cdots < a_d \leq n$, the Schubert variety $\Omega_{\mathbf{a}}^n = \Omega_{a_0, \ldots, a_d}^n \subset G(d, n)$ defined by

$$\Omega_{\mathbf{a}}^n = \{l \in G(d, n) \mid \dim(l \cap L_i) \geq a_i, i = 0, \ldots, d\}.$$

Now, let $\{\mathbf{a}^1 \subset \ldots \subset \mathbf{a}^r\}$ be a nested collection of sequences, such that the length of $\mathbf{a}^i$ is d_i, and define the Schubert variety $\Omega_{\mathbf{a}^1; \ldots; \mathbf{a}^r}^n \subset F(d_1, \ldots, d_r; \mathbb{P}^n)$ by

$$\Omega_{\mathbf{a}^1; \ldots; \mathbf{a}^r}^n = \big\{(l_1, \ldots, l_r) \in F(d_1, \ldots, d_r; \mathbb{P}^n) \mid l_i \in \Omega_{\mathbf{a}^i}^n\big\}.$$

Facts 1.2.4.9

1. The Schubert variety $\Omega^n_{\mathbf{a}^1;\dots;\mathbf{a}^r}$ is an irreducible subvariety of $F(d_1,\dots,$ $d_r;\mathbb{P}^n)$ of dimension $d(\mathbf{a}^1;\dots;\mathbf{a}^r) = \sum_{i=1}^r \sum_{j=0}^{d_i}(a^i_j - j)$. In this sum, any term $(a^i_j - j)$ is omitted if a^i_j appears in the previous $\mathbf{a}^{i-1}$.

2. The surjection $\Pi_p\big(F(d_1,\dots,d_r;\mathbb{P}^n)\big) \to \mathcal{A}^{\geq}_p\big(F(d_1,\dots,d_r;\mathbb{P}^n)\big)$ is an isomorphism. Therefore, if $\omega^n_{\mathbf{a}^1;\dots;\mathbf{a}^r}$ denotes the class of $\Omega^n_{\mathbf{a}^1;\dots;\mathbf{a}^r}$ in $\Pi_*\big(F(d_1,\dots,d_r;\mathbb{P}^n)\big)$, then the monoid $\Pi_p\big(F(d_1,\dots,d_r;\mathbb{P}^n)\big)$ is freely generated by the collection

$$\left\{\omega^n_{\mathbf{a}^1;\dots;\mathbf{a}^r} \,\middle|\, d(\mathbf{a}^1;\dots;\mathbf{a}^r) = p\right\}.$$

Remark 1.2.4.10 Let $\mathbf{a}$ be the sequence $0 < 1 < \cdots < d - 1 < d + 1$. Then, it is easy to see that the class of a 1-dimensional orbit $[\overline{\mathbb{T} \cdot x}] \in \Pi_1\big(G(d,n)\big)$, is precisely $\omega^n_{\mathbf{a}}$.

Consider the universal bundles U_d and U_{d+1} over the flag variety $F(d - 1, d; \mathbb{P}^{n-1})$, of ranks d and $d + 1$, respectively. Then, denote the quotient line bundle U_{d+1}/U_d by L, and let $\rho : \mathbb{P}(L \oplus 1) \to F(d - 1, d; \mathbb{P}^{n-1})$ be the bundle projection. The following result describes the trace map $t_p :$ $\mathcal{C}_p\big(F(d - 1, d; \mathbb{P}^{n-1})\big) \to \mathcal{C}_{p+1}\big(G(d,n)\big)$, in a very explicit and geometric manner.

Proposition 1.2.4.11

1. *The blow-up of $G(d,n)$ along the fixed point set $G(d,n)^{\mathbb{C}^*} = G(d - 1, n - 1) \amalg G(d, n - 1)$ is the variety $\mathbb{P}(L \oplus 1)$.*

2. *Let $b : \mathbb{P}(L \oplus 1) \to G(d,n)$ denote the blow-up map. Then $\phi_p : \mathcal{C}_p\big(F(d - 1, d; \mathbb{P}^{n-1})\big) \to \mathcal{C}_{p+1}\big(G(d,n)\big)$ is the composition $b_* \circ \rho^*$, where b_* denotes the proper push-forward under b, and ρ^* is the flat pull-back.*

The induced map $\varphi_p : \Pi_p\big(F(d - 1, d; \mathbb{P}^{n-1})\big) \to \Pi_{p+1}\big(G(d,n)\big)$ can be computed either from the proposition or by a direct argument.

Corollary 1.2.4.12 *Given a sequence $0 \leq a_1 < \cdots < a_d \leq n - 1$ and $0 \leq j \leq d$, one has*

$$\varphi_p\big(\omega^{n-1}_{a_0,\dots,\widehat{a_j},\dots,a_d;a_0,\dots,a_d}\big) = \omega^n_{a_0,\dots,a_{j-1},a_j+1,\dots,a_d+1}. \tag{32}$$

The flag variety $F(0, 1; \mathbb{P}^n)$

The orbit structure of the $\mathbb{C}^*$ action on $F(0, 1; \mathbb{P}^n)$ is described in the following statement. Here we denote by $Q_{d,n}$ and $S_{d,n}$ the universal quotient bundle and the tautological bundle over $G(d, n)$, respectively.

Proposition 1.2.4.13

1. *The fixed point set* $F(0, 1; \mathbb{P}^n)^{\mathbb{C}^*}$ *has three connected components* F_1, F_2 *and* F_3, *respectively isomorphic to* $F(0, 1; \mathbb{P}^{n-1})$, $\mathbb{P}^{n-1}$ *and* $\mathbb{P}^{n-1}$.
2. *There are two* $\mathbb{C}^*$-*invariant subvarieties* W_1 *and* W_2 *of* $F(0, 1; \mathbb{P}^n)$ *which are equivariantly isomorphic to the projective bundles* $\mathbb{P}(Q_{0,n-1} \oplus 1)$ *and* $\mathbb{P}(S_{0,n-1} \oplus 1)$ *over* $\mathbb{P}^{n-1}$, *respectively. Furthermore,* W_1 *and* W_2 *are Schubert cycles for an appropriate choice of coordinates, such that their Schubert classes are given by* $[W_1] = \omega^n_{n-1;n-1,n}$ *and* $[W_2] = \omega^n_{n;0,n}$.
3. *The component* F_1 *is precisely* $\mathbb{P}(Q_{0,n-1}) \subset \mathbb{P}(Q_{0,n-1} \oplus 1)$. *The strata* W_1 *and* W_2 *intersect at* F_2, *where* $F_2 \equiv \mathbb{P}(1) \subset \mathbb{P}(Q_{0,n-1} \oplus 1)$ *and* $F_2 \equiv \mathbb{P}(S_{0,n-1}) \subset \mathbb{P}(S_{0,n-1} \oplus 1)$. *The component* F_3 *is identified with* $\mathbb{P}(1) \subset \mathbb{P}(S_{0,n-1} \oplus 1)$.

Applying [Bia73, Lemma 4.1] to our situation, one concludes that the closure of a 1-dimensional orbit $\mathbb{C}^* \cdot x$ in $F(0, 1; \mathbb{P}^n)$ intersects precisely two connected components of the fixed point set. Furthermore, an application of [Bia73, Theorems 4.1 and 4.3] implies that these two components completely determine the *type* of the orbit; cf. Definition 1.2.4.1.

There are three types of 1-dimensional orbits.

Type 1: the 1-dimensional orbits contained in the stratum W_1. These are the orbits of the points in $W_1 - \{F_1 \cup F_2\}$.

Type 2: the 1-dimensional orbits contained in the stratum W_2. These are the orbits of the points in $W_2 - \{F_2 \cup F_3\}$.

Generic type: these are the orbits of points in the *generic locus* $F(0, 1; \mathbb{P}^n)^o = F(0, 1; \mathbb{P}^n) - \{W_1 \cup W_2\}$ of the action.

Proposition 1.2.4.14

1. *The Chow quotient* $W_1 /\!\!/ \mathbb{C}^*$ *is naturally isomorphic to* F_1.
2. *The Chow quotient* $W_2 /\!\!/ \mathbb{C}^*$ *is naturally isomorphic to* F_3.
3. *The Chow quotient* $F(0, 1; \mathbb{P}^n) /\!\!/ \mathbb{C}^*$ *is naturally isomorphic to* $\mathbb{P}^{n-1}[2]$, *the blow-up of* $\mathbb{P}^{n-1} \times \mathbb{P}^{n-1}$ *along the diagonal.*

We now compute the Euler–Chow series for the flag variety $F(0, 1; \mathbb{P}^2)$, using the above information. In this case one has

$$F_1 \cong W_1 /\!\!/ \mathbb{C}^* \cong F_2 \cong F_3 \cong W_2 /\!\!/ \mathbb{C}^* \cong \mathbb{P}^1,$$

and

$$F(0, 1; \mathbb{P}^2) /\!\!/ \mathbb{C}^* \cong \mathbb{P}^1[2] = \mathbb{P}^1 \times \mathbb{P}^1.$$

It follows from the isomorphisms $\Pi_1\big(F(0, 1; \mathbb{P}^2)\big) \cong \mathbb{Z}_+ \cdot \omega^2_{1;0,1} \oplus \mathbb{Z}_+ \cdot \omega^2_{0;0,2}$
$\cong \mathbb{Z}_+ \oplus \mathbb{Z}_+$, that we may denote the connected components of the Chow
monoid $\mathcal{C}_1\big(F(0, 1; \mathbb{P}^2)\big)$ by $\mathcal{C}_{1,(r,s)}\big(F(0, 1; \mathbb{P}^2)\big)$, where $(r, s) \in \mathbb{Z}_+ \oplus \mathbb{Z}_+$.

The Chow quotients described above induce inclusions:

1. $\mathbb{P}^1 \cong W_1 /\!/ \mathbb{C}^* \subset \mathcal{C}_1\big(F(0, 1; \mathbb{P}^2)\big)$,
2. $\mathbb{P}^1 \cong W_2 /\!/ \mathbb{C}^* \subset \mathcal{C}_1\big(F(0, 1; \mathbb{P}^2)\big)$, and
3. $\mathbb{P}^1 \times \mathbb{P}^1 \cong F(0, 1; \mathbb{P}^2) /\!/ \mathbb{C}^* \subset \mathcal{C}_1\big(F(0, 1; \mathbb{P}^2)\big)$,

whose associated trace maps are given as follows.

Lemma 1.2.4.15

1. $\psi_1 : \mathcal{C}_{1,d}\big(\mathbb{P}^1\big) \equiv \mathcal{C}_{1,d}\big(W_1 /\!/ \mathbb{C}^*\big) \to \mathcal{C}_{2,(d,0)}\big(F(0, 1; \mathbb{P}^2)\big)$, *sends* $d \cdot \mathbb{P}^1$ *to* $d \cdot W_1$.
2. $\psi_2 : \mathcal{C}_{1,d}\big(\mathbb{P}^1\big) \equiv \mathcal{C}_{1,d}\big(W_2 /\!/ \mathbb{C}^*\big) \to \mathcal{C}_{1,(0,d)}\big(F(0, 1; \mathbb{P}^2)\big)$, *sends* $d \cdot \mathbb{P}^1$ *to* $d \cdot W_2$.
3. *Let* $t_1 : \mathcal{C}_{1,(r,s)}\big(\mathbb{P}^1 \times \mathbb{P}^1\big) \to \mathcal{C}_{2,(r,s)}\big(F(0, 1; \mathbb{P}^2)\big)$ *be the trace map induced by the Chow quotient* $\mathbb{P}^1 \times \mathbb{P}^1 = F(0, 1; \mathbb{P}^2) /\!/ \mathbb{C}^*$, *and let Z be an irreducible subvariety of* $\mathbb{P}^1 \times \mathbb{P}^1$. *If the 2-cycle* $t_1(Z)$ *contains either* W_1 *or* W_2 *as an irreducible component, then* $Z = \Delta =$ *the diagonal of* $\mathbb{P}^1 \times \mathbb{P}^1$, *in which case* $t_1(\Delta) = W_1 + W_2$.

With the above data, we can prove the following.

Theorem 1.2.4.16 *Let x and y be variables associated to the Schubert classes* $\omega^2_{1;1,2}$ *and* $\omega^2_{2;0,2}$, *respectively. Then the second Euler–Chow series of* $F(0, 1; \mathbb{P}^2)$ *is given by the generating function*

$$E_2\big(F(0, 1; \mathbb{P}^2)\big) = \frac{1 - xy}{(1 - x)^3 (1 - y)^3}. \tag{33}$$

In other words,

$$E_2\big(F(0, 1; \mathbb{P}^2)\big)(r, s) := \chi\big(\mathcal{C}_{2,(r,s)}\big(F(0, 1; \mathbb{P}^2)\big)\big)$$

$$= \frac{1}{2}(r + 1)(s + 1)(r + s + 2).$$

We have the following corollary.

Corollary 1.2.4.17 *The 0th and 1st Euler–Chow series of* $F(0, 1; \mathbb{P}^2)$ *are given by the following generating functions:*

$$E_0\big(F(0, 1; \mathbb{P}^2)\big) = \frac{1}{(1 - t)^6}$$

and

$$E_1\big(F(0, 1; \mathbb{P}^2)\big) = \frac{1}{(1 - r)^3(1 - s)^3(1 - rs)^3},$$

where t is a variable associated to $\omega_{0;0,1}^2$ *and r, s are associated to* $\omega_{0;0,2}^2$ *and* $\omega_{1;1,2}^2$, *respectively.*

One can now use the above information to compute the *p*th Euler–Chow series of the Grassmannian $G(1, 3)$, using the prescription of Corollary 1.2.4.8.

In the next result we use the following association {variable} $\leftrightarrow$ {Schubert class}:

$$t \leftrightarrow \omega_{0,1}^2, \quad s \leftrightarrow \omega_{0,2}^2, \quad x \leftrightarrow \omega_{0,3}^2, \quad y \leftrightarrow \omega_{1,2}^2, \quad z \leftrightarrow \omega_{1,3}^2. \quad (34)$$

Proposition 1.2.4.18 *The Euler–Chow series of the Grassmannian* $G(1, 3)$ *is given by the following generating functions:*

$$E_0\big(G(1, 3)\big) = \frac{1}{(1 - t)^6}, \qquad\qquad E_1\big(G(1, 3)\big) = \frac{1}{(1 - s)^{12}},$$

$$E_2\big(G(1, 3)\big) = \frac{1}{(1 - x)^4(1 - y)^4(1 - xy)^3}, \qquad E_3\big(G(1, 3)\big) = \frac{1 + z}{(1 - z)^5},$$

where the latter coincides with the Hilbert series, as explained in Example 1.2.1.1.

Proof This follows directly from Corollaries 1.2.4.8 and 1.2.4.12, and the computations of the Euler–Chow functions of the flag varieties $F(0, 1; \mathbb{P}^2)$. $\qquad \square$

1.2.5 The case of abelian varieties

$$\chi\left(\mathcal{C}_\lambda(X)\right) = \begin{cases} \chi\left(\mathcal{C}_\beta(Y)\right) & \text{if } \alpha = \pi^*(\beta) \text{ for some } \beta \in H_\bullet(Y) \\ 0 & \text{otherwise.} \end{cases}$$

We follow the article [EH96]. Let us begin by stating the main result in this section.

Theorem 1.2.5.1 *Suppose that A is an abelian variety and that X is a projective variety, not necessarily smooth, both defined over* $\mathbb{C}$. *If*

$$\mu : A \times X \longrightarrow X$$

is an algebraic action of A on X, then $\chi(X) = \chi(X^A)$, *where* $\chi(X)$ *is the Euler characteristic of X and* $\chi(X^A)$ *is the Euler characteristic of the fixed point set* X^A.

Finally, we apply this to a Chow variety of a principal *A*-space.

Corollary 1.2.5.2 *If A is an abelian variety and X is a principal A-bundle over a projective variety Y, then*

$$\chi\left(\mathcal{C}_\lambda(X)\right) = \begin{cases} \chi\left(\mathcal{C}_\beta(Y)\right) & \text{if } \alpha = \pi^*(\beta) \text{ for some } \beta \in H_\bullet(Y), \\ 0 & \text{otherwise.} \end{cases}$$

Proof This follows directly from Theorem 1.2.5.1 using the facts that if $\alpha = \pi^*\beta$, then

$$\mathcal{C}_\alpha(X)^A = \mathcal{C}_\beta(Y),$$

and that $\mathcal{C}_\alpha(Y)^A$ is non-empty if and only if $\alpha = \pi^*(\beta)$ for some $\beta \in H_\bullet(Y)$. □

We conclude with some examples.

Example 1.2.5.3 If $X = A$, then

$$\chi\left(\mathcal{C}_\alpha(A)\right) = \begin{cases} 0 & \text{if } \alpha \neq n[A] \\ 1 & \text{if } \alpha = n[A]. \end{cases}$$

Example 1.2.5.4 If $X = A \times Y$, where *A* acts on *X* by $a : (b, y) \mapsto (b + a, y)$, then we have

$$\chi\left(\mathcal{C}_\alpha(X)\right) = \begin{cases} 0 & \text{if } \alpha \neq [A] \times \beta \\ \chi\left(\mathcal{C}_\beta(Y)\right) & \text{if } \alpha = [A] \times \beta. \end{cases}$$

Here $\beta \in H_\bullet(Y)$.

1.3 Some open problems, interesting relations for the Euler–Chow series, and the total coordinate ring

In this section we mention some open problems with the Euler–Chow series and Chow varieties, and we define and mention some properties of the total coordinate ring of a projective variety. We follow [ES02] and [EKiW02].

1.3.1 Euler–Chow series for the Néron–Severi group

Let *X* be a complex projective variety, and λ an element in the homology group $H_{2p}(X, \mathbb{Z})$. Let $\mathcal{C}_\lambda(X)$ denote the Chow variety of *X* with respect to λ, namely the space of all effective algebraic cycles in *X* of dimension *p* with homology class λ.

If X is a non-singular complex projective variety of dimension n, then using Poincaré duality to identify $H^{2p}(X, \mathbb{Z})$ with $H_{2(n-p)}(X, \mathbb{Z})$, we may similarly consider the Chow variety $\mathcal{C}_\lambda(X)$ associated to any cohomology class $\lambda \in H^{2p}(X, \mathbb{Z})$, parametrizing effective algebraic cycles on X of codimension p with cohomology class λ. In particular, if $p = 1$, $\mathcal{C}_\lambda(X)$ is a Chow variety of effective divisors in X; in this case $\mathcal{C}_\lambda(X) \neq \emptyset$ implies that $\lambda \in \mathrm{NS}(X) \subset H^2(X, \mathbb{Z})$, where $\mathrm{NS}(X)$ denotes the Néron–Severi group of algebraic (or homological) equivalence classes of divisors on X.

If X and Y are non-singular complex projective varieties, there is a natural inclusion $\mathrm{NS}(X) \oplus \mathrm{NS}(Y) \hookrightarrow \mathrm{NS}(X \times Y)$, obtained by associating to the classes of divisors D on X and E on Y the class of the divisor $X \times E + D \times Y$ on $X \times Y$.

We have the following theorem.

Theorem 1.3.1.1 *Let X and Y be non-singular complex projective varieties. Let $\mathcal{C}_\xi(X)$, $\mathcal{C}_\eta(Y)$ and $\mathcal{C}_{(\xi,\eta)}(X \times Y)$ denote the Chow varieties of X, Y and $X \times Y$ respectively, with respect to the divisor classes $\xi \in \mathrm{NS}(X)$, $\eta \in \mathrm{NS}(Y)$ and $(\xi, \eta) \in \mathrm{NS}(X) \oplus \mathrm{NS}(Y) \subset \mathrm{NS}(X \times Y)$. Then*

$$\chi\big(\mathcal{C}_{(\xi,\eta)}(X \times Y)\big) = \chi\big(\mathcal{C}_\xi(X)\big)\, \chi\big(\mathcal{C}_\eta(Y)\big).$$

In terms of Euler–Chow series we have

Corollary 1.3.1.2 *Let M be the monoid of effective divisor cycle classes on $X \times Y$ which are contained in $\mathrm{NS}(X) \oplus \mathrm{NS}(Y)$. Then we have*

$$E^1(M)(X \times Y) = E^1(X)\, E^1(Y) \in \mathbb{Z}[[M]].$$

Where E^1 is the Euler–Chow series for codimension 1.

An interesting problem that we would like to mention, is to find a generalization of Corollary 1.3.1.2.

Problem 1 Let X and Y be projective varieties of dimension n and m. We wonder under what conditions the following product formula holds:

$$E_{n+m-1}(X \times Y) = E_{n-1}(X)\, E_{m-1}(Y).$$

For example, the formula is valid for toric varieties [Eli94], or more generally if $\mathrm{Pic}^0(X) = \mathrm{Pic}^0(Y) = 0$. By Corollary 1.3.1.2, the formula is equivalent to the assertion that $\mathrm{NS}(X) \oplus \mathrm{NS}(Y) = \mathrm{NS}(X \times Y)$.

It would of course be very interesting to find criteria for the analogues of Corollary 1.3.1.2, or the validity of a product formula as stated above, for the Euler–Chow series associated to cycles of codimension > 1.

1.3.2 Open problems

Let M be the monoid of linear equivalence classes of effective divisors on a smooth proper variety X over an algebraic closed field k, such that linear and homological equivalence coincide for divisors on X, or equivalently, the Picard variety $\text{Pic}^0(X)$ is trivial.

Let us define the *total coordinate ring* of X by

$$\mathcal{TC}(X) = \bigoplus_{D \in M} H^0(X, \mathcal{O}(D)).$$

There is a series associated to it, namely the Euler–Chow series $E = E^1(X) \in \mathbb{Z}[[M]]$ for divisors,

$$E = \sum_{D \in M} \dim H^0(X, \mathcal{O}(D))\, D.$$

We can also define the *Riemann–Roch ring* $R = R(D)$ for any effective divisor (class) $D \in M$ by

$$R(D) = \bigoplus_{n \geq 0} H^0(X, \mathcal{O}(nD)),$$

and the series associated to this ring is the Riemann–Roch series

$$S_R = \sum_{n \geq 0} \dim H^0(X, \mathcal{O}(nD))\, t^n.$$

Here we identify the subring $\mathbb{Z}[[\mathbb{N}D]] \subset \mathbb{Z}[[M]]$ with the formal power series $\mathbb{Z}[[t]]$. It is well known that if either of the two rings $\mathcal{TC}(X)$ or R is a finitely generated k-algebra, then the corresponding series defines a rational function.

Proposition 1.3.2.1 *With the above notation, if $\mathcal{TC}(X)$ is a finitely generated k-algebra, then so is $R(D)$, for any effective divisor D.*

Proof The natural graded inclusion $R(D) \hookrightarrow \mathcal{TC}(X)$ of k-algebras has a splitting as $R(D)$-modules. Indeed, we note that the graded projection map from $\mathcal{TC}(X)$ to $R(D)$ with kernel

$$\bigoplus_{D' \in M \setminus \mathbb{N}D} H^0(X, \mathcal{O}(D'))$$

where D' ranges over the divisor classes in M which are not multiples of D, gives an $R(D)$-linear splitting of the inclusion map $R(D) \hookrightarrow \mathcal{TC}(X)$. The proposition now follows from [Eis95, Corollary 1.5, p. 31]. □

This gives us our first relation between the Euler–Chow series and the Riemann–Roch series. In general, we would like to know which of the implications appearing in the following diagram hold:

$$\mathcal{TC}(X) \text{ finitely generated} \underset{2}{\overset{1}{\rightleftarrows}} E \text{ rational}$$

$$4 \Big\Uparrow\Big\Downarrow 3 \qquad\qquad\qquad 8 \Big\Uparrow\Big\Downarrow 7$$

$$R(D) \text{ finitely generated } \forall\, D \underset{6}{\overset{5}{\rightleftarrows}} S_R \text{ rational } \forall\, D$$

Proposition 1.3.2.1 implies that implication 3 holds, and as we mentioned earlier, implications 1, 5 and 7 also hold.

We take a closer look at the other implications that appear in the diagram. Implication 4 is very interesting, albeit in general false; here is an example that appears in [EKiW02].

Example 1.3.2.2 Let us consider the weighted polynomial ring A in three variables with weight a, b, c to the variables x, y, z with coefficient over the complex numbers $\mathbb{C}$. Take the weighted projective plane

$$\mathbb{P}^2(a, b, c) = \text{Proj}(A)$$

and consider the blow-up $X(a, b, c)$ of $\mathbb{P}^2(a, b, c)$ at the smooth point

$$p(a, b, c) := \text{Ker}(k[x, y, z] \longrightarrow k[t^a, t^b, t^c]).$$

Then it is well known that the section ring $R(D)$ is finitely generated. However, Goto, Nishida and Watanabe proved in [GNiW94] that $\mathcal{TC}(X)$ is not a finitely generated if $a = 7n - 3$, $b = n(5n - 2)$ and $c = 8n - 3$ with $n \geq 4$.

On the other hand, if we require the variety to be smooth we do not know the answer. Assuming finite generation of M, we have the following problem.

Problem 2 Let X be a smooth projective variety. And let $R(D)$ be a finitely generated k-algebra, for all D. If $\text{Pic}^0(X) = 0$ and M is a finitely generated monoid then $\mathcal{TC}(X)$ is a finitely generated k-algebra.

Now, let us look at the remaining implications. Cutkosky and Srinivas proved that for any smooth algebraic surface in characteristic 0, the series S_R always defines a rational function, even if R is not a finitely generated k-algebra (see [CS93]). In particular, implication 6 is false. About implications 2 and 8, we have neither positive results nor counterexamples. We guess that all of these implications are false, but examples are difficult to compute at present.

The total coordinate ring of X

We have two important results for the total coordinate ring of X; the proofs can be seen in [EKiW02].

Theorem 1.3.2.3 *Let X be a normal Noetherian scheme with function field K. Let $D_1, \ldots, D_r$ be Weil divisors on X, and let $t_1, \ldots, t_r$ be variables over K. We set*

$$\mathfrak{TC}(X) = R = \bigoplus_{n_1, \ldots, n_r \in \mathbb{Z}} H^0\left(X, \mathcal{O}_X\left(\sum_i n_i D_i\right)\right) t_1^{n_1} \cdots t_r^{n_r} \subset S$$

$$= K\left[t_1^{\pm 1}, \ldots, t_r^{\pm 1}\right].$$

1. *Then, R is a Krull domain.*
2. *Assume that $Q(R) = Q(S)$, where $Q(\ )$ stands for the field of fractions. Then there is a natural surjection*

$$\varphi : \mathrm{Cl}(X)/\langle \overline{D}_1, \ldots, \overline{D}_r \rangle \longrightarrow \mathrm{Cl}(R),$$

   where $\overline{D}$ is the image in $\mathrm{Cl}(X)$ of a Weil divisor D on X.
3. *If there exist integers $n_1, \ldots, n_r$ such that $\sum_i n_i D_i$ is an ample divisor, then the map φ as above is an isomorphism.*

Remark 1.3.2.4 With notation as in the theorem above, it is easily seen that $Q(R)$ coincides with $Q(S)$ if there exist integers $n_1, \ldots, n_r$ such that $\sum_i n_i D_i$ is an ample divisor.

However, the converse is not true in general as in the following example. Let $\pi : X \to \mathbb{P}^2$ be the blow-up of $\mathbb{P}^2$ along a point p. Put $e = \pi^{-1}(p)$. Let H be a hyperplane in $\mathbb{P}^2$ and put $h = \pi^{-1}(H)$. Let K be the function field and let t be a variable over K. Set

$$R = \bigoplus_{n \in \mathbb{Z}} H^0(X, \mathcal{O}_X(nh)) t^n \subset S = K[t^{\pm 1}].$$

Using the projection formula, one can prove $Q(R) = Q(S)$. On the other hand, nh is not ample for any integer n by the Nakai–Moišezon criterion (e.g., 365p in [Har77]). In that case, $\mathrm{Cl}(X)$ is a $\mathbb{Z}$-free module generated by e and h. Since R is a polynomial ring, the map φ in the theorem as above is not an isomorphism.

Corollary 1.3.2.5 *With notation as in the theorem above, R is a unique factorization domain if the set of the images of $D_1, \ldots, D_r$ generates $\mathrm{Cl}(X)$.*

In particular, a total coordinate ring is a unique factorization domain for a normal projective variety whose divisor class group is a finitely generated torsion-free abelian group.

We remark that, in the case of a smooth projective variety whose divisor class group is a finitely generated torsion-free abelian group, the unique factorization of the total coordinate ring is proved independently by Berchtold and Hausen [BH].

Theorem 1.3.2.6 *Let* $p_1, \ldots, p_n$ *be n points in a projective line contained in* $\mathbb{P}^r$. *Then the total coordinate ring* $\mathfrak{TC}(X)$ *of the blow-up* $X = Bl_{p_1,\ldots,p_n}(\mathbb{P}^r)$ *of* $\mathbb{P}^r$ *along* $\{p_1, \ldots, p_n\}$ *is a Noetherian ring.*

Acknowledgements

I am grateful to the organizers for their kind support, and for inviting me to participate in this summer school. I am also grateful for support from DGAPA IN109800 and CONACYT 40531-F.

References

[BDP90]   Bifet, E., De Concini, C. and Procesi, C. Cohomology of regular embeddings, *Adv. Math.* **82** (1990) 1–34.

[BH]   Berchtold, F. and Hausen, J. Homogeneous coordinates for algebraic varieties, AG.0211413.

[Bia73]   Bialyniki-Birula, A. Some theorems on actions of algebraic groups, *Ann. Math.* **98**(2) (1973) 480–497.

[CS93]   Cutkosky, S. D. and Srinivas, V. On a problem of Zariski on dimensions of linerar systems, *Ann. Math.* **137** (1993) 531–559.

[Dan78]   Danilov, V. I. The geometry of toric varieties, *Russ. Math. Surv.* **33**(2) (1978) 97–154.

[EH92]   Eisenbud, D. and Harris, J. The dimension of the Chow variety of curves, *Compositio Math.* **83** (1992) 291–310.

[EH96]   Elizondo, E. J., and Hain, R. Chow varieties of abelian varieties, *Bol. Soc. Mat. Mex.* **2**(3) (1996) 95–99.

[Eis95]   Eisenbud, D. *Commutative Algebra: with a View Toward Algebraic Geometry, Graduate Texts in Mathematics* **150**, 1st edn., Springer-Verlag, New York, 1995, corrected second printing, 1996.

[EKiW02]   Elizondo, E. J. Kurano, K. and Watanabe, K.-I. The total coordinate ring of a normal projective variety, submitted for publication, 2002.

[ELF98]   Elizondo, E. J. and Lima-Filho, P. Euler–Chow series and projective bundle formulas, *J. Algebraic Geom.* **7** (1998) 695–729.

[Eli94]   Elizondo, J. The Euler series of restricted Chow varieties, *Compositio Math.* **94**(3) (1994) 297–310.

[ES02]   Elizondo, E. J. and Srinivas, V. Some remarks on Chow varieties and Euler–Chow series, *J. Pure Appl. Algebra* **166**(1–2) (2002) 67–81.

[FL92]   Friedlander, E. and Lawson, H. B. A theory of algebraic cocycles, *Ann. Math.* **136**(2) (1992) 361–428.

[FM91]   Friedlander, E. and Mazur, B. Filtrations on the homology of algebraic
            varieties, preprint (eleventh draft), 1991.
[Fri91]   Friedlander, E. Algebraic cycles, Chow varieties and Lawson homology,
            Compositio Math. **77** (1991) 55–93.
[Ful93]   Fulton, W. *Introduction to Toric Varieties, Annals of Mathematics Studies*,
            131, 1st edn., Princeton University Press, Princeton, NJ, 1993.
[Ful98]   *Intersection Theory*, 2nd edn., Springer-Verlag, Heidelberg, 1998.
[GKZ94]   Gelfand, I. M, Kapranov, M. M. and Zelevinsky, A. V. *Discriminants,
            Resultants and Multidimensional Determinants*, Birkhäuser, Boston, MA,
            1994.
[GNiW94]   Goto, S., Nishida, K. and Watanabe, K.-I. Non-Cohen–Macaulay symbolic
            blow-ups for space monomial curves and counterexamples to Cowsik's
            question, *Proc. Am. Math. Soc.* **120**(2) (1994) 383–392.
[Har75a]   Hartshorne, R. (ed.), *Algebraic Geometry, Arcata 1974, Proc. Symp. Pure
            Math.* **29**, American Mathematical Society, Providence, RI, 1975.
[Har75b]   Equivalence relations on algebraic cycles and subvarieties of small codi-
            mension. In: *Algebraic Geometry, Arcata 1974, Proc. Symp. Pure Math.*
            29, pp. 129–164, American Mathematica Society, Providence, RI, 1975.
[Har77]   *Algebraic Geometry, Graduate Texts in Mathematics* **52**, Springer-Verlag,
            New York, 1977.
[Hoy66]   Hoyt, W. On the Chow bunches of different projective embeddings of a
            projective variety, *Am. J. Math.* **88** (1966) 273–278.
[Kap93]   Kapranov, M. M. Chow quotients of Grassmannians I, *Adv. Sov. Math.*
            16(2) (1993) 29–109.
[Kol96]   Kollár, J. *Rational Curves on Algebraic Varieties, Ergebnisse der Mathe-
            matik und ihrer Grenzgebiete*, vol. 3, **32**, Springer-Verlag, 1996.
[Law95]   Lawson, B. *Spaces of Algebraic Cycles, Surveys in Differential Geometry*,
            vol. 2, pp. 137–213, International Press, 1995.
[LS75]   Lascou, A. T. and Scott, D. B. An algebraic correspondence with applica-
            tions to blowing up Chern classes, *Ann. Mat.* **102** (1975) 1–36.
[LY87]   Lawson, H. B., Jr. and Yau, S. S. T. Holomorphic symmetries, *Ann. Sci.
            Ec. Norm. Super.* **20** (1987) 557–577.
[Mac62]   MacDonald, I. G. The Poincaré polynomial of a symmetric product, *Proc.
            Cam. Philos. Soc.* **58** (1962) 563–568.
[Nee91]   Neeman, A. 0-cycles in $\mathbb{P}^n$, *Adv. Math.* **89** (1991) 217–227.
[Sam71]   Samuel, P. *Séminaire sur l'Equivalence Rationnelle, Publ. Math. Orsay*
            425, pp. 1–17, Université de Paris-Sud, Département de Mathématique,
            Orsay, 1971.
[Sha77]   Shafarevich, I. *Basic Algebraic Geometry*, revised edn., Springer-Verlag,
            New York, 1977.

2

Introduction to Lawson homology

Chris Peters and Siegmund Kosarew

Department of Mathematics, University of Grenoble I, UMR 5582 CNRS-UJF,
38402 Saint-Martin d'Hères, France

Abstract

Lawson homology has quite recently been proposed as an invariant for algebraic varieties. Various equivalent definitions have been suggested, each with its own merit. Here we discuss these for projective varieties and we also derive some basic properties for Lawson homology. For the general case we refer to Paulo Lima-Filho's lectures (Chapter 3).

Keywords: Lawson homology, cycle spaces
2000 Mathematics subject classification: 14C25, 19E15, 55Qxx

Introduction

This paper is meant to serve as a concise introduction to Lawson homology of projective varieties. For another introduction the reader should consult [14].

It is organized as follows. In the first section we recall some basic topological tools needed for a first definition of Lawson homology. Then some basic examples are discussed. In the second section we discuss the topology of the so-called 'cycle spaces' in more detail in order to understand functoriality of Lawson homology. In the third and final section we relate various equivalent definitions. Here the language of simplicial spaces is needed and we only summarize some crucial results from the vast literature on this highly technical subject.

2.1 Basic notions

2.1.1 Homotopy groups

We start by recalling the definition and the basic properties of the homotopy groups. For any two pairs of topological spaces (X, A) and (Y, B) we use

Transcendental Aspects of Algebraic Cycles ed. S. Müller-Stach and C. Peters.
© Cambridge University Press 2004.

the notation $[(X, A), (Y, B)]$ for the set of homotopy classes of maps $X \to Y$ sending A to B (any homotopy is supposed to send A to B as well). Then, fixing a point s on the k-sphere S^k, we have

$$\pi_k(X, x) = [(I^k, \partial I^k), (X, x)] = [(S^k, s), (X, x)].$$

There is a natural product structure on these sets (divide I^k in two and use the first map on one half and the second map on the other half). This makes $\pi_k(X, x)$ into a group, which turns out to be abelian for $k \geq 2$.

Homotopy and homology are related through the *Hurewicz homomorphism*

$$h_k : \pi_k(Y, y) \to H_k(Y),$$

defined by associating to the class of a map $f : S^k \to Y$ the image under f_* of a generator of $H_k(S^k)$. The following important result tells us when the Hurewicz homomorphism actually is an isomorphism:

Theorem 1 (Hurewicz' theorem) *Suppose that (X, x) is $(k - 1)$-connected, i.e. $\pi_s(X, x) = 1, s = 0, \ldots, k - 1$. Then h_k is an isomorphism.*

One can show that homotopic maps induce the same map in homology. Hurewicz' theorem tells us that any map inducing isomorphisms on the homotopy groups will also induce isomorphisms on the homology groups. This motivates the following definitions.

Definition 2

(1) A continuous map $f : X \to Y$ is a *homotopy equivalence* if there is a continuous map $g : Y \to X$ such that $f \circ g$ and $g \circ f$ are homotopic to the identity.

(2) A continuous map $f : X \to Y$ is a *weak homotopy equivalence* if the induced maps on the homotopy groups are all isomorphisms.

(3) Two topological spaces are *(weakly) homotopically equivalent* if there exists a (weak) homotopy equivalence between them.

Example 3 A space X is said to be an *Eilenberg–Mac Lane $K(\pi, k)$-space* if its only non-trivial homotopy group is $\pi_k(X) = \pi$. Hence any space homotopy equivalent to a $K(\pi, k)$-space is again a $K(\pi, k)$-space. For instance S^1 is a $K(\mathbb{Z}, 1)$, and the inductive union of projective spaces, $\mathbb{P}^\infty$, is a $K(\mathbb{Z}, 2)$.

An important class of topological spaces is the class of *CW-complexes*. Here we do not give the precise definition, but refer to [20]. Roughly speaking, a CW-complex is defined inductively by specifying its cells in a given dimension $(k + 1)$ together with the attaching maps of the cells to the union of the cells of lower dimension, the k-skeleton. In general one needs infinitely many cells, but a

CW-complex has to satisfy a certain local finiteness condition. Any topological
space admitting a triangulation gives an example. For instance (see [17]), any
differentiable manifold has the structure of a CW-complex. An important remark
is that although an Eilenberg–Mac Lane space need not be a CW-complex, it
has the homotopy type of a CW-complex.

For CW-complexes any weak homotopy equivalence is a homotopy equivalence. This result is due to Whitehead [20, p. 405]. Another important result (see
also [20]) for CW-complexes is the fact that $K(\mathbb{Z}, m)$ classifies cohomology:

$$[X, K(\mathbb{Z}, m)] \cong H^m(X; \mathbb{Z}).$$

For CW-complexes with finitely many cells, there is a homomorphism re-
fining the Hurewicz map, due to Almgren [1]. Recall that singular homology is
computed as the cohomology of the singular complex $S_\bullet(X)$. Instead, one can
also use the complex of $I_\bullet(X)$ 'integral currents', a refinement due to Federer
and Fleming [4] of the well known result that one can use the complex of currents to compute real (co)homology. The so-called 'flat-norm topology' makes
the spaces of integral currents into topological spaces so that we can speak
of the homotopy group of the cycle spaces $Z'_k = \ker(\partial : I_k(X) \to I_{k-1}(X))$.
We shall summarize this by saying that we give $Z'_k(X)$ the *Federer topology*.
Almgren's thesis tells us that any continuous map $f : S^r \to Z'_k(X)$ of the
r-sphere into this space can be seen as a $(k + r)$-cycle $f_k \in Z'_{k+r}(X)$ and so,
by the aforementioned result by Federer and Fleming, yields a class in the
singular homology group $H_{k+r}(X)$. This map is an isomorphism:

Theorem 4 *For a CW-complex X with finitely many cells (such as a projective
manifold) equip the set of k-cycles $Z'_k(X)$ with the Federer topology. The map
(as defined above)*

$$\pi_r Z'_k(X) \to H_{r+k}(X)$$

$$[f] \mapsto [f_k]$$

is an isomorphism.

Intuitively, f is a continuous family of k-cycles on X and sweeps out a
$(k + r)$-cycle. Homotopic maps give rise to homologous cycles. For $k = 0$
we can work with the usual notion of 0-cycles, obtaining a refinement of the
Hurewicz map. This gives the Dold–Thom theorem [3].

We next discuss the concept of a *(Hurewicz) fibration*. This is a continuous
surjective map between topological spaces $p : E \to B$ which has the homotopy
lifting property: given a map $g : X \to E$, every homotopy of $p \circ g$ can be lifted

to a homotopy of g. For such a fibration any two fibers are homotopy equivalent [20, p. 101] and with $e \in E$ the base point and F the typical fiber, one has the homotopy exact sequence [20, p. 377]

$$\cdots \pi_n(F, e) \to \pi_n(E, e) \to \pi_n(B, p(e)) \to \pi_{n-1}(F, e) \cdots$$

Examples include locally trivial fiber bundles such as smooth Kähler families or smooth projective families.

Although fibrations look rather special, in homotopy theory all maps are fibrations. Indeed, one can replace functorially any continuous map $f : X \to Y$ by a Hurewicz fibration. Using the path space PY of continuous paths in Y, the total space of the fibration is

$$E_f = \{(x, \gamma) \in X \times PY \mid \gamma(0) = f(x)\}$$

and the map $\pi_f : E_f \to Y$ given by sending a pair (x, γ) to the endpoint $\gamma(1)$ of γ gives it the structure of a fibration. The map $s : X \to E_f$ which sends x to the pair $(x, \text{constant path at } f(x))$ is a homotopy equivalence so that indeed $f : X \to Y$ may be replaced by the fibration $\pi_f : E_f \to Y$. The *homotopy fiber* $E_f(y)$ of f above y by definition is the fiber of π_f above y. Its homotopy type depends only on the path component to which y belongs. So, if we start with a Hurewicz fibration over a path connected space, any fiber is homotopy equivalent to the homotopy fiber.

2.1.2 Lawson homology

Let X be a complex projective variety. An *effective* (*algebraic*) *m-cycle* on X is a finite formal linear combination $Z = \sum n_V[V]$, $n_V \in \mathbb{N}$, of (irreducible) subvarieties $V \subset X$ of dimension m. The union $\cup V$ is called the *support of* Z and will be denoted supp(Z). If a projective embedding is fixed, one can speak of the *degree* of the cycle $\deg Z = \sum n_V \deg V$. The set $C_{m,d}(X)$ which parametrizes the effective m-cycles of degree d is known to be a projective variety (see Elizondo's lectures, Chapter 1). It comes with the complex topology. Now we take the the disjoint union $\mathcal{C}_m(X)$ of the Chow varieties of effective m-cycles of degree $d = 0, 1, 2, \ldots$. The cycle $0 \in \mathcal{C}_m(X)$ by definition is the cycle with empty support and serves as a natural base point. It also acts as a zero for the addition of cycles, making $\mathcal{C}_m(X)$ into a monoid. Let us put

$$\mathcal{Z}_m(X) = C_m(X) \times C_m(X)/\sim$$

$$(x, y) \sim (x', y') \Leftrightarrow x + y' = x' + y \text{ (the naïve group completion).}$$

The complex topology induces a natural topology on the monoid $C_m(X)$ and one equips $\mathcal{Z}_m(X)$ with the quotient topology. We can be more specific. Introduce the compact sets

$$K_d = \bigcup_{d_1+d_2 \leq d} C_{m,d_1}(X) \times C_{m,d_2}(X)/ \sim$$

which filter $\mathcal{Z}_m(X)$. Then $B \subset \mathcal{Z}_m(X)$ is closed if and only if its intersection with each of the K_d is closed.

In Section 2.2 we will review the proof that this topological space is independent of the chosen embedding of X into a projective space. The induced topology will be called the *Chow topology*.

An algebraic cycle on a projective variety defines an integral current (via integration over its smooth locus), and the inclusion $\mathcal{Z}_m(X) \subset Z'_{2m}(X)$ is continuous. Using Federer's theorem (Theorem 4) this yields maps

$$\pi_{\ell-2m}\mathcal{Z}_m(X) \xrightarrow{c_{m,\ell}} H_\ell(X)$$

motivating the following definition.

Definition 5 The *Lawson homology groups* of a complex projective algebraic variety X are the homotopy groups of the cycle space:

$$L_m H_\ell(X) = \begin{cases} \pi_{\ell-2m}\mathcal{Z}_m(X) & \text{if } \ell \geq 2m \\ 0 & \text{if } \ell < 2m. \end{cases}$$

Lawson homology incorporates usual homology in view of an old result of Dold and Thom (see [3]) which can be reformulated as a natural isomorphism

$$c_{o,\ell} : L_0 H_\ell(X) \xrightarrow{\sim} H_\ell(X).$$

This reformulation will be explained in Section 2.3 (see Examples 26 and 33).

As with the Chow groups, there are relatively few classes of varieties for which one can compute Lawson homology.

One of the breakthroughs was Lawson's computation [13] of these groups for projective spaces. He used a powerful tool, the *suspension theorem*. To formulate it, we introduce some notation. Let $X \subset \mathbb{P}^N$ be a projective variety and a point $v \in \mathbb{P}^{N+1}$ not in the subspace $\mathbb{P}^N$ in which X lies. Let $\Sigma X \subset \mathbb{P}^{N+1}$ be the projective cone over X with v as its vertex.

Theorem 6 (Lawson's suspension theorem) *The map which associates to an m-cycle Z of degree d on X its projective cone ΣZ, an $(m+1)$-cycle of degree d on ΣX, induces a weak homotopy equivalence*

$$\Sigma : \mathcal{Z}_m(X) \to \mathcal{Z}_{m+1}(\Sigma X).$$

In particular we have

$$L_m H_\ell(X) \xrightarrow{\sim} L_{m+1} H_{\ell+2}(X).$$

2.1.3 Examples

1. Two cycles that are algebraically equivalent belong to the same connected component of the cycle space. One can show [5] that the converse also holds. So $\pi_0(\mathcal{Z}_m(X))$ can be identified with the group of equivalence classes of cycles modulo algebraic equivalence:

$$L_{2m} H_m(X) = \pi_0(\mathcal{Z}_m(X)) = \mathrm{Ch}_m^{\mathrm{alg}}(X)$$

$$= \mathrm{Ch}_m(X)/\{\text{algebraic equivalence}\}.$$

This is the first indication that Lawson homology also reflects the algebraic nature of the variety, for a priori it could only be a topological invariant of its analytic topology.

2. Look at zero cycles on curves X. The Abel–Jacobi map

$$C_{0,d}(X) \to \mathrm{Jac}(X)$$

is surjective for d large enough with fiber a projective space $\mathbb{P}^{d-g}$. This shows that

$$\lim_{d \to \infty} C_{0,d}(X) \sim \mathrm{Jac}(X) \times \mathbb{P}^\infty$$

and so we know its homotopy type: it is a $K(\mathbb{Z}^{2g}, 1) \times K(\mathbb{Z}, 2)$. Later we shall show that the limit computes the homotopy type of the cycle space $\mathcal{Z}_0(X)$ and so we have shown:

$$\pi_k(\mathcal{Z}_0(X)) \cong H_k(X)$$

and this is a special case of the Dold–Thom theorem.

3. Look at irreducible curves of degree d in the projective plane $\mathbb{P}^2$. Choose a point p not on a given curve C and choose coordinates such that $p = (0 : 0 : 1)$. Projection from this point is then given by $(x : y : z) \mapsto (x : y : 0)$ and it fits in a one-parameter family of maps $\pi_t : \mathbb{P}^2 \to \mathbb{P}^2, t \in \mathbb{C}$, given by

$$\pi_t(x : y : z) = (x : y : tz).$$

For $t \neq 0$ this is an automorphism and for $t = 0$ we get our projection. The curve $C_t = \pi_t C$ degenerates to dL where L is the line $z = 0$. If $p \in C$, one can see that the curve degenerates into $aL + \sum_j b_j M_j$, where $a + \sum_j b = d$

and M_j are the tangents of C in p. So the space $C_{1,d}(\mathbb{P}^2)$ is connected and simply connected. It follows that $\pi_0(\mathcal{Z}_1(\mathbb{P}^2)) \cong \mathbb{Z}$ and also that $\mathcal{Z}_1(\mathbb{P}^2)$ is simply connected.

4. The suspension theorem generalizes this to arbitrary varieties. The role of L being played by an arbitrary variety X and $\mathbb{P}^2$ is seen as the projective cone ΣL over the line L. As an application of this, recall that the Dold–Thom theorem states that the mth homotopy group of cycle space $\mathcal{Z}_0(X)$ is equal to $H_m(X)$ and so $\mathcal{Z}_0(X)$ is homotopy equivalent to the product of the Eilenberg–Mac Lane spaces $K(H_m(X), m)$. If $X = \mathbb{P}^N$ the homology groups being $\mathbb{Z}$ (for $m \le 2N$ even) or 0 (otherwise) we get:

$$\mathcal{Z}_0(\mathbb{P}^{n-m}) = K(\mathbb{Z}, 0) \times K(\mathbb{Z}, 2) \times \cdots \times K(\mathbb{Z}, 2(n-m)).$$

Using the suspension theorem we thus find

$$\mathcal{Z}_m(\mathbb{P}^n) = K(\mathbb{Z}, 0) \times K(\mathbb{Z}, 2) \times \cdots \times K(\mathbb{Z}, 2(n-m))$$

and thus

$$L_m H_\ell(\mathbb{P}^n) = \pi_{\ell-2m}(\mathbb{P}^n) = \begin{cases} 0 & \text{if } \ell \text{ odd} \\ \mathbb{Z} & \text{if } \ell = 2k, \ k = m, \ldots, n. \end{cases}$$

2.2 Chow varieties and cycle spaces

2.2.1 Chow varieties

In this section we recall the definition of the Chow variety of a projective variety X in a fixed embedding

$$i : X \subset \mathbb{P}^N = \mathbb{P}.$$

This variety $C_{m,d}(X)$ parametrizes the effective degree d cycles on X. For more details see Elizondo's lectures, Chapter 1.

First of all, one constructs the Chow variety for $\mathbb{P}$ itself as follows. Let Z be an m-dimensional subvariety of $\mathbb{P}$. The $(m+1)$-tuples of hyperplanes whose intersection meets Z form a hypersurface in $(\mathbb{P}^\vee)^{(m+1)}$ of multidegree $(d, \ldots, d)$, where d is the degree of $Z \subset \mathbb{P}$. This hypersurface F_Z defines the *Chow point* in $\mathbb{P}(m, d)$, the projective space of forms of multidegree $(d, \ldots, d)$. The *Chow coordinates* of this point are the corresponding $(m+1)d$ homogeneous coordinates. The variety Z can be reconstructed from F_Z, or its Chow point since one can show that $z \in Z$ if and only if for any $(m+1)$-tuple $(l_0, \ldots, l_m)$ of hyperplanes passing through z, one has $F_Z(l_0, \ldots, l_m) = 0$. If now Z is a cycle

$\sum n_V[V]$, its Chow point is the point corresponding to the hypersurface $\prod F_V^{n_V}$. All cycles of $\mathbb{P}$ of fixed degree d and dimension m fill up the Chow variety of $\mathbb{P}$, a projective subvariety $C_{m,d}(\mathbb{P}) \subset \mathbb{P}(m, d)$. Cycles belonging to a fixed subvariety $X \subset \mathbb{P}$ belong to the Chow variety of X, a subvariety $C_{m,d}(X) \subset C_{m,d}(\mathbb{P})$.

Let us next discuss the functoriality of this construction. First of all, there is no universal family over the Chow variety. A weak approximation of this is the *incidence correspondence* $\Gamma_{n,d}(X) \subset C_{n,d} \times X$, which is the closure inside $C_{m,d}(X)$ of the variety of pairs (Chow point of Z, x) of irreducible m-dimensional subvarieties $Z \subset X$ of degree d such that x belongs to Z. The scheme-theoretic fiber over the Chow point of a possibly reducible $Z = \sum n_V[V]$ does have support in Z but the multiplicities are not necessarily the same. This causes failure of universality and is the source of many technical problems when dealing with Chow varieties.

Let us briefly describe what sort of families of m-cycles over a base T do give rise to a morphism $T \to C_{m,d}(X)$. The following concept is crucial.

Definition-Lemma 7 Let T be a projective variety. A subscheme $Z \subset T \times X$ is an *equi-dimensional effective relative m-cycle of X over T* if the projection $p_1 : Z \to T$ is surjective and for all $t \in T$ the scheme-theoretic fiber Z_t over t has dimension m. The set of these cycles forms a Zariski-open subset

$$C_{m,d}(X; T) \subset C_{m+t,d}(T \times X) \qquad t = \dim T.$$

The fact that the complement is a Zariski-closed set follows from upper-semi-continuity of the dimensions of the fibers of the incidence correspondence $\{(Z, x, y) \in C_{m+t,d}(T \times X) \times (X \times Y) \mid (x, y) \in \text{support of } Z\}$ over the first factor.

A standard example of such a situation arises when Z dominates T and $p_1 : Z \to T$ is *flat*. To any scheme like Z_t one can associate a cycle $[Z_t] = \sum n_V[V]$, where V runs over the irreducible components of Z_t and n_V is the length of the local artinian ring $\mathcal{O}_{V,Z_t}$ (see [10, 1.5]). In our situation, if Z is flat over T the degree of the cycle $[Z_t]$ is constant and one indeed does have an associated map $\varphi_Z : T \to C_{m,d}(X)$ which is a morphism as we will indicate below. In fact, one has

Lemma 8 *Let Z be a flat relative m-cycle of degree d dominating a smooth projective variety T. The map*

$$\varphi_Z : T \to C_{m,d}$$

sending t to the cycle associated to the scheme-theoretic fiber of $Z \to T$ over t is a morphism. Conversely, a morphism $\varphi : T \to C_{m,d}$ determines (by taking *the fiber product over $C_{m,d}$ of T and the incidence correspondence $\Gamma_{m,d}(X)$) a flat relative m-cycle of degree d whose associated morphism is φ.*

The following example, taken from [9, p. 29], illustrates what happens in the non-flat case.

Example 9 Take $T = \mathbb{C}^2$ with coordinates (x_1, x_2), $X = \mathbb{P}^2$ with homogeneous coordinates (X_0, X_1, X_2) and Z given by the union of the two varieties $Z_1 = \{X_1 = x_1 X_0, X_2 = x_2 X_0\}$ and $Z_2 = \{X_1 = -x_1 X_0, X_2 = -x_2 X_0\}$. The scheme-theoretic fiber $Z_{a,b}$ consists of two points $(1 : \pm a : \pm b) \in \mathbb{P}^2$ for $(a, b) \neq (0, 0)$, while the scheme-theoretic fiber over the origin is $(1 : 0 : 0) \in \mathbb{P}^2$ with multiplicity 3.

To obtain a constant degree cycle over the smooth locus, one has to replace $[Z_t]$ by the *intersection-theoretic fiber*

$$Z \cdot [t] = \sum e_V [V], \quad e_V = (e_{Z_t} Z)_V = i(V, \{t\} \cdot Z; T)$$

as in [10, Chapter 7]. Here e_V is Samuel's multiplicity of the primary ideal determined by Z_t in the local ring $\mathcal{O}_{V,Z}$. This multiplicity is bounded above by length $(\mathcal{O}_{V,Z_t})$ and is equal to it if for instance Z_t is a regular subscheme of Z, which happens for points $t \in U$, where U is the set of smooth points of T over which f is flat (use [10, Proposition 7.1] and [10, Example A.5.5]). In the above example, $Z \cdot \{0\} = 2(1 : 0 : 0)$ which has the correct multiplicity $2 < 3$.

To see that this degree is constant we recall that the notion of *degree* of an n-dimensional subvariety $X \subset \mathbb{P}$ defined as the cardinality of $X \cap L$, where L is a general linear space of codimension n in $\mathbb{P}$, can also be defined using the intersection product on Chow groups (see e.g. [10])

$$\mathrm{Ch}_p(X) \times \mathrm{Ch}_q(X) \overset{\bullet}{\to} \mathrm{Ch}_{p+q-n}$$

as $\deg(X) =$ degree of the cycle $[X] \bullet [L]$. Now one applies the 'principle of conservation of number' (see [10, Section 10.2] which says that the degree of the class of the zero cycle $(Z \cdot [t]) \cdot [L]$, $L \subset \mathbb{P}$ a subvariety of complementary dimension, is constant, say d, and is called the *degree of Z over T*. Over the Zariski-open subset $U \subset T$ consisting of smooth points over which Z is flat, this is the degree of the scheme-theoretic fiber as defined above.

The rational map

$$\varphi_Z : T \to C_{m,d}$$
$$t \mapsto \{\text{Chow point of } Z \cdot [t]\}$$

is defined over the smooth locus of T. It is a morphism over the locus where Z

is flat. To see that it extends as a morphism over the smooth locus, one needs the following continuity property of Chow varieties.

Lemma 10 *Suppose that S is a smooth variety and let Z be a relative effective m-cycle of X over S. Let $U \subset S$ be the Zariski-open subset over which Z is flat. Fix $s_0 \in S$. For any sequence $\{s_n \mid n = 1, 2, \ldots\}$ of points $s_n \in U$ converging to s_0, the limit of the Chow points of the cycles Z_{s_n} exists and is equal to the Chow point of the intersection-theoretic fiber $Z \cdot [s_0]$. In particular, this limit is independent of the chosen sequence in U.*

The relevance of this lemma in this context shows itself when one wants to prove that a rational map $f : T \to V$ defined over U is everywhere defined (but only as a continuous map in the complex topology). One considers its graph $\Gamma_f \subset T \times V$ whose points above t are in fact the limit points $(t_n, f(t_n))$ where $t_n \in U$ converges to t. So, if the limit $\lim_{n \to \infty} f(t_n)$ is independent of the chosen sequence, the rational map extends to an everywhere defined continuous map whose graph maps bijectively to V under the projection. This is the meaning of the following.

Definition 11 Let T, V be projective varieties. A (set-theoretic) map $f : T \to V$ is called a *continuous algebraic map* if its graph Γ_f is a subvariety of $T \times V$ and projection onto T induces a birational bijective morphism $\Gamma_f \to T$. More generally, if T and V are (not necessarily finite) disjoint unions of projective algebraic varieties, a map $f : T \to V$ is a continuous algebraic map if its restriction to each of the corresponding irreducible components T_α of T induces a continuous algebraic map $f \mid T_\alpha : Y_\alpha \to V_\beta$, V_β a component of V. A *bi-continuous algebraic map* is a bijective continuous algebraic map whose inverse is a continuous algebraic map.

Clearly, a continuous algebraic map is the same thing as a rational map which is everywhere defined and continuous (in the complex topology) and for Y normal, it is just a morphism (every bijective birational map from a normal variety Y to Z is a morphism). Bi-continuous algebraic maps are always homeomorphisms in the complex topology.

To understand continuous algebraic maps, one introduces

Definition-Lemma 12 The *weak normalization* $w : X^{\mathrm{wn}} \to X$ of X is the unique morphism of varieties over which the normalization $n : X^{\mathrm{n}} \to X$ factors and such that w is a homeomorphism. It is characterized by the property that $w_* \mathcal{O}_{X^{\mathrm{wn}}}$ is the sheaf of continuous meromorphic functions on X.

If $X = X^{\mathrm{wn}}$ we say that X is *weakly normal*. Weakly normal curves are precisely the unibranch curves, i.e. those that are locally irreducible. Any continuous algebraic map $f : X \to Y$ gives rise to a morphism $f \circ w : X^{wn} \to Y$ and conversely. For weakly normal varieties the continuous algebraic maps are precisely the morphisms.

Example 13

1. Let Y be a cuspidal cubic and $X \to Y$ its normalization. This is a bi-continuous algebraic map but not an isomorphism. The curve Y is unibranch (and hence weakly normal).
2. If T is smooth, the map $\varphi_Z : T \to C_{m,d}(X)$ defined in the preceding lemma is a continuous algebraic map and hence a morphism. In the case when $t_0 \in T$ is a singular point at which T is locally irreducible, one can still define a cycle Z_{t_0} of degree d, which differs from the cycle associated to the scheme theoretic fiber in such a way that for *weakly normal* T the resulting map

$$\varphi_Z : T \to C_{m,d}(X)$$

is a continuous algebraic map and hence a morphism, (see [2] and [8]).
3. The addition $C_m(X) \times C_m(X) \to C_{n+m}(X)$ is clearly an algebraic morphism. In terms of Chow points, this amounts to multiplication of Chow forms (by definition), which is an algebraic map.

2.2.2 Functoriality

The aim here is to discuss the nature of the map on Chow varieties induced by morphisms $f : X \to Y$ between projective varieties. Let us start with cycles themselves. Recall [10, p. 11] that to any variety $V \subset X$ there is associated a *push-forward cycle* f_*V which by definition is 0 if $\dim f(V) < \dim V$ and otherwise equals $f(V)$ with multiplicity $(\mathbb{C}(V) : \mathbb{C}(W))$. This map then is extended by linearity to all effective cycles on X, yielding a morphism of monoids of effective cycles

$$f_* : C_m(X) \to C_m(Y).$$

Also [10, p. 18], if $f : X \to Y$ is flat of relative dimension $c = \dim X - \dim Y$, for each variety W the scheme-theoretic inverse image $f^{-1}W$ is a pure $(c + \dim W)$-dimensional scheme whose associated cycle defines the *flat pull-back* $f^*[W]$ of the cycle $[W]$. Again, this is extended by linearity to all effective cycles of Y, yielding a morphism of monoids

$$f^* : C_m(Y) \to C_{m+c}(X).$$

We want to discuss now the maps induced on Chow varieties by a given morphism $f : X \to Y$. In particular, we want that the above maps induce *continuous algebraic* maps.

Proposition 14 *Fix some component $U \subset C_{m,d}(X)$ and consider the proper push-forward cycles $f_*(Z_u)$ where Z_u is the cycle with Chow point u. Assume that this degree is generically e. Then the degree is e for all $u \in U$ and the map*

$$\mathfrak{f} : U \to C_{e,m}(Y)$$
$$u \mapsto \text{Chow point of } f_*(u)$$

is a continuous algebraic map.

Proof Note first that $\mathfrak{f}$ is a rational map. To see this, assume for simplicity that $f : X \to Y$ is induced by a morphism $F : P \to P'$ of projective spaces. Then for an irreducible subvariety $V \subset P$ the Chow form F_W of the proper push-forward $f(V) = W$ is characterized by $F_W(l'_0, \dots, l'_m) = 0$ if and only if $l'_0 \cap \dots \cap l'_m \cap W \neq \emptyset$. This is the case if and only if $F_V(F^* l'_0, \dots, F^* l'_m) = 0$. So the rational map $P(m, d) \to P'(m, d)$ defined by $F^* l'_k = l_k$ sends the component U to $C_{m,d}(Y)$. Next, let us see that this map is defined at any point $u_0 \in U$ and is continuous there. One chooses a pointed smooth curve (C, c_0) and a morphism $(C, c_0) \xrightarrow{\varphi} (U, u_0)$ such that a pointed disc around c_0 maps entirely into the locus where $\mathfrak{f}$ is defined. Let $\{c_n\}_{n=1,\dots}$ be a sequence of points in this neighbourhood converging to c_0 and put $x_n = \mathfrak{f}c_n$.

The equi-dimensional relative m-cycle Z_φ over C yields the equi-dimensional relative m-cycle $(1 \times f)_* Z_\varphi$ with fiber $f_* Z_{u_0}$ over u_0. This cycle defines in its turn the morphism $\psi = \mathfrak{f} \circ \varphi : C \to C_{m,e}(Y)$. By Lemma 10 the limit of the Chow points $\mathfrak{f}(u_n)$ is equal to $\lim \psi(c_n)$, the Chow point of $f_* Z_{u_0}$. But this limit is independent of the chosen curve and hence the map $\mathfrak{f}$ is a continuous algebraic map. $\qquad\square$

In a similar fashion one can prove (see [5, Proposition 2.9])

Proposition 15 *Let $f : X \to Y$ be a flat morphism of relative dimension $c = \dim X - \dim Y$. Let $V \subset C_{m,d}(Y)$ be a component. If the degree of the flat pull-back $f^*[Z_v]$ is e, where $[Z_v]$ is a generic cycle with Chow point $v \in V$, the degree of all flat pull-backs of cycles with Chow point in V is e and there results a continuous algebraic map*

$$V \to C_{m+c,e}(X).$$

2.2.3 Cycle spaces and their group completions

The next step is to define

$$\mathcal{C}_m(X) = \mathcal{C}_m(X, i) = \coprod_{d \geq 0} C_{m,d}(X)$$

with the topology on each component given by a fixed projective embedding $i : X \hookrightarrow \mathbb{P}$. To show that this topology in fact does not depend on the embedding, one considers another embedding $i' : X \hookrightarrow \mathbb{P}'$ and the Segre embedding $\mathbb{P} \times \mathbb{P}' \to \mathbb{P}''$. This defines an embedding $i'' : X \to \mathbb{P}''$ dominating both embeddings. There are evident bijections $\mathcal{C}_m(X, i'') \to \mathcal{C}_m(X, i)$ and $\mathcal{C}_m(X, i'') \to \mathcal{C}_m(X, i')$ induced by the projections from $P \times P'$ onto its factors. These are continuous algebraic maps. This can be seen as follows. Let $X_i, i = 1, \ldots, N$ respectively $Y_j, j = 1, \ldots, M$ be homogeneous coordinates on $\mathbb{P}$ respectively $\mathbb{P}'$. Then $T_{ij}, i = 1, \ldots, N, j = 1, \ldots, M$ can be taken as homogeneous coordinates on $\mathbb{P}''$ with Segre embedding given by $T_{ij} = X_i Y_j$. If the Chow point of Z in the embedding i'' is given by the form $F(T_{ij}^{(\alpha)})$ separately homogeneous of the same degree d in each of the $m + 1$ sets of variables $T_{ij}^{(\alpha)}, \alpha = 0, \ldots, m$, we may write $F(X_i^{(\alpha)} Y_j^{(\alpha)})$ as a sum of products $G_s(X_i^{(\alpha)}) H_s(Y_j^{(\alpha)}), s = 1, \ldots, S$ such that G_s and H_s are bihomogeneous in each set of variables. The greatest common divisor $G(X_i^{(\alpha)})$ of $G_1, \ldots, G_s$ is a form and this form is the Chow form of Z in the embedding i. The assignment $F \mapsto G$ is algebraic in the sense that the coefficients of G depend rationally on the coefficients of F and hence the bijection $\mathcal{C}_m(X, i'') \to \mathcal{C}_m(X, i)$ is a (birational) morphism and hence a bi-continuous algebraic map. It follows that the induced bijection $\mathcal{C}_m(X, i) \to \mathcal{C}_m(X, i')$ is a bi-continuous algebraic map. It is in particular a homeomorphism and the topology on $\mathcal{C}_m(X, i)$ is therefore independent of the embedding i. In the sequel we therefore omit the reference to i.

From Propositions 14 and 15 one obtains immediately:

Proposition 16 *A morphism $f : X \to Y$ between projective varieties induces a continuous algebraic map*

$$f_* : \mathcal{C}_m(X) \to \mathcal{C}_m(Y)$$

and if f is flat, there is an induced continuous algebraic map

$$f_* : \mathcal{C}_m(Y) \to \mathcal{C}_{m+c}(Y), \qquad c = \dim X - \dim Y.$$

The last step consists in considering the group of possibly non-effective m-cycles on projective variety X.

Definition 17 The *naïve group completion* $\mathcal{Z}_m(X)$ of $\mathcal{C}_m(X)$ consists of considering the set of m-cycles as the topological quotient

$$\mu_m : \mathcal{C}_m(X) \times \mathcal{C}_m(X) \to \mathcal{Z}_m(X)$$

under the equivalence relation $(Z_1, Z_2) \sim (Z_1', Z_2')$ if $Z_1 + Z_2' = Z_1' + Z_2$. In other words $\mu_m(Z_1, Z_2) = Z_1 - Z_2$ is a well defined m-cycle and any m-cycle can be written as a unique equivalence class.

We already explained how to put a topology on the naïve group completion. As to functoriality, the preceding proposition implies:

Corollary 18 *A morphism* $f : X \to Y$ *between projective varieties induces a group homomorphism which is a continuous map*

$$f_* : \mathcal{Z}_m(X) \to \mathcal{Z}_m(Y)$$

and if f *is flat, there is an induced group homomorphism which is a continuous map*

$$f_* : \mathcal{Z}_m(Y) \to \mathcal{Z}_{m+c}(Y), \quad c = \dim X - \dim Y.$$

2.3 Defining Lawson homology

2.3.1 Simplicial stuff

Apart from the definition given in Section 2.1, there are various other equivalent definitions, each having its advantages. These definitions all use the language of *simplicial spaces*, and so we will first briefly review this.

Recall that the standard p-simplex Δ_p is the convex hull in $\mathbb{R}^{p+1}$ of the $p + 1$ standard unit vectors

$$\Delta_p = \{(x_0, \ldots, x_p) \mid x_i \geq 0, \sum_i x_i = 1\}.$$

Its boundary consists of the $(p - 1)$-simplices $\Delta_p^q = \Delta_p \cap \{x_q = 0\}$, $q = 1, \ldots, p + 1$ inducing the embeddings $\Delta^q : \Delta_{p-1} \to \Delta_p$. Its vertices, the $p + 1$ standard unit vectors, are often identified with elements from the ordered set $\{0, \ldots, p\}$ by the correspondence $i \Longleftrightarrow e_i$. The standard p-simplex will also be denoted by $[p]$, which means the *ordered* set $\{0, \ldots, p\}$. Thus, the maps Δ^q are examples of non-decreasing maps $[p - 1] \to [p]$. Other examples of non-decreasing maps are the degeneracy maps $\sigma^q : \Delta_p \to \Delta_{p-1}, q = 0, \ldots, p$ defined by $\sigma^q e_0 = e_0, \ldots, \sigma^q e_q = \sigma^q e_{q+1} = e_q, \sigma^q e_{q+2} = e_{q+1}, \ldots, \sigma^q e_p = e_{p-1}$. All non-decreasing maps are obtained upon composing face and degeneracy maps.

The standard simplices and all their face and degeneracy maps form a *co-simplicial set*

$$\Delta_0 \rightrightarrows \Delta_1 \Rrightarrow \Delta_2 \Rrightarrow \Delta_3 \cdots$$

Non-decreasing maps form the morphisms of a category Δ whose objects are the standard simplices Δ_n. All information of this category is given by the corresponding co-simplicial set. Dually, a *simplicial set* $K_\bullet$ is a collection of sets $K_0, K_1, \ldots$ together with mappings $K(f) : K_p \to K_q$, one for each non-decreasing map $f : [q] \to [p]$ such that

$$K(\mathrm{id}) = \mathrm{id}, \qquad K(g \circ f) = K(f) \circ K(g).$$

In other words, if considering the collection of standard simplices with non-decreasing maps as a category, a simplicial set is just a contravariant functor of this category to the category of sets. So, a simplicial set can be given as a diagram as before, but by reversing the arrows.

$$\cdots K_3 \Rrightarrow K_2 \Rrightarrow K_1 \rightrightarrows K_0$$

The arrows correspond to the face and degeneracy maps, so we have face maps $d_j : K_n \to K_{n-1}$ for $j = 0, \ldots, n$ and degeneracy maps $s_j : K_n \to K_{n+1}$, $j = 0, \ldots, n$. These satisfy certain compatibility relations (the *simplicial identities* to be found in [16, Section 1] they do not play a role here) which can be used to define a simplicial space directly. Using these, one may write any non-decreasing map in a unique way in the form $s_{j_t} \circ \cdots \circ s_{j_1} \circ d_{i_s} \circ \cdots \circ d_{i_1}$, which makes precise how the simplicial set is determined completely by the data of face and degeneracy maps.

Example 19

1. The complex of singular simplices $S_\bullet(X)$ (with the usual face and degeneracy maps).
2. Any simplicial complex can be viewed as a simplicial set by considering its simplices as non-degenerate simplices and by adding to these the degenerate simplices obtained by letting the face and degeneracy maps act on these. The simplicial unit interval I, or more generally the ordinary n-simplices Δ^n, thus define simplicial sets denoted $\Delta[n]$. The standard boundary maps $\Delta_j : \Delta^{n-1} \to \Delta^n$ (inclusion of the jth face) and degeneracy maps $\sigma_j : \Delta^{n+1} \to \Delta^n$ (collapsing by leaving out the jth vertex) induce simplicial maps $(\Delta_j)_* : \Delta[n-1] \to \Delta[n]$ and $(s_j)_* : \Delta[n+1] \to \Delta[n]$.

Using this language two degree-preserving simplicial maps $f, g : K_\bullet \to$ $L_\bullet$ are *homotopic* if there exists a simplicial map $h : K_\bullet \times \Delta[1] \to L_\bullet$ such that $d_0 h = f$ and $d_1 h = g$.

3. Let SS be the category of simplicial sets. Then for any two simplicial sets $K_\bullet$ and $L_\bullet$ the set

$$\text{Hom}_{SS}(K_\bullet, L_\bullet)$$

is also a simplicial set. Here we define an n-simplex as a simplicial map $f : K_\bullet \times \Delta[n] \to L_\bullet$, the boundary maps are defined by $d_i f(k, t) = f(k, (\sigma_i)_* t)$ and the degeneracy maps by $s_i f(k, t) = f(k, (\Delta_i)_* t), k \in K_q,$ $t \in (\Delta[n])_q$.

4. A *Kan complex* is a simplicial set satisfying the extension property: given exactly $n + 1$ simplices $\sigma_0, \sigma_1, \ldots, \sigma_{k-1}, \sigma_{k+1}, \ldots, \sigma_{n+1}$ whose boundaries match $(d_i \sigma_j = d_{j-1} \sigma_i, i < j, i \neq k, j \neq k)$, there exists an $(n + 1)$-simplex σ, whose ith boundary is σ_i. In other words, the simplicial set contains a simplex, if all but one of its faces are already in it. A standard example of a Kan complex is the singular complex (any continuous map defined on all but one of the n-dimensional faces of Δ^{n+1} extends to Δ^{n+1}).

A *simplicial map* from $K_\bullet$ to $L_\bullet$ is a degree-preserving map which commutes with face and degeneracy operators. Equivalently, it is a natural transformation of functors.

One can define a *pointed simplicial set* as a simplicial set $K_\bullet$ together with a map $[0] \to K_\bullet$. Standard examples are obtained from subcomplexes $K_\bullet \subset L_\bullet$ by considering for each n the equivalence classes L_n / K_n, where all elements of K_n are made equivalent. This yields the *quotient simplicial complex* $(L/K)_\bullet$ with natural base point the equivalence class of points in $K_\bullet$. The standard example is the *simplicial sphere* $S[n]$ obtained by identifying all $(n - 1)$-faces except the last in $\Delta[n]$.

It should be clear what is meant by a *homotopy* between two simplicial maps between (pointed) simplicial sets. This however is *not* an equivalence relation in general. For this reason Kan complexes have been introduced, since homotopy is an equivalence on $\text{Hom}_{SS}(K_\bullet, L_\bullet)$ whenever $L_\bullet$ is a Kan complex (see [16, Section 6]) and so for these we can introduce the set

$$[K_\bullet, L_\bullet]$$

of equivalence classes of pointed simplicial maps from $K_\bullet$ to $L_\bullet$ under the homotopy relation. So for a Kan complex $L_\bullet$, there are *homotopy groups*

$$\pi_n(L_\bullet) = [S[n], L_\bullet]$$
$$= \{\ell \in L_n \mid d_i \ell = *, \ \forall i\} / \sim,$$

where $\ell \sim \ell'$ if there exists some $z \in L_{n+1}$ whose boundary components $dz = (d_0z, \ldots, d_{n+1}z)$ are given by $dz = (*, \ldots, *, \ell, \ell', *, \ldots, *)$. In fact, one can introduce a product in these sets as follows. If $[x], [y] \in \pi_n(L)$, by the Kan property, there exists $v \in L_{n+1}$ such that $dv = (x, ?, y, *, \ldots, *)$ and one sets $[xy] = [?]$. This does yield a group structure for $n \geq 1$, which is abelian if $n > 1$ (see [16, Section 4]).

The functor which to any topological space X associates the simplicial set $S_\bullet(X)$ of singular simplices has a natural adjoint functor, the geometric realization functor. This functor associates to any simplicial set $K_\bullet$ the topological space

$$|K| = \left(\coprod_{p=0}^{\infty} \Delta_p \times K_p \right) \Big/ R,$$

where the equivalence relation R is generated by identifying $(s, x) \in \Delta_q \times K_q$ and $(f(s), y) \in \Delta_p \times K_p$ if $x = K(f)y$ and $f : \Delta_q \to \Delta_p$ is any non-decreasing map. The topology on $|K|$ is the quotient topology under R obtained from the direct product topology, where the sets K_p are given the discrete topology. Observe that $|K|$ has a natural structure as a CW-complex.

Let us make explicit that the geometric realization functor is adjoint to the functor of singular simplices. Given a simplicial complex $K_\bullet$ and a topological space X, there are natural bijections

$$(*) \begin{cases} \mathrm{Hom}_{SS}(K, S_\bullet X) \xrightarrow{\phi} \{\text{Continuous maps } |K| \to X\} \\ \{\text{Continuous maps } |K| \to X\} \xrightarrow{\psi} \mathrm{Hom}_{SS}(K, S_\bullet X) \end{cases}$$

given by $\phi(f)(t, k) = f(k)t$ and $\psi(g)(k)t = g(t, k)$. These preserve homotopies and so in particular, taking for K the simplicial n-sphere $S[n]$, one has

Lemma 20 *For any topological space X, the bijection $(*)$ induces an isomorphism*

$$\pi_n(X) \cong \pi_n(S_\bullet X).$$

While this assertion is quite straightforward (see [16]), the fact that for a Kan complex $K_\bullet$ the analogous isomorphism

$$\pi_n(K_\bullet) \cong \pi_n(|K|)$$

holds, is more difficult, see [16].

Even more is true. The above bijections induce adjunction morphisms

$$K_\bullet \to S_\bullet|K|$$
$$X \to |S_\bullet X|$$

(take $K = S_\bullet X$ and the identity, respectively $X = |K|$ and the identity in (∗) above). These induce isomorphisms on the level of homotopy (see [19, Section 8.6]. We have seen that a weak homotopy equivalence between CW-complexes is a homotopy equivalence. In particular, any Kan complex K has the same homotopy type as the singular complex of its geometric realization (i.e. $K_\bullet$ is homotopy equivalent to $S_\bullet(|K|)$), and any CW-complex X has the same homotopy type as the geometric realization of its associated singular complex (i.e. X is homotopy equivalent to $|S_\bullet X|$). So, in homotopy theory CW-complexes can be replaced by Kan complexes.

Simplicial sets are very flexible. For instance the K_p could have extra structure, i.e. they could be topological spaces, complex varieties, groups, monoids etc. In fact, a *simplicial object in a category* $\mathfrak{C}$ is a contravariant functor from the category Δ of standard simplices to the category $\mathfrak{C}$. We then speak of a *simplicial topological space, simplicial complex variety, simplicial group, simplicial monoid* etc.

A simplicial group is a Kan complex [19, Theorem 8.3.1]. Its homotopy groups π_i are all abelian for $i \geq 1$; in fact this is true for a Kan monoid complex [16, Proposition 17.3]. The homotopy groups can be calculated as the homology groups of a (not necessary abelian) chain complex, the *normalized chain complex*

$$NG_\bullet = \{\cdots \to NG_n \xrightarrow{d_0} NG_{n-1} \xrightarrow{d_0} \cdots NG_2 \xrightarrow{d_0} NG_1\},$$

where $NG_n \triangleleft G_n$ is the intersection of the kernel of all face maps except d_0. Since $\mathrm{Im}\,d_0$ is a normal subgroup of $\ker d_0$ one can form the quotient group $H_n(NG_\bullet, d_0)$. One has [19, Theorem 8.3.2]

$$\pi_n(G_\bullet) = H_n(NG_\bullet, d_0).$$

In the case when $G_\bullet$ is abelian, the normalized chain complex is a subcomplex of a complex, naturally associated to the simplicial set, and denoted by the same letter:

$$G_\bullet = \{\cdots G_p \xrightarrow{\partial_p} G_{p-1} \xrightarrow{\partial_{p-1}} G_{p-2} \cdots \xrightarrow{\partial_1} G_0.\}$$

where $\partial_p = \sum_{q=0}^{p}(-1)^q G(i_q)$. The inclusion of $NG_\bullet \subset G_\bullet$ is a chain homotopy [19, Theorem 8.5.1], and so, for *abelian* groups

$$H_n(G_\bullet, \partial) \cong H_n(NG_\bullet, d_0) \cong \pi_n(G_\bullet).$$

This is useful if one wants to define long exact sequences of homotopy groups.

Lemma 21 *Let $H_\bullet \subset G_\bullet$ be an inclusion of simplicial abelian groups. The quotients G_n/H_n then form a simplicial abelian group $(G/H)_\bullet$ and there is a long exact sequence*

$$\cdots \pi_p(H_\bullet) \to \pi_p(G_\bullet) \to \pi_p((G/H)_\bullet) \to \pi_{p-1}(H_\bullet) \to \cdots$$

Proof The exact sequence of simplicial abelian groups

$$0 \to H_\bullet \to G_\bullet \to (G/H)_\bullet \to 0$$

induces a short exact sequence for the normalized chain complexes. The long exact sequence in homology then gives the result.   □

Example 22 To any simplicial set $K_\bullet$ one associates the simplicial abelian group $\mathbb{Z}K_\bullet$, obtained upon replacing K_n by the free abelian group on K_n (and the naturally induced face and degeneracy maps). This makes it possible to define *homology groups* for simplicial sets:

$$H_n(K_\bullet) = H_n(\mathbb{Z}K_\bullet).$$

Clearly, the complex of singular homology on a space X is just the complex defined by the simplicial abelian group associated to the simplicial set $S_\bullet(X)$ of singular simplices. So its homology is singular homology of X. One can show further that the adjunction morphism $K_\bullet \to S_\bullet|K|$ and its inverse $|S_\bullet X| \to X$ induce homology isomorphisms [19, Theorem 8.5.5] so that there is no ambiguity when speaking of homology of simplicial sets and topological spaces.

Let $\mathfrak{C}$ be any category. Its *classifying space* is the simplicial space given by

$$B\mathfrak{C}_n = \text{set of strings of morphisms}\{a_0 \to a_1 \to \cdots a_n\}$$

with face map d_i defined by leaving out a_i, replacing $a_{i-1} \to a_{i+1}$ by the composition of the arrows $a_{i-1} \to a_i$ and $a_i \to a_{i+1}$; the degeneracy map s_j is defined by inserting a_j and id_{a_j} between a_j and a_{j+1}; see [18].

This can be applied to obtain the classifying space for any group G by regarding it as a category with one object $*$ and whose morphisms are given by the group elements, multiplication defining composition of morphisms. This yields BG where

$$(BG)_p = \underbrace{G \times \cdots \times G}_{p \text{ times}}$$

and for any non-decreasing $f : [q] \to [p]$ one has

$$BG(f)(g_1, \ldots, g_p) = (h_1, \ldots, h_q), \qquad h_i = \prod_{j=f(i-1)+1}^{f(i)} g_j$$

$$(= e \text{ if } f(i-1) = f(i)).$$

In terms of face and degeneracy maps this comes indeed down to

$$d_i(g_1, \ldots, g_p) = (g_1, \ldots, g_{i-1}, g_i g_{i+1}, g_{i+2}, \ldots, g_p)$$
$$s_i(g_1, \ldots, g_p) = (g_1, \ldots, g_{i-1}, e, g_{i+1}, \ldots, g_{p-1}).$$

This pointed simplicial set has indeed the desired property that $\pi_1(BG) = G$ and $\pi_i(BG) = 0$ as in the case of the ordinary classifying space.

If instead we take a simplicial group, we have to modify this as follows:

$$(BG)_p = G_{p-1} \times \cdots \times G_0, \qquad p \geq 1$$
$$(BG)_0 = *, \qquad \text{one point.}$$

We leave the determination of face and degeneracy operators as an exercise for the reader.

Note that the notation $(BG)_\bullet$ is consistent in that it gives back the old construction for a simplicial group associated to a group. It can be proved that this simplicial set plays the role of the classifying space in the set-up of Kan complexes; see [16, Section 21].

Note also that the simplicial complexes $K_\bullet = (BG)_\bullet$ thus obtained are *reduced*, meaning that K_0 is a single point. There is an adjoint functor from reduced complexes $K_\bullet$ to group complexes which plays the role of the loop-space functor and which is defined as follows. One sets

$$(\Omega K)_n = (\text{free group on } K_{n+1}) / \sim,$$
$$s_0 x \sim 1, \qquad \forall x \in K_n$$

and one defines face and degeneracy maps on the generators as follows (here $[-]$ denotes the class of $- \in K_{n+1}$ in $(\Omega K)_n$ and $x \in K_{n+1}$)

$$d_0[x] = [d_0 x]^{-1}[d_1 x]$$
$$d_i[x] = [d_{i+1} x], \quad i \geq 1$$
$$s_i[x] = [s_{i+1} x], \quad i \geq 0.$$

Of course $(\Omega K)_n$ is a free group, and the maps above extend uniquely to group homomorphisms, making $(\Omega K)_\bullet$ into a simplicial group.

See [16, Section 27] for a proof that Ω and B are adjoint functors. Kan has shown (see [12]):

Lemma 23 *The adjunction morphism*

$$\Psi(G_\bullet) : \Omega B G_\bullet \to G_\bullet$$

is a homotopy equivalence.

Note also that the B-construction makes sense for any simplicial monoid $M_\bullet$, since one only needs multiplication and a unity for the formulas to make sense. In the sequel, we will assume that the monoid law is *abelian* and we will write it additively.

Definition 24 Let $M_\bullet$ be a simplicial abelian monoid. Its *homotopy theoretic group completion* is $(\Omega B M)_\bullet$. Its *naïve group completion* $M_\bullet^+$ is built from the naïve group completions M_n^+ of the constituent monoids M_n.

The naïve group completion imitates the construction from $\mathbb{Z}$ out of the natural numbers. So M_n^+ consists of pairs $(x, y) \in M_n^+$ modulo the equivalence relation $(x, y) \sim (x', y')$ if $x + y' = x' + y$. Clearly, the monoid operation induces one on M_n^+, making it into an abelian group. Also, face and degeneracy maps extend uniquely to give $M_\bullet^+$ the structure of a simplicial abelian group. Moreover, the natural injective monoid morphisms $M_n \to M_n^+$ given by $x \mapsto (x, 0)$ extend to

$$i : M_\bullet \to M_\bullet^+ \qquad \text{(the plus morphism).}$$

The plus morphism induces a natural homomorphism of simplicial abelian groups

$$u : \Omega B M_\bullet \xrightarrow{\Omega B(i)} \Omega B M_\bullet^+ \xrightarrow{\Psi} M_\bullet^+.$$

Quillen has shown (see [9, Appendix Q]) that the map

$$B M_\bullet \xrightarrow{B(i)} B(M_\bullet^+)$$

is a homotopy equivalence. Since $\Omega B(i)$ is then also a homotopy equivalence, and one knows already that Ψ is a homotopy equivalence, this holds likewise for the composition u. So in homotopy theory there is no difference between the naïve group completion and the homotopy theoretic group completion. Summarizing, one has

Proposition 25 *Let $M_\bullet$ be a simplicial abelian monoid. There is a natural homotopy equivalence $\Omega B M_\bullet \to M_\bullet^+$ between the the homotopy theoretic group completion and the naïve group completion. This holds in particular for abelian monoids themselves and for the simplicial monoid of singular simplices $S_\bullet X$ of a topological space X.*

2.3.2 Base systems

The fundamental idea behind this is that one wants to glue together the components of a topological abelian monoid M by choosing base points in each connected component and using the addition for the gluing procedure. Formally, a *base system* for M is a pair (I, b) consisting of a set I and a map $b : I \to M$ such that every connected component of M contains at least one point in the image of b. The free group $F(I)$ has a natural partial ordering given by $\lambda \leq \mu$ if $\exists v \in F(I)$ with $\mu = \lambda + v$. This ordering can be used to define the *associated directed monoid* $\overrightarrow{M}_b$ by taking one copy $M = M_\lambda$ for each element $\lambda \in F(I)$ and by defining $b_{\lambda\mu} : M_\lambda \to M_\mu$ by $x \mapsto x + b(v)$. This map is a continuous base point preserving map (but it does not preserve the addition). The topological space

$$\varinjlim_{b} \overrightarrow{M}_b$$

is defined to be the infinite mapping telescope obtained by taking the disjoint union of the $M_\lambda \times I$ and identifying $(x, 1)$ with $(b_{\lambda\mu}(x), 0)$. It can be viewed as the limit space associated to the directed monoid.

Identifying M with M_0, where $0 \in F(I)$ is the zero element, there is a natural map

$$i_b : M \to \overrightarrow{M}_b$$

which induces a map to the mapping telescope and which will be denoted by the same letter.

Example 26 Let X be a connected topological space and let $X^{(d)}$ be its dth symmetric power. Let $*$ be a base point in X and use $*^d$ as the base point in $X^{(d)}$. One takes $I = \mathbb{N}$ and one defines $b(d) = *^d$. Using the inclusions $X^{(e)} \hookrightarrow X^{(d+e)}$ defined by $[x] \mapsto [(x, *^d)]$ one builds an inductive system whose limit is $X^{(\infty)}$. The disjoint union

$$X^{[\infty]} := \coprod_{d \geq 0} X^{(d)}$$

is an abelian monoid and the choice of base points induces the structure of a directed monoid $\overrightarrow{X}_*$ whose limit is exactly $X^{(\infty)}$.

The classical Dold–Thom theorem [3] states:

Theorem 27 *Let X be a CW-complex X. There are natural isomorphisms*

$$\pi_k(X^{(\infty)}) \xrightarrow{\sim} H_k(X)$$

such that the Hurewicz map $\pi_k(X) \to H_k(X)$ is obtained after composing this *isomorphism with the homomorphism $\pi_k(X) \to \pi_k(X^{(\infty)})$ induced by the obvious inclusion $X \to X^{(\infty)}$.*

Let us next define the *homotopy groups of a directed monoid $\vec{M}_b$* by first applying the homotopy functor to the directed monoid and then taking the direct limit of the associated direct system

$$\pi_k(\vec{M}_b) := \varinjlim_\lambda \pi_k(M_\lambda).$$

Of course this equals the homotopy group $\pi_k(\varinjlim_b \vec{M}_b)$ of the corresponding mapping telescope, but one rarely uses this.

Let us next compare $\vec{M}_b$ with the associated singular simplicial directed set $\vec{S_\bullet M}_b$. Again, it is clear that the functor S applied to the directed monoid yields a direct system of simplicial sets and one may form

$$\varinjlim_b \vec{S_\bullet M}_b,$$

its direct limit. Also, the map $i_b : M \to \vec{M}_b$ induces the monoid homomorphism

$$S(i_b) : S_\bullet(M) \to \varinjlim_b \vec{S_\bullet M}_b$$

and the result from [9, Section 2.7] states that in fact up to homotopy this is the $+$-construction (naïve group completion). Note that the monoid structure on M induces one on $S_\bullet M$ and it is this monoid structure that is meant to be completed. To define the comparison map, let $b_n(\lambda) \in S_n(M), \lambda \in F(I)$, be the totally degenerate n-simplex at the point $b(\lambda)$. Then define

$$S_n(M_\lambda) \to (S_n M)^+$$
$$s \mapsto s - b_n(\lambda).$$

This induces

$$h : \varinjlim_b \vec{S_\bullet M}_b \to (S_\bullet M)^+$$

and one has

Proposition 28 *In the following commutative diagram*

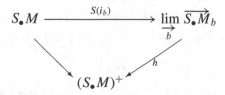

the map h is a homotopy equivalence.

The proof of this result uses the full strength of Quillen's result reproduced in [9, Appendix Q].

2.3.3 Tractable monoids

One final comparison has to be made; one needs to know in what sense the plus construction and the simplicial functor commute. To explain this, one has to remark that the naïve group completion of an abelian monoid M is the quotient of $M \times M$ under the diagonal action of M and there is a natural map

$$S_M : S_\bullet(M \times M)/S_\bullet M \to S_\bullet((M \times M)/M).$$

Suppose that one can prove that this is a homotopy equivalence. Then one can combine it with the next easy lemma to conclude that the natural map $(S_\bullet M)^+ \to S_\bullet(M^+)$ is a homotopy equivalence as well.

Lemma 29 $S_\bullet(M \times M)/S_\bullet(M)$ *is naturally isomorphic to* $(S_\bullet M)^+$.

Proof Let us map $S_\bullet(M \times M)$ to $S_\bullet M \times S_\bullet M$ by sending $u \in S_n(M \times M)$ to $(p_1 u, p_2 u) \in S_n(M) \times S_n(M)$, where p_1 and p_2 are induced by the projections. Since the diagonal action of $S_\bullet M$ on both sides is equivariant with respect to this map, there is an induced map between the quotients. This is a homomorphism between simplicial groups. The inverse is induced from the map $S_\bullet M \to S_\bullet(M \times M)$ which sends s to $s \times 0$. This map can be extended in a natural way to a map $(S_\bullet M)^+ \to S_\bullet(M \times M)/S_\bullet M$ by sending (s, t) to $s \times t$. That this is the sought after inverse follows from the observation that letting 0_n, the totally degenerate n-simplex at $0 \in M$, act on a singular n-simplex $s : \Delta_n \to M \times M$, yields the (degenerate) $2n$-simplex $\pi_1 s \times \pi_2 t : \Delta_{2n} \to M \times M$. $\square$

· So, more generally, for a monoid M acting on a topological space T, one needs to compare the singular simplicial set of a quotient space T/M and the quotient $S_\bullet T/S_\bullet M$. The notion of *tractable monoid action* of M on a topological space T is conceived just so that the natural map

$$S_\bullet T/S_\bullet M \to S_\bullet(T/M)$$

induces isomorphisms on homotopy groups. To explain tractability we recall an auxiliary notion, that of a particular kind of inclusion, a *cofibration* $i : A \hookrightarrow X$. This means by definition that given a map $\tilde{f} : X \to Y$ whose restriction $f = \tilde{f}|A$ fits into a homotopy $H : A \times I \to Y$ (i.e. $f(x) = H(x, 0)$), fits itself in a homotopy $\tilde{H} : X \times I \to Y$ extending H.

Definition-Lemma 30 An abelian monoid M *acts tractably* on a topological space T, if there is a filtration $T_0 \subset T_1 \cdots \subset T_n \subset$ whose topological union is T

and such that there are cofibrations $R_n \subset S_n$ such that for $n \geq 1$ the inclusions $T_{n-1} \subset T_n$ fit in a push-out square (or fiber coproduct)

$$
\begin{array}{ccc}
R_n \times M & \hookrightarrow & S_n \times M \\
\downarrow & & \downarrow \\
T_{n-1} & \hookrightarrow & T_n.
\end{array}
$$

If the cofibrations $R_n \subset S_n$ are relative CW-complexes, we say that (T, M) is a *tractable CW-space*. For these T/M is a CW-complex.

If the diagonal action of M on $M \times M$ is tractable, one says that *the monoid M is tractable*.

If M acts tractably on T and if M has the 'cancellation property' ($sx = sy$ implies $x = y$), the natural map

$$
S_{\bullet}T/S_{\bullet}M \to S_{\bullet}(T/M)
$$

induces isomorphisms on homotopy, and so since a tractable monoid M has the cancellation property, the natural map

$$
(S_{\bullet}M)^+ \to S_{\bullet}(M^+)
$$

induces isomorphisms on homotopy.

The proof of this is not hard. It can be found in [7, Theorem 1.4]; see also [6, Lemma 1.3].

Remark The terminology tractable monoids is just another name for certain properties of monoids introduced in [15].

Example 31 Let X be a projective variety and $M = \mathcal{C}_m(X)$ be the monoid of effective algebraic m-cycles with the Chow topology. Let us verify that this is a tractable monoid. So one considers $T = M \times M$ with the diagonal action of M. One takes M_d to be the degree d cycles (with respect to a fixed projective embedding). Now fix some bijection $v : \mathbb{N} \times \mathbb{N} \to \mathbb{N}$ and set

$$
T_n := \left[\bigcup_{v(a,b) \leq n} M_a \times M_b \right] \cdot M
$$
$$
S_n := M_{a_n} \times M_{b_n}, \qquad v(a_n, b_n) = n
$$
$$
R_n := \mathrm{Im} \left[\bigcup_{c>0} M_{a_n - c} \times M_{b_n - c} \times M_c \to M_{a_n} \times M_{b_n} \right].
$$

It is easily verified that these fit into a push-out diagram as above. One can now inductively provide S_n with a semi-algebraic triangulation so that $R_n \subset S_n$ is a subcomplex. Here one uses that any projective algebraic variety can be triangulated by semi-algebraic simplices in such a way that any given finite union of semi-algebraic closed subsets figures as a subcomplex (see [11]) together with the fact that the image of a semi-algebraic map under a continuous algebraic map (such as the multiplication maps which define R_n) stays semi-algebraic. This also shows that $R_n \subset S_n$ is a relative CW-complex and so $\mathcal{Z}_m(Y)$ is a tractable CW-space.

2.3.4 Application to Lawson homology

Let us apply the results of the previous two sections to the topological monoid $\mathcal{C}_m(X)$ of algebraic m-cycles on a projective variety X, where one puts the Chow topology on $\mathcal{C}_m(X)$. One chooses any base system $b : I \to \mathcal{C}_m(X)$ (see Section 2.3.2).

Collecting the results from the previous sections, one has

Theorem 32

(1) *There is a natural homotopy equivalence*

$$\Omega B \mathcal{C}_m X \to \mathcal{Z}_m X$$

from the homotopy theoretic group completion $\Omega B \mathcal{C}_m X$ of $\mathcal{C}_m X$ to the group $\mathcal{Z}_m X$ of algebraic m-cycles with the Chow topology.
(2) *There are natural homotopy equivalences*

$$\varinjlim_b \overrightarrow{S_\bullet \mathcal{C}_m X}_b \to (S_\bullet (\mathcal{C}_m X))^+ \leftarrow S_\bullet(\mathcal{Z}_m X).$$

(3) *There are natural isomorphisms*

$$\pi_k \overrightarrow{S_\bullet (\mathcal{C}_m X)}_b \tilde{\to} \pi_k (S_\bullet (\mathcal{C}_m X))^+$$

and hence natural isomorphisms

$$\pi_k \overrightarrow{\mathcal{C}_m X}_b \tilde{\to} \pi_k (S_\bullet (\mathcal{C}_m X))^+ \tilde{\leftarrow} \pi_k(\mathcal{Z}_m X) \tilde{\leftarrow} \pi_k(\Omega B \mathcal{C}_m X).$$

Proof

(1) This follows directly from Proposition 25. It is stated explicitly in [15].
(2) The first homotopy equivalence is Proposition 28. The second homotopy equivalence follows from the fact that $\mathcal{C}_m X$ is a tractable monoid (Example 31) and from Definition-Lemma 30.
(3) This follows from the fact that direct limits commute with homotopy groups.

The last assertion follows from the previous assertions together with the fact that the homotopy groups of a topological space are isomorphic to those of the associated simplicial set of singular simplices (Lemma 20). □

Example 33 Continuing with Example 26, let us look at a complex projective variety X. The group of zero cycles $\mathcal{Z}_0(X)$ is the naïve group completion of the abelian monoid $X^{[\infty]}$ of effective zero cycles. Since $\pi_k(\mathcal{Z}_0(X)) \cong \pi_k(\overrightarrow{\mathcal{C}_0 X}_b) \cong \pi_k(X^{(\infty)})$, the classical Dold–Thom theorem can thus be reinterpreted as the existence of a canonical isomorphism

$$\pi_k(\mathcal{Z}_0(X)) \xrightarrow{\sim} H_k(X).$$

Acknowledgement

We wish to thank the referee for suggestions on improving the exposition.

References

[1] Almgren, F. Homotopy groups of the integral cycle groups, *Topology* **1** (1962) 257–299.

[2] Barlet, D. Espace analytique réduit des cycles analytiques complexes compacts d'un espace analytique complex de dimension finie. In *Fonctions de Plusieurs Variables Complexes II, Lecture Notes in Mathematics*, **482** pp. 1–158, 1975.

[3] Dold, A. and Thom, R. Quasifaserungen und unendliche symmetrische Produkte, *Ann. Math.* **67** (1958) 239–281.

[4] Federer, H. and Fleming, W. Normal and integral currents, *Ann. Math.* **72** (1960) 458–520.

[5] Friedlander, E. Algebraic cycles, Chow varieties, and Lawson homology, *Compositio Math.* **77** (1991) 55–93.

[6] ——— Algebraic cocycles on normal, quasi-projective varieties, *Compositio Math.* **110** (1998) 127–162.

[7] Friedlander, E. and Gabber O. Cycle spaces and intersection theory. *Topological Methods in Modern Mathematics*, pp. 325–370, Publish or Perish, 1993.

[8] Friedlander, E. and Lawson B. A theory of algebraic cocycles, *Ann. Math.* **136** (1992) 361–428.

[9] Friedlander, E. and Mazur B. Filtration on the homology of algebraic varieties, *Mem. Am. Math. Soc.* **529** (1994).

[10] Fulton, W. *Intersection Theory*, Springer Verlag, 1984.

[11] Hironaka, H. Triangulation of algebraic sets, *Proc. Symp. Pure Math.* **29** (1976) 165–185.

[12] Kan, D. On homotopy theory and c.s.s. groups, *Ann. Math.* **68** (1958) 282–312.

[13] Lawson, B. Algebraic cycles and homotopy theory, *Ann. Math.* **129** (1989) 253–291.

[14] ——— Spaces of algebraic cycles. In *Surveys in Differential Geometry* **2**, pp. 137–213, International Press, 1995.

[15] Lima-Filho, P. Completions and fibrations for topological monoids, *Trans. Am. Math. Soc.* **340** (1993) 127–147.

[16] May, P. *Simplicial Objects in Algebraic Topology*, Van Nostrand, 1967.

[17] Munkres, E. *Elementary Differential Topology, Ann. Math. Studies.* **54**, Princeton University Press, Princeton, NJ, 1963.

[18] Segal, G. Classifying spaces and spectral sequences, *Publ. Math., Inst. Hautes Etudes Sci.* **34** (1968) 105–112.

[19] Selick, P. *Introduction to Homotopy Theory*, Fields Institute Monographs, AMS, Providence, RI, 1997.

[20] Spanier, E. *Algebraic Topology*, Springer Verlag, 1966.

Part II

Lawson (co)homology

3
Topological properties of the algebraic cycles functor

Paulo Lima-Filho
Department of Mathematics, Texas A&M University, College Station, TX 77840, USA

3.1 Introduction

In this survey, we present various applications of algebraic cycles, considered as a *topological group functor*, to the study of complex algebraic varieties. In the classical literature in algebraic geometry, one can already find a vast number of results and techniques that can be considered as precursors of the approach presented here. For example, the study of divisors on Riemann surfaces, Picard groups and Albanese varieties [GH78], Abel–Jacobi maps and generalized Jacobians [Gri68], [Kin83], all contain manifestations of many phenomena that occur when one studies algebraic cycles varying *continuously* on a family. In the theory presented here, many of these classical ideas are combined with homotopy theory techniques to create the appropriate framework for our study.

The approach used here was introduced in the pioneering work of Lawson [Law89], whose main objective was to prove his *complex suspension theorem* (see Theorem 3.3.11) and, in particular, to compute the homotopy type of algebraic cycles on projective spaces. However, the techniques and ideas introduced in this paper went far beyond these results. In fact, the 'complex suspension theorem' itself was a disguised form of a homotopy invariance (cf. Section 3.3.2). It was due to the insight of Friedlander [Fri91] that the techniques used in [Law89] were shown to yield highly non-trivial homology-like functors on projective varieties.

The basic premise is rather simple. Given an algebraic variety X, the group $\mathcal{Z}_p(X)$ of algebraic p-cycles on X is the free abelian group generated by the irreducible p-dimensional subvarieties of X. The starting point is the fact that one can endow $\mathcal{Z}_p(X)$ with a well behaved topology, in a functorial fashion, and

Transcendental Aspects of Algebraic Cycles ed. S. Müller-Stach and C. Peters.
© Cambridge University Press 2004.

hence one can derive various invariants for the variety X in terms of homotopy invariants for $\mathcal{Z}_p(X)$. The primary invariants obtained in this way are called the *Lawson homology* and the *morphic cohomology* of X, respectively.

We do not follow the historical development of the subject. Instead, we try to present the theory in a more structured way, providing a natural perspective on some of its fundamental features. We hope that this perspective, along with the various exercises included in the text, will have a sound pedagogical value for non-specialists. In a few places we provide completely new and simpler proofs of key results, and prove generalizations of many others. However, we emphasize that this is not an extensive survey, but rather an incursion into various facets of the theory. For a more comprehensive historical account of the subject, we refer the reader to the excellent survey [Law95].

We compiled results from various sources, notably the works of Lawson [Law89]; Lawson and Michelsohn [LM88]; Friedlander [Fri91], [Fri95]; Fried-lander and Mazur [FM94]; Friedlander and Lawson [FL92], [FL97], [FL98]; Friedlander and Gabber [FG93]; Lima-Filho [LF92], [LF93a], [LF93b], [LF94]; Lawson, Lima-Filho and Michelsohn [LLFM], [LLFM96], [LLFM98]; dos Santos [dS]; Lima-Filho and dos Santos [dSLF]. This is not a complete list, and the author hopes that any unintentional omission will be forgiven.

Each individual section of the survey contains an introduction to its contents, hence we will only give a brief outline of the material here, leaving most citations to the main text.

We start Section 3.2 with a presentation of three approaches to introducing a topology on $\mathcal{Z}_p(X)$, each one having its own special features. Historically, the first approach used Chow varieties [Law89], [Fri91] on projective varieties, and was subsequently extended to quasi-projective varieties in [LF92]. However, we first introduce topologies using suitable families of cycles, in an approach closer to subsequent works [FL92], [LF94]. In many ways, this goes back to the aforementioned techniques in classical algebraic geometry. Ultimately, these three approaches are shown to coincide. Among their main properties one shows that the assignment $X \mapsto \mathcal{Z}_p(X)$ is *covariantly functorial for proper maps, contravariantly functorial for flat maps, and transforms closed inclusions into principal fibrations.*

In the case where X is projective, the Chow monoid $\mathcal{C}_p(X)$ of effective p-cycles can be written as a disjoint union of Chow varieties, thus becoming an abelian topological monoid. The inclusion $\mathcal{C}_p(X) \hookrightarrow \mathcal{Z}_p(X)$ is the universal (naïve) group completion of $\mathcal{C}_p(X)$ in the category of topological monoids. It is certainly desirable that this group-completion, in the level of spaces, also have suitable homotopy theoretic properties. We conclude the section with a discussion of the homotopy theoretic properties of $\mathcal{Z}_p(X)$, in the projective case. This is done by exhibiting a simplicial space, a triple bar construction $B(\mathcal{C}_p(X) \times \mathcal{C}_p(X), \mathcal{C}_p(X), *)$, whose geometric realization is homotopy equivalent to the homotopy group-completion $\Omega B \mathcal{C}_p(X)$. It turns out that $\mathcal{Z}_p(X)$ is also homotopy equivalent to the geometric realization of this simplicial space, and hence $\mathcal{Z}_p(X) \cong \Omega B \mathcal{C}_p(X)$. We describe this simplicial variety in detail, for we will use this model subsequently, when we discuss mixed Hodge structures in Section 3.5.

In Section 3.3 we introduce the Lawson homology $L_* H_*(X)$ of a complex variety X, a bigraded group where $L_p H_n(X)$ is defined as the homotopy groups $\pi_{n-2p}(\mathcal{Z}_p(X))$. The topological properties of $\mathcal{Z}_p(X)$ described in Section 3.2 immediately yield the functorial properties of Lawson homology, including localization sequences. We end the section by discussing Lawson's suspension theorem and its main consequence, the *homotopy property*, stating that the flat pull-back map $p^* : \mathcal{Z}_p(X) \to \mathcal{Z}_{p+e}(E)$ induced by a vector bundle projection $p : E \to X$ is a homotopy equivalence, inducing, in particular, an isomorphism in Lawson homology.

The first applications of the properties above are described in Section 3.4, where we present Friedlander and Mazur's [FM94] *s-map* $s : L_p H_n(X) \to L_{p-1} H_n(X)$ and its associated *cycle map* $s^p : L_p H_n(X) \to H_n(X; \mathbb{Z})$ from Lawson homology to the Borel–Moore homology of X. We extend the definition of the s-map to arbitrary varieties and provide a novel presentation of the map and its functoriality. One of the most interesting applications of this construction is Friedlander–Mazur's filtration on the Griffiths groups, interpolating between the filtrations introduced by Nori in [Nor93] and Bloch and Ogus in [BO84].

In Section 3.5 we use the simplicial space model for $\mathcal{Z}_p(X)$ described in Section 3.2.3 to provide a structure of colimits of mixed Hodge structures on the homology of cycle spaces $\mathcal{Z}_p(X)$, for arbitrary complex varieties X. We show that this structure coincides with Friedlander–Mazur's in the case where X is projective. Using the Hurewicz map we also endow Lawson homology with a colimit of mixed Hodge structures.

The intersection theory for Lawson homology, developed by Friedlander and Gabber [FG93], is explained in Section 3.6. Here we work directly on the homotopy category, instead of the original derived category approach, transforming the *deformation to the normal cone* technique into a homotopy lifting-extension problem. This gives Gysin (homotopy class of) maps $j^! : \mathcal{Z}_p(X) \to \mathcal{Z}_{p-e}$ associated to regular embeddings $j : Y \hookrightarrow X$ of codimension e. The intersection product for the Lawson homology of smooth varieties is then defined in the standard way using the Gysin map associated to the diagonal embedding $X \hookrightarrow X \times X$.

In the last section we briefly present the cohomological counterpart of Lawson homology, the morphic cohomology group $L^*H^*(X)$, introduced by Friedlander and Lawson in [FL92]. A more thorough description of its properties and the body of work associated with morphic cohomology would require an equally lengthy survey, beyond the scope of this one. We simply present the natural relation between morphic and ordinary cohomology, as a motivation to its definition. Then we conclude by displaying the surprisingly natural duality map from morphic cohomology to Lawson homology. In the deep and beautiful papers [FL97] and [FL98], it is shown, among many other results, that the duality map gives an isomorphism when the varieties are smooth.

3.2 Topological properties of algebraic cycles

Throughout these lectures an *algebraic variety* is a reduced scheme of finite type over $\mathbb{C}$. To avoid excessive notation, we sometimes use the same letter X to denote either the variety X or the space of complex points $X(\mathbb{C})$ of X with the analytic topology. We hope that the context will suffice to determine the meaning of the notation used.

Our goal here is to describe a topology on the group $\mathcal{Z}_p(X)$ of algebraic p-cycles on an arbitrary complex variety X. We present this topology in detail below and prove some of the following properties. These properties are a compilation of results proven in [Law89], [Fri91], [LF92] and [LF94].

Property 1 $\mathcal{Z}_p(X)$ is a Hausdorff topological group of the homotopy type of a CW-complex.

Property 2 The connected component $\mathcal{Z}_p(X)_o$ of the identity element $0 \in \mathcal{Z}_p(X)$ is the group $\mathcal{Z}_p(X)_{\mathrm{alg}}$ of p-cycles algebraically equivalent to zero.

Property 3 If $f : X \to Y$ is a proper morphism, then the *push-forward* map $f_* : \mathcal{Z}_p(X) \to \mathcal{Z}_p(Y)$ is a continuous homomorphism. If $f : X \to Y$ is a flat morphism of relative dimension k, then the *flat pull-back* $f^* : \mathcal{Z}_p(Y) \to \mathcal{Z}_{p+k}(X)$ is a continuous homomorphism.

Property 4 If $Y \subseteq X$ is a closed subvariety, then $\mathcal{Z}_p(Y) \hookrightarrow \mathcal{Z}_p(X)$ is a closed embedding and the sequence $\mathcal{Z}_p(Y) \hookrightarrow \mathcal{Z}_p(X) \to \mathcal{Z}_p(X - Y)$ is a principal fibration.

Remark 3.2.1 The fibration described in **Property 4** is functorial with respect to maps of pairs satisfying the conditions of **Property 3**.

The following additional properties are instrumental in many situations.

Property 5 The topology on $\mathcal{Z}_p(X)$ is given as the direct limit topology of a filtering sequence

$$\mathcal{Z}_p(X)_{\leq 1} \subseteq \mathcal{Z}_p(X)_{\leq 2} \subseteq \cdots \subseteq \mathcal{Z}_p(X)_{\leq d} \subseteq \cdots$$

of closed, compact subsets of $\mathcal{Z}_p(X)$, satisfying:
1. the inclusions $\mathcal{Z}_p(X)_{\leq d} \subseteq \mathcal{Z}_p(X)_{\leq d+1}$ are closed cofibrations;
2. if X is quasi-projective, the successive differences $\mathcal{Z}_p(X)_{\leq d+1} - \mathcal{Z}_p(X)_{\leq d}$ are homeomorphic to quasi-projective varieties;
3. the filtration is compatible with the group operation, in other words,

$$\mathcal{Z}_p(X)_{\leq d} + \mathcal{Z}_p(X)_{\leq e} \subseteq \mathcal{Z}_p(X)_{\leq d+e}.$$

We will present three different approaches to introducing such a topology on $\mathcal{Z}_p(X)$, each one having its own merits and natural properties. At the end, one can show that the three approaches produce the same topology, cf. [LF94].

3.2.1 The flat and equidimensional topologies

The first two approaches are introduced à la Bourbaki, using algebraic families of cycles.

Definition 3.2.2 Let $\mathcal{F} = \{i_\lambda : S_\lambda \to T\}_{\lambda \in \Lambda}$ be a family of maps from topological spaces S_λ into a set T. The collection $\tau_{\mathcal{F}}$ of subsets $U \subset T$ such that $i_\lambda^{-1}(U)$ is open in S_λ, for all $\lambda \in \Lambda$, defines a topology on T, the finest topology making all the maps i_λ continuous. Denote by $T^{\mathcal{F}}$ or $(T, \tau_{\mathcal{F}})$ the resulting topological space.

Remark 3.2.3 Given $(T, \tau_{\mathcal{F}})$ as above, one can show that a map $f : T \to Y$ from T to a space Y is continuous if and only if for each $i_\lambda : S_\lambda \to T$ in $\mathcal{F}$ the composition $f \circ i_\lambda : S_\lambda \to Y$ is continuous.

Our two primary examples are the following.

Example 3.2.4 (Flat families) Consider an algebraic variety X, and let (S, σ) be a pair consisting of an algebraic variety S of pure dimension k, and an algebraic cycle $\sigma = \sum_i n_i \Gamma^i$ on $S \times X$ which is flat of relative dimension p over S. One can define a map $\pi_\sigma : S(\mathbb{C}) \to \mathcal{Z}_p(X)$ defined on a closed point $s \in S(\mathbb{C})$ by $\pi_\sigma(s) = \sum_i n_i [\Gamma_s^i]$, where $[\Gamma_s^i]$ is the cycle associated to the scheme-theoretic fiber Γ_s^i, cf. [Ful84, Section 1.5].

Let $\mathcal{Z}_p(X/S)^{\text{fl}}$ denote the group of all cycles in $S \times X$ which are flat of relative dimension p over S, and define

$$\mathcal{F}_p(X)^{\text{fl}} := \{\pi_\sigma : S(\mathbb{C}) \to \mathcal{Z}_p(X) \mid \sigma \in \mathcal{Z}_p(X/S)^{\text{fl}}$$

$$\text{and } S \text{ is pure dimensional}\}. \quad (1)$$

If one gives $S(\mathbb{C})$ the analytic topology, then $\mathcal{F}_p(X)^{\text{fl}}$ becomes a family of maps from a collection of topological spaces to the set $\mathcal{Z}_p(X)$. Using Definition 3.2.2 one obtains a unique *flat topology* on $\mathcal{Z}_p(X)$, which we denote by $\mathcal{Z}_p(X)^{\text{fl}}$.

In a similar fashion, we can replace flat families over arbitrary base spaces by equidimensional families over smooth base spaces.

Example 3.2.5 (Equidimensional families) Let $\mathcal{Z}_p(X/S)^{\text{eq}}$ denote the group of all cycles σ in $S \times X$ which are *equidimensional* of relative dimension p over a smooth base S. For a closed point $s \in S(\mathbb{C})$ define $\pi_\sigma(s) = \sum_i n_i[\Gamma_s^i] \in \mathcal{Z}_p(X)$, where $[\Gamma_s^i]$ is the *intersection-theoretic* fiber $\Gamma^i \cdot (s \times X)$ over s. Define

$$\mathcal{F}_p(X)^{\text{eq}} := \{\pi_\sigma : S(\mathbb{C}) \to \mathcal{Z}_p(X) \mid \sigma \in \mathcal{Z}_p(X/S)^{\text{eq}} \text{ and } S \text{ is smooth}\}.$$

$$(2)$$

If one gives $S(\mathbb{C})$ the analytic topology, then $\mathcal{F}_p(X)^{\text{fl}}$ becomes a family of maps from a collection of topological spaces to the set $\mathcal{Z}_p(X)$. Using Definition 3.2.2 one obtains a unique *equidimensional topology* on $\mathcal{Z}_p(X)$, which we denote by $\mathcal{Z}_p(X)^{\text{fl}}$.

Exercise 3.2.1

1. Prove that, given an algebraic variety X, the flat topology is completely determined by the subfamily

$$\mathcal{F}_p(X)^{\text{fl}}_{\text{smooth}} := \{\pi_\sigma : S(\mathbb{C}) \to \mathcal{Z}_p(X) \mid \sigma \in \mathcal{Z}_p(X/S)^{\text{fl}} \text{ and } S \text{ is smooth}\}$$
$$\subset \mathcal{F}_p(X)^{\text{fl}}.$$

2. Let $f : X \to Y$ be a flat morphism of relative dimension k. Prove that the flat pull-back $f^* : \mathcal{Z}_p(Y) \to \mathcal{Z}_{p+k}(Y)$ is a continuous map.

The behaviour of these topologies under *proper push-forward* is a bit more subtle. Consider a proper map $f : X \to Y$, and let $\Gamma \subset S \times X$ be an irreducible subvariety, flat over a smooth variety S. Define $\Gamma' = (1 \times f)(\Gamma) \subset S \times Y$. The projection $\Gamma \to S$ factors as a composition $\Gamma \xrightarrow{p} \Gamma' \xrightarrow{g} S$, where p is surjective and $g \circ p$ is flat. In particular, $g \circ p$ is universally open, and so is g, cf. [GD66, Proposition 14.3.4(i)]. On the other hand, since S is normal and Γ' is irreducible, these conditions show that $\Gamma' \to S$ is equidimensional over S, cf. [GD66, Corollary 14.4.9(b)]. Now, given $s \in S(\mathbb{C})$, it follows from [Ful84, Theorem 6.2(a)] that

$$\pi_{\Gamma'}(s) := \pi_{(1 \times f)(\Gamma)}(s) = f_*(\pi_\Gamma(s)),$$

where the map $\pi_{\Gamma'}$ is defined as in Example 3.2.5 and π_Γ is defined in Example 3.2.4. This observation, along with Remark 3.2.3, proves the following.

Lemma 3.2.6 *Given a proper morphism $f : X \to Y$, the push-forward homomorphism $f_* : \mathcal{Z}_p(X)^{\mathrm{fl}} \to \mathcal{Z}_p(Y)^{\mathrm{eq}}$ is a continuous map.*

This lemma provides the first step in comparing the two topologies. The next step uses the following *flatification/extension* result.

Theorem 3.2.7 *Let $j : U \hookrightarrow V$ be a dense open subvariety of an algebraic variety V, and let S be a smooth k-dimensional affine variety. Given a subvariety $\Gamma \subset S \times U$, which is equidimensional of relative dimension p over S, there are a blow-up $b : \tilde{S} \to S$ and a subvariety $\tilde{\Gamma}$ of $\tilde{S} \times V$ flat over $\tilde{S}$, with relative dimension P and satisfying the following:*

(i) *$\tilde{\Gamma}$ is the proper transform of the closure $\overline{\Gamma}$ of Γ in $S \times V$;*
(ii) *the diagram*

$$
\begin{array}{ccc}
\tilde{S}(\mathbb{C}) & \xrightarrow{\pi_{\tilde{\Gamma}}} & \mathcal{Z}_p(V) \\
\downarrow{\scriptstyle b} & & \downarrow{\scriptstyle j^*} \\
S(\mathbb{C}) & \xrightarrow{\pi_\Gamma} & \mathcal{Z}_p(U)
\end{array}
$$

commutes, where $\pi_{\tilde{\Gamma}} \in \mathcal{F}_p(X)^{\mathrm{fl}}$ and $\pi_\Gamma \in \mathcal{F}_p(X)^{\mathrm{eq}}$.

The next result follows from this proposition.

Theorem 3.2.8 [LF94, Corollary 3.5] *Let $j : U \hookrightarrow V$ be the inclusion of an open dense subvariety of V. Then the restriction homomorphism $j^* : \mathcal{Z}_p(V)^{\mathrm{fl}} \to \mathcal{Z}_p(U)^{\mathrm{eq}}$ is a quotient map.*

Corollary 3.2.9 *Given an algebraic variety X, the identity map induces a homeomorphism $i_* : \mathcal{Z}_p(X)^{\mathrm{fl}} \to \mathcal{Z}_p(X)^{\mathrm{eq}}$.*

Proof It follows from the theorem that the identity $i : X \to X$ induces a closed map $i_* : \mathcal{Z}_p(X)^{\mathrm{fl}} \to \mathcal{Z}_p(X)^{\mathrm{eq}}$, and Lemma 3.2.6 shows that this map is continuous. $\square$

Corollary 3.2.10 *Proper push-forwards and flat pull-backs are continuous maps in the flat ($=$ equidimensional) topology on groups of algebraic cycles.*

3.2.2 The Chow topology

One can introduce yet a third topology on $\mathcal{Z}_p(X)$, the *Chow topology* $\mathcal{Z}_p(X)^{\text{ch}}$. This approach allows one to prove various point-set topological properties of $\mathcal{Z}_p(X)$. The definition, however, requires a few more steps than the *flat* and *equidimensional* topologies.

Case 1: projective varieties

Given a projective embedding $X \subseteq \mathbb{P}^n$, let

$$\mathcal{C}_p(X) := \amalg_{\alpha \in \Pi_p(X)} \, \mathcal{C}_{p,\alpha}(X) \tag{3}$$

denote the *Chow monoid* of X. This is the free abelian monoid generated by the irreducible p-dimensional subvarieties of X. One can write $\mathcal{C}_p(X)$ as a disjoint union of *Chow varieties* $\mathcal{C}_{p,\alpha}(X)$. These are connected projective varieties, indexed by the monoid $\Pi_p(X)$ of *effective algebraic equivalence classes of effective algebraic p-cycles*. There is a monoid morphism deg $: \Pi_p(X) \to \mathbb{Z}_{\geq 0}$ such that $\deg^{-1}(d)$ is finite and

$$\mathcal{C}_{p,d}(X) := \amalg_{\alpha \in \deg^{-1}(d)} \mathcal{C}_{p,\alpha}(X)$$

is a finite union of closed connected subvarieties of the classical Chow variety $\mathcal{C}_{p,d}(\mathbb{P}^n)$ of effective cycles of degree d in $\mathbb{P}^n$.

Remark 3.2.11 The addition map $+ : \mathcal{C}_p(X) \times \mathcal{C}_p(X) \to \mathcal{C}_p(X)$ induces algebraic maps $+ : \mathcal{C}_{p,d}(X) \times \mathcal{C}_{p,e}(X) \to \mathcal{C}_{p,d+e}(X)$.

The group $\mathcal{Z}_p(X)$ is the Grothendieck group of the monoid $\mathcal{C}_p(X)$, which can be described as the quotient of $\mathcal{C}_p(X) \times \mathcal{C}_p(X)$ under the relation $(a, b) \sim (a', b')$ if and only if $a + b' = a' + b$. Now, define $\mathcal{Z}_p(X)^{\text{ch}}$ as the quotient topology induced by the quotient map $\rho : \mathcal{C}_p(X) \times \mathcal{C}_p(X) \to \mathcal{Z}_p(X)$.

Several properties follow from this definition. If one defines

$$\mathcal{Z}_p(X)_{\leq d} := \rho(\amalg_{r+s \leq d} \, \mathcal{C}_{p,r}(X) \times \mathcal{C}_{p,s}(X)) \subset \mathcal{Z}_p(X), \tag{4}$$

then one can show that the topology on $\mathcal{Z}_p(X)$ is the direct limit topology induced by this filtration, as described in **Property 5**. In fact, in this context we can prove **Properties 1, 3** and **5** for the Chow topology, in the case of projective varieties, by using basic results from [Ste67], and certain cofibration properties resulting from triangulations of complex algebraic varieties, cf. [Hir75]

Case 2: quasi-projective varieties

Given a closed subvariety Y of a projective variety X, it is easy to show that $\mathcal{Z}_p(Y)^{\text{ch}}$ is a closed subgroup of $\mathcal{Z}_p(X)^{\text{ch}}$, cf. [LF92]. The following result allows one to extend the Chow topology to arbitrary varieties, and has many deep consequences.

Theorem 3.2.12 [LF92] *If $f : (X, Y) \to (X', Y')$ is a relative isomorphism of pairs of projective varieties, then it induces a topological group isomorphism*

$$f_* : \mathcal{Z}_p(X)^{\mathrm{ch}}/\mathcal{Z}_p(Y)^{\mathrm{ch}} \to \mathcal{Z}_p(X')^{\mathrm{ch}}/\mathcal{Z}_p(Y')^{\mathrm{ch}}.$$

With this result, one shows that the following definition is independent of projective embeddings and compactifications.

Definition 3.2.13 [LF92] Let $U \subset \mathbb{P}^n$ be a quasi-projective variety. Define $\mathcal{Z}_p(U)^{\mathrm{ch}}$ as the topological group quotient $\mathcal{Z}_p(\overline{U})^{\mathrm{ch}}/\mathcal{Z}_p(\overline{U} - U)^{\mathrm{ch}}$.

Case 3: arbitrary varieties

An *envelope* of a variety X is a proper morphism $p : X' \to X$ such that for every closed irreducible subvariety V of X there is a subvariety V' of X' such that p maps V' birationally onto V. We call p a *Chow envelope* if, in addition, X' is a quasi-projective variety. One can show that any variety of finite type X has a Chow envelope, and that envelopes are preserved under base-extension, cf. [FG83]. Observe that if $p : X' \to X$ is an envelope, then the induced (proper push-forward) homomorphism $p_* : \mathcal{Z}_p(X') \to \mathcal{Z}_p(X)$ is surjective.

Definition 3.2.14 Given a variety X and a Chow envelope $p : X' \to X$, define the *Chow topology* $\mathcal{Z}_p(X)^{\mathrm{ch}}$ to be the quotient topology on $\mathcal{Z}_p(X)$ induced by the surjection $p_* : \mathcal{Z}_p(X')^{\mathrm{ch}} \to \mathcal{Z}_p(X)$. It is easy to see that this is independent of the Chow envelope.

Remark 3.2.15 In this case one defines $\mathcal{Z}_p(X)_{\leq d} = p_*(j^*\mathcal{Z}_p(\overline{X'})_{\leq d})$, where $j : X \hookrightarrow \overline{X'}$ is a projective compactification of X'. This filtration expresses the topology on $\mathcal{Z}_p(X)$ as a colimit of compact subsets, and is the basis of various inductive arguments used in proving **Properties 1, 3** and **5**.

It is easy to compare the Chow and flat topologies in the case of a projective variety $X \subset \mathbb{P}^n$, for one can use finitely many Hilbert schemes $Hilb_X^P$ parametrizing subschemes of $\mathbb{P}^n$, whose Hilbert polynomial has leading coefficients $\frac{d}{p!}t^p$, to produce a surjective proper map

$$\rho_d : \amalg_P Hilb_X^P \to \mathcal{C}_{p,d}(X)$$

for all d. These maps, along with the universal flat families over Hilbert schemes, provide the desired comparison.

Theorem 3.2.16 *If X is a projective variety then the identity map induces a homeomorphism $i : \mathcal{Z}_p(X)^{\mathrm{fl}} \to \mathcal{Z}_p(X)^{\mathrm{ch}}$.*

Corollary 3.2.17 *The conclusion of the theorem still holds for quasi-projective varieties.*

Proof Let $j : U \hookrightarrow \overline{U} \subset \mathbb{P}^n$ be a projective compactification of U. One has a commutative diagram

$$
\begin{array}{ccc}
\mathcal{Z}_p(\overline{U})^{\mathrm{fl}} & = & \mathcal{Z}_p(\overline{U})^{\mathrm{ch}} \\
{\scriptstyle j^*} \downarrow & & \downarrow {\scriptstyle j^*} \\
\mathcal{Z}_p(U)^{\mathrm{fl}} & \longrightarrow & \mathcal{Z}_p(U)^{\mathrm{ch}}
\end{array}
$$

where the left vertical arrow is a quotient map, according to Theorem 3.2.8, and the right vertical arrow is a quotient map by definition. This suffices to prove the corollary.   □

Remark 3.2.18 The result holds for arbitrary varieties, and the proof follows from the same flatification/extension result as Theorem 3.2.8 does. Roughly speaking, one just needs to show that any flat family of cycles in X can be lifted to a Chow envelope $p : X' \to X$ after a proper base change.

Once we have proven equality of all three topologies, we simply denote the resulting topological group by $\mathcal{Z}_p(X)$.

Exercises 3.2.2

1. Given an algebraic variety X, prove that the flat topology is determined completely by the subfamily

$$
\mathcal{F}_p(X)^{\mathrm{fl}}_{\mathrm{smooth}} := \{ \pi_\sigma : S(\mathbb{C}) \to \mathcal{Z}_p(X) \mid \sigma \in \mathcal{Z}_p(X/S)^{\mathrm{fl}}
$$

$$
\text{and } S \text{ is smooth} \} \subset \mathcal{F}_p(X)^{\mathrm{fl}}.
$$

2. Let $f : X \to Y$ be a flat morphism of relative dimension k. Prove that the flat pull-back $f^* : \mathcal{Z}_p(Y) \to \mathcal{Z}_{p+k}(Y)$ is a continuous map.

3.2.3 On group completions

In this section we give a brief discussion of homotopy theoretic properties of the group $\mathcal{Z}_p(X)$ when X is projective. A more extensive study, for a broader class of monoids, can be found in [LF93a].

Consider an arbitrary topological monoid M. If (A, M, B) is a triple where A is a right M-space and B is a left M-space, then one can construct a simplicial space $\mathcal{B}_*(A, M, B)$, called the *triple bar construction*, as follows, cf. [May75].

The space of n-simplices is defined as

$$\mathcal{B}_n(A, M, B) := A \times M^{\times n} \times B,$$

with faces defined by

$$d_i(a; m_1, \ldots, m_n; b) := \begin{cases} (a * m_1; m_2, \ldots, m_n; b) & \text{for } i = 0 \\ (a; m_1, \ldots, m_i m_{i+1}, \ldots, m_n; b) & \text{for } 1 \leq i < n \\ (a; m_1, \ldots, m_{n-1}; m_n * b) & \text{for } i = n \end{cases}$$

and degeneracies $s_i(a; m_1, \ldots, m_n; b) = (a; m_1, \ldots, m_{i-1}, 0, m_i, \ldots, m_n; b)$, where 0 is the identity element of the monoid.

This construction is functorial on triples and its *geometric realization* $\mathcal{B}(A, M, B)$ satisfies the following properties. We refer the reader to [May75] for details of this construction.

1. If (A, M, B) is such a triple and M acts trivially on C, then $\mathcal{B}(C \times A, M, B) = C \times \mathcal{B}(A, M, B)$, where M acts diagonally on $C \times A$, cf. [May75].
2. $\mathcal{B}(*, M, *) = BM$ is the classifying space of M and the map $EM := \mathcal{B}(M, M, *) \to BM$ induced by the obvious map of triples $(M, M, *) \to (*, M, *)$ is the universal quasi-fibration for M, cf. [May75].
3. Given a triple of the form $(A, M, *)$, the bar construction $\mathcal{B}(A, M, *)$ is the homotopy quotient of A under the action of M.
4. If M is abelian, then BM is an abelian monoid and so is ΩBM under pointwise addition.

The following result seems to be well known, according to the observation in [Seg74, p. 305], but the first proof known to the author is found in [LF93a].

Proposition 3.2.19 *If M is abelian, and $(M \times M, M, *)$ is the triple where M acts diagonally on $M \times M$, then ΩBM is naturally homotopy equivalent to $\mathcal{B}(M \times M, M, *)$, which is the homotopy quotient of $M \times M$ by the diagonal action.*

Remark 3.2.20 1. In [Seg74, p. 305], Segal mentions that the result still holds if the monoid is 'sufficiently abelian', by which we presume that this means up to sufficiently high coherent homotopies. We only present the proof in the case where M is actually abelian.
2. The involution $M \times M \to M \times M$ sending (m, n) to (n, m) induces an involution $\iota_M : \mathcal{B}(M \times M, M, *) \to \mathcal{B}(M \times M, M, *)$ which is natural on M and corresponds to giving the 'inverse' of an element. In other words, $\mathrm{id} + \iota_M$ is naturally homotopic to zero.

The proposition allows one to use the model $\mathcal{B}(M \times M, M, *)$ for the homotopy theoretic group-completion ΩBM of an abelian monoid M. Note that one has a map of triples $(M \times M, M, *) \to (M^+, *, *)$, where M^+ is the Grothendieck group of the monoid, with the quotient topology from $M \times M$. It turns out that, under mild conditions on the monoid M, the induced map on triple bar constructions is a homotopy equivalence, as we explain below.

Definition 3.2.21 Let M be a monoid whose topology is given by a filtration $\cdots \subseteq M_n \subset M_{n+1} \subset \cdots$. Define $(M \times M)_d = \cup_{r+s \leq d} (M_r \times M_s) \subset M \times M$, and

$$\Delta(M \times M)_d := \{(m, m') \in (M \times M)_d \mid (m + a, m' + a) = (n + b, n' + b)$$
$$\text{for some } (n, n') \in (M \times M)_{d-1} \text{ and } a, b \in M\}.$$

Define M to be properly c-filtered if M_d is compact for all d and the inclusion $\Delta(M \times M)_d \subseteq (M \times M)_d$ is a cofibration.

Example 3.2.22 Given a projective variety X, every Chow monoid $\mathcal{C}_p(X)$ is properly c-filtered. Actually, it is c-graded in the sense that the filtration comes from a grading.

Theorem 3.2.23 [LF93a] *Let M be a properly c-graded abelian topological monoid. Then the map*

$$B(M \times M, M, *) \to M^+$$

is a homotopy equivalence.

Corollary 3.2.24 *For every projective variety X, one has a homotopy equivalence*

$$\mathcal{Z}_p(X) \cong \Omega B(\mathcal{C}_p(X)).$$

Remark 3.2.25 The definition above can be extended to actions of M on a space A, mimicking the properties of the action of the diagonal on $M \times M$, as defined above. This condition was called *a tractable action of M on A* in [FG93]. The proof of the theorem yields the fact that under appropriate cofibrant filtration conditions, the homotopy quotient $\mathcal{B}(A, M, *)$ is homotopy equivalent to the actual quotient A/M of A by the action of the monoid.

3.3 Lawson homology

The *Lawson homology groups* of an algebraic variety X are defined in terms of the homotopy groups of the various topological groups $\mathcal{Z}_p(X)$. They form a family of invariants for the variety X that encodes hybrid properties of X, interpolating from purely topological invariants at one end to purely algebraic at the other. The term *Lawson homology* for projective varieties was coined by Friedlander in [Fri91] after Lawson's work [Law89]. In [Fri91] an ℓ-adic version of Lawson homology is developed for *projective varieties* over fields of characteristic $p \neq \ell$. Subsequently, the theory was extended to arbitrary (complex) varieties in [LF92] and [LF94].

Definition 3.3.1 Define the Lawson homology groups of X as

$$L_p H_n(X) := \pi_{n-2p}(\mathcal{Z}_p(X)),$$

for $n \geq 2p$ and $0 \leq p \leq \dim X$.

3.3.1 Basic properties

The topological properties of the functors $X \mapsto \mathcal{Z}_p(X)$, described in the previous section, along with basic properties of homotopy groups, yield the following basic properties.

Theorem 3.3.2 *The Lawson homology functor satisfies the following properties.*

Proper push-forward *It is a covariant functor for proper morphisms. In other words, a proper morphism $f : X \to Y$ induces group homomorphisms $f_* : L_p H_n(X) \to L_p H_n(Y)$, and $(f \circ g)_* = f_* \circ g_*$ for proper morphisms f, g.*

Flat pull-back *It is contravariant for flat morphisms. In other words, a flat morphism $f : X \to Y$ of relative dimension k induces group homomorphisms $f^* : L_p H_n(Y) \to L_{p+k} H_{n+2k}(X)$, and $(f \circ g)^* = g^* \circ f^*$ for flat morphisms f, g.*

Localization sequence *Given a closed subvariety $Y \subset X$ one has a long exact sequence*

$$\cdots \to L_p H_{n+1}(X - Y) \to L_p H_n(Y) \to L_p H_n(X)$$
$$\to L_p H_n(X - Y) \to \cdots .$$

In special cases one can recover classical invariants out of Lawson homology:

Theorem 3.3.3 *Let X be an algebraic variety.*

(i) [Fri91] $L_p H_{2p}(X) := \pi_0(\mathcal{Z}_p(X)) \cong \mathcal{A}_p(X)$, *where $\mathcal{A}_p(X)$ is the group of algebraic p-cycles on X modulo algebraic equivalence.*

(ii) [DT56] $L_0 H_n(X) := \pi_n(\mathcal{Z}_0(X)) \cong H_n^{BM}(X(\mathbb{C}); \mathbb{Z})$, *where the latter denotes the Borel–Moore homology of the analytic space $X(\mathbb{C})$ with coefficients in $\mathbb{Z}$.*

(iii) [Fri91] *Given a non-singular projective variety X of dimension n, there are isomorphisms:*

$$L_{n-1} H_{2n}(X) \cong \mathbb{Z}$$
$$L_{n-1} H_{2n-1}(X) \cong H_{2n-1}(X; \mathbb{Z})$$
$$L_{n-1} H_{2n-2}(X) \cong NS(X),$$

where the latter is the Néron–Severi group $H_{n-1,n-1}(X; \mathbb{Z})$.

3.3.2 The homotopy property

The fundamental result that triggered the development of this theory was B. Lawson's seminal work [Law89]. In order to present his *complex suspension theorem*,[1] instead of following the historical development of the subject we first introduce the *join pairing* of cycles, a construction that will appear on multiple occasions hereafter.

The join pairing

Given a finite-dimensional complex vector space V, let $\mathbb{P}(V)$ denote the projective space of complex lines through the origin in V. In other words, $\mathbb{P}(V) := \mathrm{Proj}(Sym_*(\check{V}))$, where $\check{V}$ denotes the dual of V. Let $Z \subset \mathbb{P}(V)$ and $Z' \subset \mathbb{P}(W)$ be irreducible subvarieties of dimensions r and s and degrees d and e, respectively. Under the natural embeddings $\mathbb{P}(V) \subset \mathbb{P}(V \oplus W)$ and $\mathbb{P}(W) \subset \mathbb{P}(V \oplus W)$ as two disjoint linear subspaces, one can consider both Z and Z' as subvarieties of $\mathbb{P}(V \oplus W)$. Define the *algebraic join $Z\#Z'$* of Z and Z' as the subvariety of $\mathbb{P}(V \oplus W)$ consisting of all projective lines joining points in Z to points in Z'.

Remark 3.3.4

1. Observe that if $CZ \subset V$ denotes the affine cone over Z and CZ' is the corresponding cone for Z' then $Z\#Z'$ is the subvariety of $\mathbb{P}(V \oplus W)$ whose

---

[1] This was Lawson's original terminology.

affine cone is $CZ \times CZ'$. This shows that $Z\#Z'$ is indeed an irreducible subvariety of dimension $r + s + 1$, where $r = \dim Z$ and $s = \dim S$.

2. It can also be shown that if $\deg Z = d$ and $\deg Z' = e$, then $\deg(Z\#Z') = d \cdot e$.

Definition 3.3.5 Let X and X' be projective varieties, with respective embeddings $j : X \hookrightarrow \mathbb{P}(V)$ and $j' : X' \hookrightarrow \mathbb{P}(W)$. The *join pairing*

$$\# : \mathcal{Z}_r(X) \times \mathcal{Z}_{r'}(X) \to \mathcal{Z}_{r+s}(X\#X')$$

is the bilinear extension of the join of subvarieties of X and X'. That is, for cycles $\sigma = \sum_i n_i S_i \in \mathcal{Z}_r(X)$ and $\tau = \sum_j m_j T_j \in \mathcal{Z}_s(X')$, one has $\sigma\#\tau = \sum_{i,j} n_i m_j S_i \# T_j$ as described above.

Proposition 3.3.6 *For projective subvarieties $X \subset \mathbb{P}(V)$ and $X' \subset \mathbb{P}(W)$, the join pairing $\# : \mathcal{Z}_r(X) \times \mathcal{Z}_{r'}(X) \to \mathcal{Z}_{r+s}(X\#X')$ is a continuous map.*

Proof Since X, X' and $X\#X'$ are closed subvarieties of their corresponding projective spaces, it follows from **Property 4** in Section 3.2 that one only needs to prove continuity of the pairing $\# : \mathcal{Z}_r(\mathbb{P}(V)) \times \mathcal{Z}_{r'}(\mathbb{P}(W)) \to \mathcal{Z}_{r+s+1}(\mathbb{P}(V \oplus W))$. In this case, consider the blow-up B of $\mathbb{P}(V \oplus W)$ along $\mathbb{P}(V) \amalg \mathbb{P}(W)$. It is easy to see that $B = \mathbb{P}\left(p_1^*(\mathcal{O}_{\mathbb{P}(V)}(-1)) \oplus p_2^*(\mathcal{O}_{\mathbb{P}(W)}(-1))\right)$, where $p_1 : \mathbb{P}(V) \times \mathbb{P}(W) \to \mathbb{P}(V)$ and $p_2 : \mathbb{P}(V) \times \mathbb{P}(W) \to \mathbb{P}(V)$ are the respective projections, and $\mathcal{O}(-1)$ denotes tautological bundles. Hence we have maps

$$\mathbb{P}(V \oplus W) \xleftarrow{b} B \xrightarrow{p} \mathbb{P}(V) \times \mathbb{P}(W),$$

where b is proper and p is flat of relative dimension 1. $\qquad\qquad\square$

Claim 3.3.7 *Let $\mu : \mathcal{Z}_r(\mathbb{P}(V)) \times \mathcal{Z}_s(\mathbb{P}(W)) \to \mathcal{Z}_{r+s}(\mathbb{P}(V) \times \mathbb{P}(W))$ denote the bilinear extension of the map that sends irreducible subvarieties $Z \subset \mathbb{P}(V)$ and $Z' \subset \mathbb{P}(W)$ to $Z \times Z'$. Then μ is a continuous map.*

Proof We use the 'flat families' description of the cycle spaces, cf. Section 3.2. If $\Gamma \in \mathcal{Z}_r(\mathbb{P}(V)/S)^{\mathrm{fl}}$ and $\Gamma' \in \mathcal{Z}_s(\mathbb{P}(W)/S')^{\mathrm{fl}}$ are flat cycles over smooth varieties S and S', then the product cycle $\Gamma \times \Gamma'$ is flat over $S \times S'$. Since $\mu \circ (\pi_\Gamma \times \pi_{\Gamma'}) = \pi_{\Gamma \times \Gamma'}$, where the π_- are defined in Example 3.2.4, one concludes that μ is continuous. $\qquad\qquad\square$

It is easy to show that

$$\sigma\#\sigma' = b_* \circ p^* \circ \mu(\sigma, \sigma'). \tag{5}$$

This expresses the join pairing as the composition of a proper push-forward, a flat pull-back and a continuous pairing, as shown in Claim 3.7 and **Property 3.3.7**.   □

Remark 3.3.8 The proof of continuity of the join pairing, used in [FM94] for example, uses Chow varieties and the notion of algebraic continuous maps. Subsequently, Barlet and (independently) Plümer showed that the join of cycles induces an algebraic map on the level of Chow varieties. The description of the join pairing given here appeared in [LF93b].

Lawson's suspension theorem

Using the join of algebraic cycles one can define Lawson's *complex suspension homomorphism* as follows.

Definition 3.3.9 Given a projective variety $X \subset \mathbb{P}(V)$, let $\Sigma X \subset \mathbb{P}(V \oplus \mathbb{C})$ denote the projective cone $X \# x_\infty$ over X, where $x_\infty := \mathbb{P}(0 \oplus \mathbb{C}) \in \mathbb{P}(V \oplus \mathbb{C})$. Extending this construction linearly to cycles supported on X, one obtains a continuous homomorphism:

$$\Sigma \ : \ \mathcal{Z}_p(X) \to \mathcal{Z}_{p+1}(\Sigma X),$$

called the (complex) algebraic join homomorphism.

Remark 3.3.10 Observe that[2] ΣX and $L := \Sigma X - \{x_\infty\}$ are, respectively, the Thom space and the total space of the hyperplane bundle associated to $\mathcal{O}_X(1)$ over X. Furthermore, since a $(p+1)$-cycle cannot be supported on the vertex x_∞ of the cone, one has identifications

$$\mathcal{Z}_{p+1}(\Sigma X) \equiv \mathcal{Z}_{p+1}(\Sigma X)/\mathcal{Z}_{p+1}(\{x_\infty\})$$

$$\equiv \mathcal{Z}_{p+1}(\Sigma X - \{x_\infty\}) \equiv \mathcal{Z}_{p+1}(L). \tag{6}$$

Using the 'Chow varieties description' of the topology on cycles (cf. Definition 3.2.13) one sees that the identities above are topological group isomorphisms. Under this identification, the 'suspension homomorphism' is simply the flat pull-back $\rho^* : \mathcal{Z}_p(X) \to \mathcal{Z}_{p+1}(L)$ under the bundle projection $\rho : L \to X$. We must point out, however, that Lawson's work [Law89] preceded the introduction of the topological structure of cycles on quasi-projective varieties [LF92].

---

[2] In this assertion we are actually identifying ΣX and $\Sigma X(\mathbb{C})$ with the analytic topology.

Theorem 3.3.11 [Law89] *For any projective variety X, the algebraic suspension homomorphism $\Sigma : \mathcal{Z}_p(X) \to \mathcal{Z}_{p+1}(\Sigma X)$ is a homotopy equivalence. Equivalently, pull-back induces a homotopy equivalence $\mathcal{Z}_p(X) \to \mathcal{Z}_{p+1}(L)$.*

Corollary 3.3.12 *For any projective variety $X \subset \mathbb{P}(W)$, and complex vector space V of dimension v, the join homomorphism $\mathbb{P}(V)\#(-) : \mathcal{Z}_p(X) \to \mathcal{Z}_{p+v}(\mathbb{P}(V)\#X)$ is a homotopy equivalence.*

Using **Property 4** one can prove the following *homotopy property* for Lawson homology.

Corollary 3.3.13 (Homotopy property) [FG93] *Let $\rho : E \to X$ be an algebraic vector bundle of rank e over an algebraic variety X. Then the flat pull-back $\rho^* : \mathcal{Z}_p(X) \to \mathcal{Z}_{p+e}(E)$ is a homotopy equivalence.*

Proof (Sketch) The first step is to show that the corollary holds for the trivial line bundle over any affine variety Y. This follows from Theorem 3.3.11 along with Noetherian induction and localization exact sequences.

Once this is shown, one can use induction on rank of the bundle and prove the corollary for any trivial vector bundle over an affine variety Y. Using localization sequences and Noetherian induction once again, one then proves the corollary for an arbitrary variety X. $\qquad\square$

Example 3.3.14 As an immediate consequence of Lawson's theorem, one can compute the Lawson homology of projective spaces $\mathbb{P}^N = \mathbb{P}(\mathbb{C}^{N+1}) = \mathbb{P}(\mathbb{C}^p \oplus \mathbb{C}^{N-p+1}) = \mathbb{P}(\mathbb{C}^p)\#\mathbb{P}^{N-p}$, for $0 \le p \le N$. Corollary 3.3.12 gives a homotopy equivalence $\mathcal{Z}_0(\mathbb{P}^{N-p}) \cong \mathcal{Z}_p(\mathbb{P}^N)$. This yields isomorphisms

$$L_p H_n(\mathbb{P}^N) := \pi_{n-2p}(\mathcal{Z}_p(\mathbb{P}^N)) \cong \pi_{n-2p}(\mathcal{Z}_0(\mathbb{P}^{N-p})) \cong H_{n-2p}(\mathbb{P}^{N-p}; \mathbb{Z}),$$

where the latter comes from Theorem 3.3.3(ii).

Exercises 3.3.2

1. Prove the assertions in Remark 3.3.4.
2. Prove that the algebraic join is given by the composition $\sigma\#\sigma' = b_* \circ p^* \circ \mu(\sigma, \sigma')$, as described in equation (5).
3. Prove Corollary 3.3.12.
4. Write the details of the proof of Corollary 3.3.13.
5. Given an algebraic variety X and $x_\infty \in \mathbb{P}^1$, one has two maps $i_\infty : \mathcal{Z}_{r+1}(X) \to \mathcal{Z}_{r+1}(X \times \mathbb{P}^1)$ and $p_1^* : \mathcal{Z}_r(X) \to \mathcal{Z}_{r+1}(X \times \mathbb{P}^1)$, where the first

is induced by the inclusion and the latter is a flat pull-back by the proper projection $p_1 : X \times \mathbb{P}^1 \to X$. Prove that the map

$$\psi : \mathcal{Z}_r(X) \times \mathcal{Z}_{r+1}(X) \to \mathcal{Z}_{r+1}(X \times \mathbb{P}^1)$$
$$(\sigma, \tau) \mapsto i_\infty(\sigma) + p_1^* \tau$$

is a homotopy equivalence. (This is a baby version of the projective bundle formula proven in [FG93].)

Hint: Consider the commutative diagram

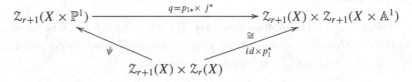

where $j^* : \mathcal{Z}_{r+1}(X \times \mathbb{P}^1) \to \mathcal{Z}_{r+1}(X \times \mathbb{A}^1)$ is the restriction map induced by the inclusion $j : \mathbb{A}^1 = \mathbb{P}^1 - \{x_\infty\} \hookrightarrow \mathbb{P}^1$. Then, use the localization sequence for the pair $(X \times \mathbb{P}^1, X \times \{x_\infty\})$ to show that the map q in the diagram is a homotopy equivalence.

Remark 3.3.15 Note that this exercise shows the following:

1. $L_{r+1}H_n(X \times \mathbb{P}^1) \cong L_{r+1}H_n(X) \oplus L_r H_{n-2}(X)$;
2. the kernel of the homomorphism $L_{r+1}H_n(X \times \mathbb{P}^1) \to L_{r+1}H_n(X)$ is the direct summand of $L_{r+1}H_n(X \times \mathbb{P}^1)$ given as the image of the flat pull-back homomorphism $L_r H_{n-2}(X) \to L_{r+1}H_{n-2}(X \times \mathbb{P}^1)$.

3.4 First applications

In this section we present a quick overview of some applications that can be derived immediately from the basic properties of algebraic cycle functors and Lawson homology.

3.4.1 The s-map and cycle maps

Using the join pairing and Lawson's suspension theorem, Friedlander and Mazur constructed a set of operations on the Lawson homology of a projective variety X that yields interesting filtrations both on the ordinary homology of X and on the groups $\mathcal{Z}_p(X)$. Here we present only the operation called the *s-map*.

Consider the composition $\mathbb{P}^1 \times \mathbb{Z}_p(X) \xrightarrow{u \times id} \mathbb{Z}_0(\mathbb{P}^1)_o \times \mathbb{Z}_p(X) \xrightarrow{\#} \mathbb{Z}_{p+1}(\mathbb{P}^1 \# X)_o$, where $u : \mathbb{P}^1 \to \mathbb{Z}_0(\mathbb{P}^1)_o$ is the inclusion sending $x \in \mathbb{P}^1$ to $x - x_\infty$, where $x_\infty = \mathbb{P}(0 \oplus \mathbb{C})$ is a fixed base point in $\mathbb{P}^1 = \mathbb{P}(\mathbb{C} \oplus \mathbb{C})$. It is clear that this map factors through $\mathbb{P}^1 \wedge \mathbb{Z}_p(X)$ and fits into the following diagram.

$$
\begin{array}{ccc}
\mathbb{P}^1 \times \mathbb{Z}_p(X) & \xrightarrow{\ u \times id\ } & \mathbb{Z}_0(\mathbb{P}^1)_o \times \mathbb{Z}_p(X) \\
\downarrow & & \downarrow{\scriptstyle \#} \\
\mathbb{P}^1 \wedge \mathbb{Z}_p(X) & \xrightarrow[\ \bar{u}\ ]{} & \mathbb{Z}_{p+1}(\mathbb{P}^1 \# X)_o \\
& {\scriptstyle (\Sigma^2)^{-1} \circ \bar{u}} \searrow & \ \ \Sigma^2 \uparrow \ \Big) (\Sigma^2)^{-1} \\
& & \mathbb{Z}_{p-1}(X)_o.
\end{array}
\tag{7}
$$

Let

$$
\sigma \ : \ \mathbb{Z}_p(X) \to \Omega^2 \mathbb{Z}_{p-1}(X)_o
\tag{8}
$$

be the unique homotopy class of maps given as the adjoint of the composition $(\Sigma^2)^{-1} \circ \bar{u}$, where $(\Sigma^2)^{-1}$ is a homotopy inverse for Σ^2. Therefore, σ induces homomorphisms $\sigma_* : \pi_r(\mathbb{Z}_p(X)) \to \pi_{r+2}(\mathbb{Z}_{p-1}(X))$, and for $r = n - 2p$ this yields the following.

Definition 3.4.1 [FM94] Let X be a projective variety.

1. The s-map $\quad s : L_p H_n(X) \to L_{p-1} H_n(X) \quad$ is the map induced by $\sigma : \mathbb{Z}_p(X) \to \Omega^2 \mathbb{Z}_{p-1}(X)_o$ on Lawson homology.
2. The pth successive composition induces the *cycle map* for projective varieties:

$$
s^p \ : \ L_p H_n(X) \to L_0 H_n(X) \equiv H_n(X; \mathbb{Z}).
$$

See Theorem 3.3.3(ii).

Remark 3.4.2 One must note that both the s-map and the cycle map depend a priori on the projective embedding. We shall see shortly that this is not the case.

Digression on the cycle map Using geometric measure theory (Lawson's initial approach to the subject), one can give an alternative description of the cycle map, which proves its functorial nature and consequent independence from embeddings. Let $\mathcal{I}_k(X)$ denote the group of integral k-currents [Fed69] on the analytic space X, with the flat-norm topology. Generalizing the classical Dold–Thom theorem, Almgren [Alm62] exhibited a natural isomorphism (holds for any compact ANR)

$$\Phi : \pi_r(\mathfrak{I}_k(X)) \cong H_{r+k}(X;\mathbb{Z}).$$

For a projective variety, Lawson showed that the group $\mathcal{Z}_p(X)$ (Chow topology) sits inside $\mathfrak{I}_{2p}(X)$ as a closed subgroup. The following theorem provided the first proof that the cycle map is natural, in particular that it is independent from the embedding.

Theorem 3.4.3 [LF93b] *The composition*

$$L_p H_n(X) := \pi_{n-2p}(\mathcal{Z}_p(X)) \to \pi_{n-2p}(\mathfrak{I}_{2p}(X)) \xrightarrow{\Phi} H_n(X;\mathbb{Z})$$

coincides with Friedlander–Mazur cycle maps s^p.

This result also shows that the cycle map is compatible with localization exact sequences.

The s-map has a somewhat simpler description, once one uses the various properties described in previous sections. The following result, for quasi-projective varieties, was shown in [FG93] and [Fri95]. The proof presented here avoids the intersection theory machinery used in the original papers.

Proposition 3.4.4 *Given a projective algebraic variety X and $p \geq 1$, the homotopy class of the map $\sigma : \mathcal{Z}_p(X) \to \Omega^2 \mathcal{Z}_{p-1}(X)_o$ is given by the adjoint of the composition*

$$\mathbb{P}^1 \wedge \mathcal{Z}_p(X) \xrightarrow{\bar{u}} \mathcal{Z}_0(\mathbb{P}^1)_o \times \mathcal{Z}_p(X) \xrightarrow{\mu} \mathcal{Z}_p(X \times \mathbb{P}^1)_o$$

$$\xrightarrow{j^*} \mathcal{Z}_p(X \times \mathbb{A}^1)_o \xrightarrow{\rho^{*-1}} \mathcal{Z}_{p-1}(X)_o,$$

where $\bar{u}$ is defined in diagram (7), μ is the product of cycles, $j : X \times \mathbb{A}^1 \hookrightarrow \mathbb{P}^1$ is the inclusion and $\rho : X \times \mathbb{A}^1 \to X$ is the projection.

Proof We denote by E the total space of the bundle associated to $p_1^* \mathcal{O}_{\mathbb{P}^1}(-1) \oplus p_2^* \mathcal{O}_X(-1)$, and refer the reader to Exercises 3.2(5) and diagram (7) for additional notation. The following diagram summarizes the maps involved in the proof, where pr_1, p_1, i_1 and ι_1 denote the evident projections and inclusions, respectively.

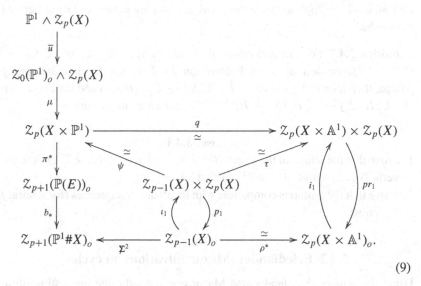

$$(9)$$

We want to show that the composition $\left(\Sigma^2\right)^{-1} \circ b_* \circ \pi^* \circ \mu \circ \overline{u}$ is homotopic to $(\rho^*)^{-1} \circ j^* \circ \mu \circ \overline{u}$. Since $j^* = pr_1 \circ q$, we will show:

$$b_* \circ \pi^* \circ \mu \circ \overline{u} \simeq \Sigma^2 \circ \left(\rho^*\right)^{-1} \circ pr_1 \circ q \circ \mu \circ \overline{u}. \qquad (10)$$

First we write

$$\Sigma^2 = b_* \circ \pi^* \circ \psi \circ \iota_1, \qquad (11)$$

and hence the second term of (10) can be written as

$$b_* \circ \pi^* \circ \psi \circ \left(\iota_1 \circ \left(\rho^*\right)^{-1}\right) \circ pr_1 \circ q \circ \mu \circ \overline{u}$$

$$= b_* \circ \pi^* \circ \psi \circ \left(\tau^{-1} \circ i_1\right) \circ pr_1 \circ q \circ \mu \circ \overline{u}, \qquad (12)$$

where the equality comes from the fact that $i_1 \circ \rho^* = \tau \circ \iota_1$; see diagram (9) above.

The main observation now is the fact that the image of $\mu : Z_0(X)_o \times Z_p(X) \to Z_p(X \times \mathbb{P}^1)$ lies in the kernel of the map $Z_p(X \times \mathbb{P}^1) \to Z_p(X)$. Hence the image of the composition $q \circ \mu \circ \overline{u}$ lies in the kernel of $pr_2 : Z_p(X \times \mathbb{A}^1) \times Z_p(X) \to Z_p(X)$. It follows that $q \circ \mu \circ \overline{u} = i_1 \circ pr_1 \circ q \circ \mu \circ \overline{u}$. Applying this identity to (12), along with the fact that $\psi \circ \tau^{-1} \circ q \simeq \mathrm{id}$, one obtains (10). $\qquad \square$

Corollary 3.4.5 *The s-map is a functorial construction, and hence it is independent of the embedding.*

Remark 3.4.6 Friedlander and Gabber [FG93] used their intersection theory for Lawson homology to prove the above proposition in the context of derived categories.

Definition 3.4.7 For an *arbitrary* algebraic variety X, let $\sigma : \mathcal{Z}_p(X) \to \Omega^2 \mathcal{Z}_{p-1}(X)_o$ be defined as in Proposition 3.4.4. In the level of homotopy groups, this defines the s-map $s : L_p H_n(X) \to L_{p-1} H_n(X)$ and the cycle map $s^p : L_p H_n(X) \to L_0 H_n(X) \cong H_n^{BM}(X; \mathbb{Z})$, for an arbitrary variety X.

Exercises 3.4.1

1. Prove that the image of the map $\mathcal{Z}_0(\mathbb{P}^1)_o \times \mathcal{Z}_p(X) \xrightarrow{\mu} \mathcal{Z}_p(X \times \mathbb{P}^1)$ lies in the kernel of the map $\mathcal{Z}_p(X \times \mathbb{P}^1) \to \mathcal{Z}_p(X)$.
2. Prove that the s-map is compatible with localization sequences (for arbitrary varieties).

3.4.2 Friedlander–Mazur filtrations on cycles

Using the s-map, Friedlander and Mazur constructed quite interesting filtrations on cycle spaces of projective varieties in [FM94]. Subsequently, these filtrations were extended to quasi-projective varieties in [Fri95]. We describe these filtrations below, where we also show that they can be defined for arbitrary complex varieties.

Recall that

$$L_p H_{2p}(X) = \pi_0(\mathcal{Z}_p(X)) \cong \mathcal{Z}_p(X)/\mathcal{Z}_p(X)_o = \mathcal{A}_p(X),$$

cf. Theorem 3.3.3(ii). In particular one has a composition

$$\mathcal{Z}_p(X) \xrightarrow{\pi} \mathcal{Z}_p(X)/\mathcal{Z}_p(X)_o \cong L_p H_{2p}(X) \xrightarrow{s^j} L_{p-j} H_{2p}(X), \quad (13)$$

where π denotes the projection and s^j is the jth iteration of the cycle map.

Definition 3.4.8 Define the jth stage of the *topological filtration* on $\mathcal{Z}_p(X)$ as $S_j \mathcal{Z}_p(X) := \ker s^j \circ \pi$. Note that this forms a filtration

$$\mathcal{Z}_p(X)_{\text{alg}} \equiv S_0 \mathcal{Z}_p(X) \subseteq S_1 \mathcal{Z}_p(X) \subseteq \cdots \subseteq S_p \mathcal{Z}_p(X) \equiv \mathcal{Z}_p(X)_{\text{hom}},$$

interpolating between the group of cycles algebraically equivalent to zero, $\mathcal{Z}_p(X)_{\text{alg}}$, and $\mathcal{Z}_p(X)_{\text{hom}}$, the group of cycles homologically equivalent to zero, i.e. the kernel of the cycle map.

Remark 3.4.9 Taking quotients by $\mathcal{Z}_p(X)_{\text{alg}}$ one obtains a filtration of the Griffiths group $\mathcal{Z}_p(X)_{\text{hom}}/\mathcal{Z}_p(X)_{\text{alg}}$.

This filtration is given a different formulation in [Fri95], using the notion of *Chow correspondences*, a generalization of the usual notion of correspondences. Roughly speaking, a Chow correspondence $f : Y \to \mathcal{C}_{p,\alpha}(X)$ is a morphism between the weak normalization of a variety Y and the weak normalization of a Chow variety $\mathcal{C}_{p,\alpha}(X)$ of p-cycles on a projective variety X. (This is also called an *algebraic continuous map*.) One can show that such a correspondence induces a continuous *graphing map* $\Gamma_f : \mathcal{Z}_r(X) \to \mathcal{Z}_{p+r}(X)$; see [Fri95] for details.

Theorem 3.4.10 [Fri95] *For any projective algebraic variety X, $S_j \mathcal{Z}_p(X) \subseteq \mathcal{Z}_p(X)$ is the subgroup generated by cycles Z of the following form: there exists a projective variety Y of dimension $2j + 1$, a Chow correspondence $f : Y \to \mathcal{C}_{p-j}(X)$, and a j-cycle W on Y homologically equivalent to 0 in Y and such that Z is rationally equivalent to $\Gamma_f(W)$.*

Among the many interesting properties of this filtration is the fact that, when X is smooth, it interpolates between Nori's filtration [Nor93] and Bloch–Ogus' filtration [BO84]. These filtrations are roughly defined as follows.

The jth level $A_j C H_p(X)$ of Nori's filtration is defined as the subgroup generated by those algebraic cycles that are rationally equivalent to cycles of the form $pr_{X*}(pr_Y{}^*X \cdot Z)$, where Y is a smooth projective variety, $W \in \mathcal{Z}_j(Y)_{\text{hom}}$ and $Z \in \mathcal{Z}_{p+\dim Y - j}(Y \times X)$.

Similarly, the jth level $B_j C H_p(X)$ of the Bloch–Ogus filtration is the subgroup generated by the p-cycles c on X such that c is supported on a subvariety V of X of dimension $p + j + 1$ and c is homologically equivalent to zero in V.

Theorem 3.4.11 [Fri95] *If X is a smooth projective variety, then one has inclusions*

$$A_j C H_p(X) \subseteq S_j \mathcal{Z}_p(X) \subseteq B_j C H_p(X)$$

where $A_j C H_p(X)$ is Nori's filtration on $\mathcal{Z}_p(X)$ and $B_j C H_p(X)$ is Bloch–Ogus' filtration.

Question 3.4.12 Let (V, σ) be a pair consisting of a complex variety V and a p-cycle $\sigma \in \mathcal{Z}_p(V)_{\text{hom}}$, homologically equivalent to zero. Under which conditions can one find a smooth variety $\widehat{V}$ along with a p-cycle $\widehat{\sigma}$ on $\widehat{V}$ and a proper map $f : \widehat{V} \to V$ such that $f_*(\widehat{\sigma}) = \sigma$ *and* $\widehat{\sigma}$ is also homologically equivalent to zero in $\widehat{V}$? If this question has a positive answer under general circumstances, then Nori's and Friedlander's filtrations coincide.

3.5 (Colimits of) mixed Hodge structures on Lawson homology

In [FM94], Friedlander and Mazur showed that one can naturally endow the Lawson homology of projective varieties with the structure of a 'colimit' of mixed Hodge structures. Here we provide an alternative construction that can also be applied to arbitrary varieties, and we show that this construction coincides with the one in [FM94] in the projective case.

3.5.1 A brief review of mixed Hodge structures

A (pure) *Hodge structure of weight m* on a finite-dimensional real vector space V is a decreasing filtration

$$\cdots \subset F^{p+1} V_{\mathbb{C}} \subset F^p V_{\mathbb{C}} \subset \cdots$$

of the complexified vector space $V_{\mathbb{C}} = V \otimes_{\mathbb{R}} \mathbb{C}$ satisfying the Hodge decomposition

$$V_{\mathbb{C}} = \bigoplus_{p+q=m} V^{p,q},$$

where $V^{p+q} = F^p V_{\mathbb{C}} \cap \overline{F^q V_{\mathbb{C}}}$ and $\overline{F^* V_{\mathbb{C}}}$ is the conjugate filtration.

If A is a subring of $\mathbb{R}$, an *A-Hodge structure of weight m* is an A-module V_A of finite type together with a Hodge structure of weight m on $V_{\mathbb{R}} = V_A \otimes_A \mathbb{R}$.

Exercises 3.5.1

1. Show that if V and W carry Hodge structures of weights m, n respectively, then $V \otimes W$ carries a Hodge structure of weight $m + n$.
2. (*Tate Hodge structure*) Prove that $\mathbb{Z}(1) := (2\pi i)\mathbb{Z} \subset \mathbb{C}$ has a unique $\mathbb{Z}$-Hodge structure of weight -2. Denote by $\mathbb{Z}(n)$ the $\mathbb{Z}$-Hodge structure $\mathbb{Z}(n) := \mathbb{Z}(1) \otimes \cdots \otimes \mathbb{Z}(1)$ (of weight $-2n$).

Remark 3.5.1 In the ground-breaking work [Hod41] on the theory of harmonic integrals, Hodge showed that the cohomology group $H^m(X; \mathbb{R})$ of a compact Kähler manifold X carries a Hodge structure of weight m.

Definition 3.5.2 Given a subring $A \subset \mathbb{R}$, an *A-mixed Hodge structure* (MHS) consists of:

1. an A-module of finite type V_A;
2. an *increasing filtration*

$$\cdots \subset W_n \subset W_{n+1} \subset \cdots$$

of the $A \otimes \mathbb{Q}$ module $V_A \otimes \mathbb{Q}$;

3. a *decreasing filtration*

$$\cdots \subset F^{p+1} \subset F^p \subset \cdots$$

of $V_{\mathbb{C}} = V_A \otimes \mathbb{C}$.

These filtrations must satisfy the following property. If one defines $Gr_j^W(V_{\mathbb{C}}) = (W_j \otimes \mathbb{C})/(W_{j-1} \otimes \mathbb{C}) = (W_j/W_{j-1}) \otimes \mathbb{C}$, then the filtrations F and $\overline{F}$ on $V_{\mathbb{C}}$ induce an A-Hodge structure of weight j on $Gr_j^W(V_A) := W_j/W_{j-1}$.

A morphism $f: V_A \to V_{A'}$ of an A-MHS is an A-module homomorphism which is compatible with both filtrations W and F. With this notion, one has a category of A-mixed Hodge structures.

Theorem 3.5.3 [Del71, Theorem 2.3.5]

(i) *The category of A-MHS is abelian, with kernels and cokernels of morphisms endowed with the induced filtrations.*
(ii) *The functor $V \mapsto Gr_j^W(V)$ is an exact functor from the category of A-MHS to the category of A-HS of weight j.*
(iii) *The functor $V \mapsto Gr_F^p(V)$ is exact.*

In [Del71] Deligne generalized Hodge's result to include arbitrary smooth complex algebraic varieties.

Theorem 3.5.4 [Del71, Theorem 3.2.5] *There are filtrations $W[p]$ and F which define on $H^p(X; \mathbb{Z})$ a mixed Hodge structure. This MHS is functorial with respect to algebraic maps.*

Subsequently, in [Del74], Deligne went even further and proved the existence of MHS on the cohomology of arbitrary complex varieties. In order to construct this MHS, one needs to resort to simplicial varieties. In fact, for any simplicial scheme $X_\bullet$ over $\mathbb{C}$, one can construct [Del74, Section 8.3] a structure of MHS on the cohomology groups $H^*(X_\bullet; \mathbb{Z})$, and this structure is functorial on $X_\bullet$.

Recall that a *simplicial variety* over $\mathbb{C}$ consists of a collection $\{X_n; d_i, s_i\}$ of complex algebraic varieties X_n, $n = 0, 1, 2, \ldots$, along with morphisms $\delta_i: X_n \to X_{n-1}$, $i = 0, \ldots, n$ (the 'face maps') and morphisms $s_i: X_n \to X_{n+1}$, $i = 0, \ldots, n$ satisfying certain compatibility relations; see [May82] for basics on simplicial objects.

Example 3.5.5 Let G be a complex algebraic group acting algebraically on a variety X. Then the *bar construction* (see Section 3.2.3) $\mathcal{B}_\bullet(X, G, *)$ with nth variety $\mathcal{B}_n(X, G, *) = X \times \underbrace{G \times \cdots \times G}_{n\text{-times}}$ (and whose face and degeneracy maps are defined in Section 3.2.3) is an example of a simplicial variety.

The particular cases $E_\bullet G := \mathcal{B}_\bullet(G, G, *)$ and $B_\bullet G := \mathcal{B}(*, G, *)$ give simplicial varieties so that the natural map $E_\bullet G \to B_\bullet G$ is (after geometric realization) the universal principal G-bundle over the classifying space BG of G. It is easy to see that $\mathcal{B}(X; G; *) \cong X \times_G EG$ is the usual *Borel construction* on X.

3.5.2 Colimits of MHS and Lawson homology

Definition 3.5.6 Let A be a subring of $\mathbb{R}$. An A-module M_A is *a colimit of A-mixed Hodge structures* if there is a directed system $\{M_{A,\lambda}\}_{\lambda \in \Lambda}$ of A-mixed Hodge structures (i.e. all maps $\varphi_\mu^\lambda \colon M_{A,\lambda} \to M_{A,\mu}$ are maps of A-MHS) such that

$$M_A = \operatorname*{colim}_\lambda M_{A,\lambda}.$$

The following result is proven in [FM94] in the case of *projective varieties*. The general case presented here is a novel result. Also, we use a rather different approach, avoiding the use of the (non-canonical) notion of *base systems* as in [FM94].

Theorem 3.5.7 *Let X be a complex algebraic variety and let $\mathcal{Z}_p(X)$ denote the group of algebraic p-cycles on X, with the topology defined in Section 3.2.*

(i) *The homology groups $H_m(\mathcal{Z}_p(X); \mathbb{Z})$ admit the structure of a colimit of $\mathbb{Z}$-MHS.*

(ii) *This structure is covariantly functorial for proper morphisms and contravariant for flat morphisms.*

(iii) *The Hurewicz homomorphism*

$$h \colon L_p H_n(X) = \pi_{n-2p}(\mathcal{Z}_p(X)) \longrightarrow H_{n-2p}(\mathcal{Z}_p(X); \mathbb{Z})$$

is injective, hence it gives a functorial structure of colimit of $\mathbb{Z}$-MHS on Lawson homology.

(iv) *If X is projective, then this structure coincides with the one in [FM94].*

Before sketching the proof, let us establish some conventions.

Notation If Y is a projective variety and $\mathcal{C}_{p,\alpha}(Y)$ is a Chow variety, let $v \colon \mathcal{C}_{p,\alpha}(Y)^w \to \mathcal{C}_{p,\alpha}(Y)$ denote its weak normalization. This is a homeomorphism of underlying topological spaces and hence it induces isomorphism between the MHSs in their cohomology. The advantage of using $\mathcal{C}_{p,\alpha}(Y)^w$ instead of $\mathcal{C}_{p,\alpha}(Y)$ lies in the fact that if $f \colon X \to Y$ is a morphism of projective varieties, then $f_* \colon \mathcal{C}_{p,\alpha}(X)^w \to \mathcal{C}_{p,f_*(\alpha)}(Y)^w$ is a morphism between Chow varieties. From now on, we denote $\mathcal{C}_{p,\alpha}(X)^w$ simply by $\mathcal{C}_{p,\alpha}(X)$; see [FM94] or [Kol96] for details.

Sketch of proof of Theorem 3.5.7 We first consider the case X quasi-projective. Let $j: \ X \hookrightarrow \overline{X}$ be a projective compactification and denote $D = \overline{X} - X$.

Consider $M := \mathcal{C}_p(\overline{X}) \times \mathcal{C}_p(\overline{X})$, where $\mathcal{C}_p(\overline{X})$ is the Chow monoid of effective p-cycles in $\overline{X}$, and denote by $N \subset M$ *the image* of the map

$$\mathcal{C}_p(D) \times \mathcal{C}_p(D) \times \mathcal{C}_p(\overline{X}) \longrightarrow \mathcal{C}_p(\overline{X}) \times \mathcal{C}_p(\overline{X})$$

$$(a\ b;\ \lambda) \longmapsto (a + \lambda, b + \lambda). \tag{14}$$

Exercises 3.5.2

1. Show that N is a submonoid of M.
2. Show that $N \cap (\mathcal{C}_{p,\alpha}(\overline{X}) \times \mathcal{C}_{p,\beta}(\overline{X}))$ is a subvariety of $\mathcal{C}_{p,\alpha}(\overline{X}) \times \mathcal{C}_{p,\beta}(\overline{X})$ for all α, β.
3. (Main point) Show that the monoid quotient M/N is homeomorphic to $\mathcal{Z}_p(\overline{X})/\mathcal{Z}_p(D) \equiv \mathcal{Z}_p(X)$.

Lemma 3.5.8 *The map* $\mathcal{B}(M, N, *) \longrightarrow M/N \equiv \mathcal{Z}_p(X)$ *is a quasi-fibration with contractible fibers homeomorphic to* $\mathcal{B}(N, N, *)$. *In particular,* $\phi: \ \mathcal{B}(M, N, *) \longrightarrow \mathcal{Z}_p(X)$ *is a homotopy equivalence.*

Proof This follows directly from [LF93a]. $\qquad\qquad\square$

Now, define

$$\mathcal{B}_k(M, N, *)_{\leq d} = \left\{ (a, b; n_1, \ldots, n_k; *) \in M \times N^{\times k} \mid \right.$$

$$\left. \deg a + \deg b + 2 \sum_i \deg n_i \leq d \right\}.$$

It is easy to see that

(a) $\mathcal{B}_\bullet(M, N, *)_{\leq d} \subset \mathcal{B}_\bullet(M, N, *)$ is a simplicial subspace,
(b) each $\mathcal{B}_\bullet(M, N.*)_{\leq d}$ is a simplicial variety and the inclusion

$$\mathcal{B}_\bullet(M, N, *)_{\leq d} \hookrightarrow \mathcal{B}_\bullet(M, N, *)_{\leq d+1}$$

is a morphism of simplicial varieties.

Since each $\mathcal{B}_\bullet(M, N, *)_{\leq d}$ is a simplicial variety whose geometric realization is a finite CW-complex, its integral homology admits the structure of a $\mathbb{Z}$-MHS via the Kronecker duality pairing $H_{*,\mathbb{Q}} \otimes H_\mathbb{Q}^* \to \mathbb{Q}$. This gives H_* a unique $\mathbb{Z}$-MHS so that the pairing is a map of MHS when $\mathbb{Q}$ is given the Hodge structure of weight 0.

We now have $\mathcal{B}_\bullet(M, N, *) = \operatorname*{colim}_d \mathcal{B}(M, N, \cdot)_{\leq d}$ and, since homology commutes with colimits, one obtains

$$H_n(\mathcal{Z}_p(X); \mathbb{Z}) \cong H_n(\mathcal{B}(M, N, *); \mathbb{Z})$$
$$= \operatorname*{colim}_d H_n(\mathcal{B}(M, N, *)_{\leq d}; \mathbb{Z}),$$

where the latter is a colimit of $\mathbb{Z}$-MHS. This proves statement (i) of the theorem.

Lemma 3.5.9 *Let G be an abelian topological monoid whose topology is given by an increasing sequence $\{0\} \subset G_1 \subset \cdots \subset G_d \subset G_{d+1} \subset \cdots$ of compact subsets. Then the Hurewicz map $h\colon \pi_k(G) \to H_k(G; \mathbb{Z})$ is injective.*

Proof This result is classical. (The filtration condition is not really needed.) Let $1 \in G$ denote the unity of G, whose operation we denote multiplicatively. Since G is abelian, the map

$$G_d \times \cdots \times G_d \xrightarrow{\;\bullet\;} G$$
$$g_1, \ldots, g_n \longmapsto g_1 \cdot \ldots \cdot g_n$$

descends to a map

$$SP_n(G_d) \longrightarrow G,$$

where $SP_n(G_d) = (G_d \times \cdots \times G_d)/S_n$ is the n-fold symmetric product. This is compatible with the inclusion

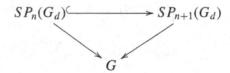

As a result, one gets a map $SP_\infty(G_d) \to G$ having the property that $G_d \hookrightarrow SP_\infty(G_d) \to G$ is simply the inclusion $G_d \hookrightarrow G$. Now, the Dold–Thom theorem gives an isomorphism $\pi_k(SP_\infty(G_d)) \cong H_k(G_d, \mathbb{Z})$ such that the composition

$$\pi_k(G_d) \longrightarrow \pi_k(SP_\infty(G_d)) \cong H_k(G_d; \mathbb{Z})$$

is the Hurewicz map. We obtain a map

$$r \;:\; H_k(G; \mathbb{Z}) = \operatorname*{colim}_d H_k(G_d; \mathbb{Z}) \cong \operatorname*{colim}_d \pi_k(SP_\infty(G_d)) \longrightarrow \pi_k(G)$$

so that the composition

$$\pi_k(G) = \operatorname*{colim}_d \pi_k(G_d) \xrightarrow{h} \operatorname*{colim}_d H_k(G_d; \mathbb{Z}) \xrightarrow{r} \pi_k(G)$$

is the identity. It follows that h is injective. $\qquad\square$

Proof of (iii) As a consequence of Lemma 3.5.9 one has a natural inclusion

$$h : L_p H_n(X) := \pi_{n-2p}(\mathcal{Z}_p(X)) \hookrightarrow H_{n-2p}(\mathcal{Z}_p(X); \mathbb{Z}).$$

Since the category of colimits of MHS is abelian, we conclude that $L_p H_n(X)$ inherits a natural structure of colimits of MHS which is clearly functorial.

Proof of (iv) If X is projective then $M = \mathcal{C}_p(X) \times \mathcal{C}_p(X)$ and $N = \Delta_{\mathcal{C}_p(X)}$ is the diagonal, cf. (14). Choose a base system $\{\mathcal{C}_{p,\alpha}(X), z_\alpha^0\}$, $z_\alpha^0 \in \mathcal{C}_{p,\alpha}(X)$, where α runs over all connected components of the Chow monoid $\mathcal{C}_p(X)$. The map

$$\iota_\alpha : \mathcal{C}_{p,\alpha}(X) \longrightarrow \mathcal{Z}_p(X)_0$$

$$\sigma \longmapsto \sigma - z_\alpha^0 \tag{15}$$

fits into a commutative diagram

$$\begin{array}{ccc}
\mathcal{C}_{p,\alpha}(X) & \longrightarrow & \mathcal{B}_\bullet(\mathcal{C}_p(X) \times \mathcal{C}_p(X), \Delta_{\mathcal{C}_p}(X), *) \\
\iota_\alpha \downarrow & \nearrow & \downarrow \\
\mathcal{Z}_p(X)_0 & \longrightarrow & \mathcal{Z}_p(X).
\end{array}$$

Hence $H_k(\mathcal{C}_{p,\alpha}(X); \mathbb{Z}) \longrightarrow H_k(\mathcal{Z}_p(X)_0; \mathbb{Z})$ is a map of (co)limits of MHS. On the other hand, general properties of homotopy theoretic group completions give an isomorphism

$$\operatorname*{colim}_\alpha H_k(\mathcal{C}_{p,\alpha}(X), \mathbb{Z}) \cong H_k(\mathcal{Z}_p(X)_0; \mathbb{Z})$$

cf. [FM94]. The previous colimit is used to define the colimit of MHS structure in [FM94].

Exercise Using Chow covers, as in Section 3.2.2, extend the proof above to include the case of arbitrary (not necessarily quasi-projective) varieties.

This concludes the (fairly complete) sketch of the proof of Theorem 3.5.7.

The constructions presented here satisfy various other properties that we leave aside in these notes, see [FM94]. Here are some additional results.

Proposition 3.5.10 *Let X be an arbitrary complex variety.*

(i) *The Pontrjagin ring structure on the homology $H_*(\mathcal{Z}_p(X); \mathbb{Z})$ becomes a Hopf algebra in the category of colimits of $\mathbb{Z}$-mixed Hodge structures.*

(ii) *The s-map $s: L_pH_n(X) \to L_{p-1}H_n(X)$ is in fact a map of colimits of $\mathbb{Z}$-MHS:*

$$S: L_pH_n(X) \to L_{p-1}H_n(X)(-1) := L_pH_n(X) \otimes \mathbb{Z}(-1).$$

Remark 3.5.11 A slight modification of Milnor–Moore's theorem allows one to identify the *rational* homotopy groups $\pi_*(\mathcal{Z}_p(X)) \otimes \mathbb{Q}$ with the primitives of the Hopf algebra $H_*(\mathcal{Z}_p(X), \mathbb{Q})$, via the Hurewicz homomorphism, cf. [FM94]. This would give $L_pH_n(X) \otimes \mathbb{Q}$ colimit of $\mathbb{Q}$-mixed Hodge structures.

Proof (i) Follows from the fact that in the level of bar constructions the various operations that induce the Hopf algebra structure are given by algebraic maps.

According to Proposition 3.4.4, the s-map is induced by the sequence of maps

$$\mathbb{P}^1 \wedge \mathcal{Z}_p(X) \xrightarrow{\bar{u}} \mathcal{Z}_0(\mathbb{P}^1)_o \times \mathcal{Z}_p(X) \xrightarrow{\mu} \mathcal{Z}_p(X \times \mathbb{P}^1)_o$$

$$\xrightarrow{j^*} \mathcal{Z}_p(X \times \mathbb{A}^1)_o \xrightarrow{\rho^{*-1}} \mathcal{Z}_{p-1}(X)_o.$$

Since π_1^* is a flat pull-back (and homotopy equivalence) it induces an isomorphism of colimits of $\mathbb{Z}$-MHS in the level of homotopy groups. The maps u and j are also derived from algebraic maps and can be easily seen to induce maps of co-MHS. The composition above then gives a pairing of co-MHS:

$$
\begin{array}{ccc}
\pi_2(\mathcal{Z}_0(\mathbb{P}^1)_0) \otimes \pi_{n-2p}(\mathcal{Z}_p(X)) & \longrightarrow & \pi_{n-2p+2}(\mathcal{Z}_{p-1}(X)) \\
\cong \downarrow & & \| \\
H_2(\mathbb{P}^1; \mathbb{Z}) \otimes \pi_{n-2p}(Z_p(X)) & & \text{def} \| \\
\text{def} \| & & \\
\mathbb{Z}(1) \otimes L_pH_n(X) & \longrightarrow & L_{p-1}H_n(X)
\end{array}
$$

whose adjoint $L_pH_n(X) \longrightarrow L_{p-1}H_n(X) \otimes \mathbb{Z}(-1)$ is precisely the s-map. $\quad\square$

3.5.3 A note on the Abel–Jacobi map

Let X be a smooth projective variety over $\mathbb{C}$. Recall that the intermediate Jacobian $\mathcal{J}^k(X)$ is the complex torus defined by

$$\mathcal{J}^k(X) := \frac{H^{2k-1}(X; \mathbb{C})}{F^kH^{2k-1}(X; \mathbb{C}) + H^{2k-1}(X; \mathbb{Z})} \cong \frac{F^{n-k+1}H^{2n-2k+1}(X, \mathbb{C})^*}{H^{2k-1}(X; \mathbb{Z})^*}.$$

The Abel–Jacobi map

$$\nu: \mathcal{Z}^k(X)_{\text{hom}} \longrightarrow \mathcal{J}^k(X)$$

sends $\sigma \in Z^k_{\text{hom}}(X)$ to

$$\nu(\sigma) = \left[\int_T \sigma \right],$$

where $\sigma = \partial T$ and T is an *integral current* of dimension $2n - 2k + 1$. Here $\int_T$ is the functional on $F^{n-k+1} H^{2n-2k+1}(X, \mathbb{C})$ defined by $\int_T[\varphi] = T(\varphi)$ and $[-]$ denotes equivalence classes in the appropriate quotient groups. It is easy to see that ν is well defined.

Proposition 3.5.12 *Let X be smooth and projective.*

(i) $\mathcal{Z}^k(X)_{\text{hom}}$ *is a closed subgroup of* $\mathcal{Z}^k(X)$;
(ii) $\mathcal{Z}^k(X)_{\text{hom}}$ *is a union of connected components of* $\mathcal{Z}^k(X)$ *and*

$$\pi_0(\mathcal{Z}^k(X)_{\text{hom}}) \cong \frac{\mathcal{Z}^k(X)_{\text{hom}}}{\mathcal{Z}^k(X)_{\text{alg}}} \equiv \text{Griffiths group};$$

(iii) *the Abel–Jacobi map* $\nu: \mathcal{Z}^k(X)_{\text{hom}} \longrightarrow \mathcal{J}^k(X)$ *is continuous.*

Proof Assertions (i) and (ii) are quite easy to show and we leave them as an exercise. Assertion (iii) follows from the continuity (holomorphicity) of normal functions [Kin83] and the description of the topology on $\mathcal{Z}^k(X)$ in terms of flat families, cf. Section 3.2. □

Speculations and remarks

Consider the restriction of the Abel–Jacobi map to the connected component of the identity $\mathcal{Z}^k_1(X)_{\text{alg}} = \mathcal{Z}^k(X)_0$ of $\mathcal{Z}^k(X)$. The image of this restriction is an abelian variety and the kernel is a closed subgroup

$$F^1 \mathcal{Z}^k(X)_0 \subset F^0 \mathcal{Z}^k(X)_0 \equiv \mathcal{Z}^k(X)_{\text{alg}}.$$

In particular, F^0/F^1 is a smooth group. On the other hand, the Abel–Jacobi map ν vanishes on the subgroup $\mathcal{Z}^k(X)_{\text{rat}} \subset \mathcal{Z}^k(X)_{\text{alg}}$ consisting of the cycles rationally equivalent to zero. Since ν is continuous, it vanishes on the closure

$$\overline{\mathcal{Z}^k(X)_{\text{rat}}} \subset \mathcal{Z}^k(X)_{\text{alg}}.$$

Considering generalized intermediate Jacobians $\mathcal{J}^k_{\Phi_p}(X)$ with supports (not exactly the same family of supports as in King's work [Kin83]), it is conceivable that one can find a filtration $\mathcal{Z}^k(X)_{\text{alg}}$ as

$$\mathcal{Z}^k(X)_{\text{rat}} \subset \overline{\mathcal{Z}}^k(X)_{\text{rat}} \overset{?}{\equiv} F^{k+1} \subset \cdots \subset F^1 \subset F^0 = Z^k_{\text{alg}}.$$

Ideally this should satisfy:

1. F^j / F^{j+1} are (possibly infinite-dimensional) smooth groups,
2. the filtration is functorial.

In particular, the most complicated portion of the topological behaviour (i.e. homotopy invariants) of the group of algebraic cycles would lie in the *closure of rational equivalence*. This would expand Rojtman's work in [Roj71] and [Roj72].

3.6 Intersection theory

In this section we present the intersection pairing in Lawson homology, as introduced for quasi-projective varieties by Friedlander and Gabber in [FG93]. Our presentation works for arbitrary varieties, not necessarily quasi-projective. The main ideas are quite universal, and can be applied in many different contexts, relying heavily on the process of *deformation to the normal cone*. This is a technique introduced by Fulton and MacPherson, and used by Fulton in his beautiful presentation of intersection theory [Ful84].

The main ingredients are the following.

1. Given algebraic varieties X, Y, the product of cycles

$$\mathcal{Z}_p(X) \times \mathcal{Z}_q(Y) \to \mathcal{Z}_{p+q}(X \times Y)$$

defines a continuous bilinear pairing, cf. Claim 3.3.7.
2. Given a closed inclusion $Y \hookrightarrow X$, the sequence

$$\mathcal{Z}_p(Y) \hookrightarrow \mathcal{Z}_p(X) \to \mathcal{Z}_p(X - Y)$$

is a locally trivial fibration. (This is **Property 4** in Section 3.2.)
3. Given an algebraic vector bundle $E \overset{p}{\longrightarrow} X$, the flat pull-back $p^*\colon \mathcal{Z}_p(X) \to \mathcal{Z}_{p+e}(E)$ is a homotopy equivalence, cf. Corollary 3.3.13.

Remark 3.6.1

1. In Fulton's intersection theory, for the Chow groups $CH_p(X) := \mathcal{Z}_p(X)/\mathcal{Z}_p(X)_{\text{rat}})$, these conditions have the following counterparts.
   (i) Multiplication of cycles preserves rational equivalence.

(ii) Given a closed subvariety $Y \subset X$ one has an exact sequence

$$CH_p(Y) \to CH_p(X) \to CH_p(X - Y) \to 0.$$

(iii) Flat-pull back $p^*\colon CH_p(X) \to CH_{p+e}(E)$ is an isomorphism for any vector bundle $E \to X$.

2. More generally, for the higher Chow groups [Blo86], the following conditions hold. Here $\mathfrak{Z}^p(X, *)$ denotes Bloch's complexes.

(i) Multiplication of cycles induces a pairing

$$\mathfrak{Z}^p(X, *) \otimes \mathfrak{Z}^q(Y, *) \to \mathfrak{Z}^{p+q}(X \times Y, *)$$

in the derived category.

(ii) A closed inclusion $Y \hookrightarrow X$ gives a distinguished triangle

$$\mathfrak{Z}^p(Y, *) \to \mathfrak{Z}^{p+e}(X, *) \to \mathfrak{Z}^{p+e}(X - Y, *)$$

in the derived category ($e = \dim X - \dim Y$). This yields the localization sequences.

(iii) The homotopy property

$$\mathfrak{Z}^p(X, *) \to \mathfrak{Z}^p(E, *)$$

is a quasi-isomorphism.

3.6.1 Deformation to the normal cone and Gysin homomorphism

Let $Y \underset{j}{\hookrightarrow} X$ be a regular embedding of codimension e. In other words, the normal cone $C_Y X$ of Y in X is in fact a vector bundle of rank e. Recall that the *normal cone* $C_Y X$ is the cone over X defined by the sheaf of graded algebras $\bigoplus_{n \geq 0} \mathfrak{J}^n/\mathfrak{J}^{n+1}$, where $\mathfrak{J}$ is the ideal sheaf of Y in X. In other words, $C_Y X = \mathrm{Spec}(\bigoplus_{n \geq 0} \mathfrak{J}^n/\mathfrak{J}^{n+1})$. When $j\colon Y \hookrightarrow X$ is a regular embedding, denote $C_Y X$ by $N_Y X$. The sheaf of sections of $N_Y X$ is $(\mathfrak{J}/\mathfrak{J}^2)^*$.

The main construction is the following. Let M denote the blow-up of $X \times \mathbb{P}^1$ along $Y \times \{\infty\}$. The normal bundle to $Y \times \infty$ in $X \times \mathbb{P}^1$ is $N \oplus 1$, where $N = N_Y X$ and 1 is the trivial line bundle. Hence, the exceptional divisor of the blow-up is $\mathbb{P}(N \oplus 1)$. Note that $\mathbb{P}(N \oplus 1)$ has two 'sections' (subvarieties):

$$\mathbb{P}(N) \equiv \mathbb{P}(N \oplus 0) \hookrightarrow \mathbb{P}(N \oplus 1) \hookleftarrow \mathbb{P}(0 \oplus 1) \equiv Y.$$

The composition $M \xrightarrow{b} X \times \mathbb{P}^1 \xrightarrow[P_1^r]{} \mathbb{P}^1$ gives a flat map $M \xrightarrow{p} \mathbb{P}^1$ satisfying:

1. The fiber above any $x \neq \infty$ is X.
2. The fiber $p^{-1}(\infty)$ above ∞ is a sum of two Cartier divisors: $\rho^{-1}(\infty) = \widetilde{X} + \mathbb{P}(N \oplus 1)$, where $\widetilde{X} = B|_Y X$ is the blow-up of X along Y.

Here is a picture of the situation:

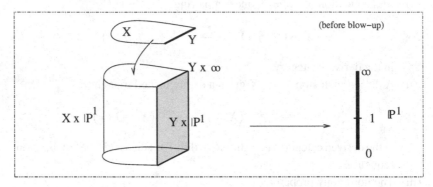

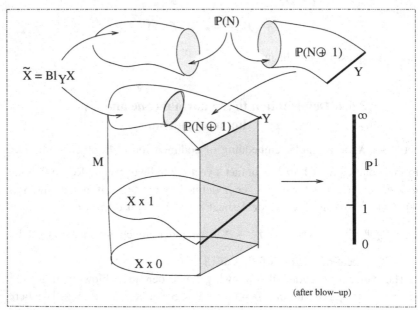

We now introduce the *deformation space*

$$M^0 := M - \widetilde{X}, \tag{16}$$

and note that $p \colon M^0 \to \mathbb{P}^1$ is still a flat map whose fiber above ∞ is

$$p^{-1}(\infty) = \mathbb{P}(N \oplus 1) - \mathbb{P}(N) \equiv N = N_Y X,$$

and $p^{-1}(t) \equiv X$ for all $t \neq \infty$.

Remark 3.6.2 A relevant observation is the fact that the normal cone $N_Y X$ is a closed subvariety of M^0 satisfying:

$$M^0 - N_Y X = X \times \mathbb{A}^1.$$

Remark 3.6.3 Contrary to the situation in differential geometry, in algebraic geometry one *does not* have a *'tubular neighbourhood'* of Y in X that can be identified with the normal bundle $N_Y X$. In many ways, this deformation space is the best replacement for tubular neighbourhoods, allowing one to 'localize' information from X to Y.

In order to use the ingredients described in the beginning of this section, we start with the following simple observation.

Lemma 3.6.4 *Given an algebraic variety X, let $i_1 \colon X \hookrightarrow X \times \mathbb{A}^1$ denote the inclusion $i_1(x) = (x, 1)$ of X as a 'slice' $X \times \{1\}$, $1 \in \mathbb{A}^1$. Then, the multiplication $\mathbb{A}^1 \times \mathbb{A}^1 \to \mathbb{A}^1$ sending (t, s) to ts induces a natural homotopy:*

$$\psi_X \colon \mathcal{Z}_p(X) \times I \longrightarrow \mathcal{Z}_p(X \times \mathbb{A}^1)$$

such that

$$\psi_X(\sigma, 0) = i_{1*}(\sigma)$$

$$\psi_X(\sigma, 1) = 0.$$

In other words, the inclusion i_{1*} is naturally homotopic to zero.

Proof Let $\gamma \colon [0, 1] \to \mathbb{P}^1(\mathbb{C})$ denote any path joining $1 \in \mathbb{A}^1 \subset \mathbb{P}^1$ to $\infty \in \mathbb{P}^1$. Consider the composition:

$$\psi_X \colon \mathcal{Z}_p(X) \times I \longrightarrow \mathcal{Z}_p(X \times \mathbb{P}^1) \xrightarrow{\quad q \quad} \mathcal{Z}_p(X \times \mathbb{A}^1)$$

$$\equiv \mathcal{Z}_p(X \times \mathbb{P}^1)/\mathcal{Z}_p(X \times \mathbb{A}^1)(\sigma_1, t) \longmapsto i_{\gamma(t)*}(\sigma)$$

where the continuity of $(\sigma, t) \mapsto i_{\gamma(t)*}(\sigma) \equiv \sigma \times \gamma(t)$ is given by Claim 3.3.7, and q is the quotient map.

Exercise 3.6.1

Verify that ψ_X is the correct homotopy. $\qquad\qquad\square$

The whole construction can now be summarized in the following diagram:

$$\mathcal{Z}_p(N_Y X)$$

$$i \downarrow$$

$$(\{\underline{0}\} \times I) \cup (\mathcal{Z}_p(X) \times 0) \longrightarrow \mathcal{Z}_p(M^0)$$

$$\downarrow \quad \overline{\psi}_X \quad \downarrow p$$

$$\mathcal{Z}_p(X) \times I \xrightarrow{\quad \psi_X \quad} \mathcal{Z}_p(X \times \mathbb{A}^1).$$

It follows that the map p is a locally trivial fibration with fiber $\mathcal{Z}_p(N_Y X)$, according to Remark 3.6.2 and **Property 4** of Section 3.2. Also, **Property 5** gives a cofibration $\{\underline{0}\} = Z_p(X)_{\leq 0} \hookrightarrow Z_p(X)$, i.e. $\underline{0} \in \mathcal{Z}_p(X)$ is a non-degenerate base point. Hence, we have assembled in the diagram above a classical homotopy-lifting problem, whose solution $\overline{\psi}_X \colon \mathcal{Z}_p(X) \times I \to \mathcal{Z}_p(M^0)$ satisfies:

1. $p \circ \overline{\psi}_X = \psi_X$,
2. $\overline{\psi}(\underline{0}, t) = \underline{0} \quad \forall t$,
3. $\overline{\psi}(\sigma, 0) = i_{1*}(\sigma)$.

Since $p \circ \overline{\psi}_X(\sigma, 1) = \psi_X(\sigma, 1) = \underline{0}$, then $\overline{\psi}_X(\sigma, 1) \in p^{-1}(0) = \mathcal{Z}_p(N_Y X) \hookrightarrow \mathcal{Z}_p(X)$. This gives the following:

Definition 3.6.5 The specialization map

$$\tau \colon \mathcal{Z}_p(X) \longrightarrow \mathcal{Z}_p(N_Y X)$$

is defined as $\tau(\sigma) = \overline{\psi}(\sigma, 1)$.

Remark 3.6.6 One can show that σ is unique up to homotopy. Observe that, up to now, one does not need the regularity of the embedding $Y \underset{j}{\hookrightarrow} X$. Hence, all constructions could have been made for any embedding, with the normal cone $C_Y X$ instead of $N_Y X$.

At this point, assume that $j \colon Y \hookrightarrow X$ is a regular embedding, i.e. $N_Y X \underset{\pi}{\longrightarrow} Y$ is a vector bundle of rank e. Hence, $\pi^* \colon \mathcal{Z}_{p-e}(Y) \to \mathcal{Z}_p(N_X Y)$ is a homotopy equivalence, by the homotopy property.

Definition 3.6.7 Define the *Gysin map*

$$j^! \colon \mathcal{Z}_p(X) \longrightarrow \mathcal{Z}_{p-e}(Y)$$

as the unique homotopy class of maps: $j^! = (\pi^*)^{-1} \circ \overline{h}$, where $(\pi^*)^{-1}$ is the homotopy inverse of π^*, and τ is the specialization map defined above.

The following properties hold, see [FG93].

Theorem 3.6.8 *Let* $j: Y \hookrightarrow X$ *be a regular embedding of codimension e. The Gysin map* $j^!: \mathcal{Z}_p(X) \to \mathcal{Z}_{p-e}(Y)$ *satisfies the following.*

(i) *If* $\mathcal{Z}_p(X)^0_{\cap Y}$ *is the subgroup of* $\mathcal{Z}_p(X)$ *consisting of those cycles intersecting Y properly, then the restriction of* $j^!$ *to* $\mathcal{Z}_p(X)^0_{\cap Y}$ *is homotopic to the usual intersection theoretic map:* $\sigma \mapsto \sigma \cdot Y$.

(ii) *If* $W \xrightarrow{i} Y$ *is another embedding of codimension d, then one has a homotopy*

$$(j \circ i)^! \simeq i^! \circ j^!: \mathcal{Z}_p(X) \to \mathcal{Z}_{p-d-e}(Y).$$

(iii) *Suppose* $g: X' \to X$ *is flat of relative dimension r and form the pull-back diagram*

$$
\begin{array}{ccc}
Y' & \xrightarrow{j'} & X' \\
{\scriptstyle g'}\downarrow & & \downarrow{\scriptstyle g} \\
Y & \xrightarrow{j} & X.
\end{array}
$$

Then one has a homotopy $j'^! \circ g^* \simeq g'^* \circ j^!$. *Similarly, if g is proper then* $j^! \circ g_* \simeq g'_* \circ j'^!$.

3.6.2 Intersection product in Lawson homology

Using the Gysin maps above, one can prove intersection theoretic properties of Lawson homology. The same formalism would prove similar properties for higher Chow groups, cf. Remark 3.6.1.

Proposition 3.6.9 *Let* $f: X \to Y$ *be a morphism between smooth varieties, where Y is equidimensional,* $\dim Y = e$. *The graph* Γ_f *induces a pairing:*

$$\Gamma_f^!: L_p H_n(X) \otimes L_q H_m(Y) \longrightarrow L_{p+q-e} H_{m+n-2e}(X). \tag{17}$$

Proof The map $(1 \times f): X \to X \times Y$ is a regular embedding of codimension $= \dim Y = e$. Consider the composition $\mathcal{Z}_p(X) \times \mathcal{Z}_q(Y) \xrightarrow{\mu} \mathcal{Z}_{p+q}(X \times Y) \xrightarrow{(1 \times f)^!} \mathcal{Z}_{p+q-e}(X)$, where $(1 \times f)^!$ is the Gysin map from Definition 3.6.7 and μ denotes the multiplication of cycles; see Claim 3.3.7. This composition satisfies $(1 \times f)^! \circ \mu(\sigma, \underline{0}) = (1 \times f)^! \circ \mu(\underline{0}, \tau) = \underline{0}$, and hence it factors through $\mathcal{Z}_p(X) \wedge \mathcal{Z}_q(Y)$:

$$\begin{array}{ccc}
\mathcal{Z}_p(X) \times \mathcal{Z}_q(Y) & \xrightarrow{\;\mu\;} & \mathcal{Z}_{p+q}(X \times Y) \\
\downarrow & & \downarrow{\scriptstyle (1 \times f)^!} \\
\mathcal{Z}_p(X) \wedge \mathcal{Z}_q(Y) & \xrightarrow[\Gamma_f^!]{} & \mathcal{Z}_{p+q-e}(X).
\end{array}$$

In the level of homotopy groups one obtains a pairing:

$$\pi_{n-2p}(\mathcal{Z}_p(X)) \otimes \pi_{m-2q}(\mathcal{Z}_q(Y)) \xrightarrow[\Gamma_f^!]{} \pi_{n+m-2(p+q)}(\mathcal{Z}_{p+q-e}(X)),$$

giving (17).   □

As a particular case, consider the identity map $\mathrm{id}: X \to X$, hence $\Gamma_{\mathrm{id}} = \Delta: X \to X \times X$ is the diagonal map. It follows that Δ induces an intersection pairing:

$$\Delta^!: L_p H_n(X) \otimes L_q H_m(X) \to L_{p+q-e} H_{m+n-2e}(X), \qquad (18)$$

where $e = \dim X$ and $p + q \geq e$.

The next result is a direct consequence of the constructions above and Theorem 3.6.8.

Theorem 3.6.10 *If X is a smooth variety of dimension e and $p + q \geq e$, then the intersection pairing (18) satisfies the following.*

(i) *It is associative and commutative.*

(ii) *At the level of π_0 it induces the usual intersection pairing:*

$$\begin{array}{ccc}
L_p H_{2p}(X) \otimes L_q H_{2q}(X) & \longrightarrow & L_{p+q-e} H_{2(p+q-e)}(X) \\
\| & & \| \\
\mathcal{A}_p(X) \otimes \mathcal{A}_q(X) & \longrightarrow & \mathcal{A}_{p+q-e}(X),
\end{array}$$

where $\mathcal{A}_p(X) := \mathcal{Z}_p(X)/\mathcal{Z}_p(X)_{\mathrm{alg}}$.

3.7 Morphic cohomology

3.7.1 Basic properties

Morphic cohomology is the cohomological counterpart of Lawson homology. It was introduced by Friedlander and Lawson in [FL92].

In order to motivate its definition, we first present basic computations. Throughout this section, we switch the notation of cycle spaces and use upperscripts to denote the codimension of cycles. The *homotopy property* gives homotopy equivalences:

$$\mathcal{Z}^q(\mathbb{A}^n) \xleftarrow{\simeq} \mathcal{Z}^q(\mathbb{A}^{n-1}) \xleftarrow{\simeq} \cdots \xleftarrow{\simeq} \mathcal{Z}^q(\mathbb{A}^q).$$

Hence

$$\pi_k(\mathcal{Z}^q(\mathbb{A}^n)) \cong \pi_k(\mathcal{Z}^q(\mathbb{A}^q)) = \pi_k(\mathcal{Z}_0(\mathbb{A}^q))$$

$$\cong H_k^{BM}(\mathbb{A}^q; \mathbb{Z}) \cong \begin{cases} 0 & k \neq 2q \\ \mathbb{Z} & k = 2q \end{cases}$$

This shows that $\mathcal{Z}^q(\mathbb{A}^n)$ is an Eilenberg–Mac Lane space of type $K(\mathbb{Z}, 2q)$. In particular, $[X, K(\mathbb{Z}, 2q)] \cong H^{2q}(X; \mathbb{Z})$ for every finite CW-complex X.

Remark 3.7.1 If one keeps track of Hodge structures, $\pi_{2q}(\mathcal{Z}_0(\mathbb{A}^q))$ with its Hodge structure, induced from the Borel–Moore homology of $\mathbb{A}^q$, is in fact $\mathbb{Z}(q)$. Hence $\mathcal{Z}^q(\mathbb{A}^n)$ is better identified with $K(\mathbb{Z}(q), 2q)$ when $n \geq q$. Furthermore, iterations of the *cycle map* give

$$\pi_{2q}(\mathcal{Z}^q(\mathbb{A}^n)) = L_{n-q}H_{2n}(\mathbb{A}^n) \xrightarrow{s^{n-q}} L_0 H_{2n}(\mathbb{A}^n) \otimes \mathbb{Z}(q-n)$$

$$\cong \mathbb{Z}(n) \otimes \mathbb{Z}(q-n) \cong \mathbb{Z}(q),$$

and this identification of pure Hodge structures holds for every $n \geq q$.

For simplicity we assume that all varieties are weakly normal, and the term 'Chow varieties' refers to the weak normalization of the usual Chow varieties.

Definition 3.7.2 [FL92] An *effective algebraic s-cocycle* on a variety X with values in a projective variety Y is a morphism $\varphi: X \longrightarrow \mathcal{C}^s(Y)$, where $\mathcal{C}^s(Y)$ is the Chow monoid of cycles of codimensions in Y. Denote by $\mathcal{C}^s(X; Y)$ the set of all such s-cocycles.

Examples 3.7.3

1. Any cycle $\Gamma \in \mathcal{C}^s(X \times Y)$ which is flat over X gives an s-cocycle $\pi_\Gamma: X \to \mathcal{C}^s(Y)$; see Section 3.2.
2. $X \subset \mathbb{P}^N$ smooth hypersurface and $Y \subset \mathbb{P}^N$ is projective subvariety. Define

$$\varphi: X \to \mathcal{C}^1(Y)$$

by $\varphi(x) = (T_p X) \cdot Y$, where the latter denotes the intersection theoretic product.

Construction 3.7.4 Any effective cocycle $\varphi \in \mathcal{C}^s(X; Y)$ can be transformed, via a *'graphing construction'*, into a cycle $\Gamma_\varphi \in \mathcal{C}^s(X \times Y)$ which is equidimensional over X, cf. [FL92].

Up to appropriate modifications in the resulting topology, we can think of the set $\mathcal{C}^s(X; Y)$ as a subspace of Map $(X(\mathbb{C}); \mathcal{C}^s(Y))$, with the compact-open topology.

Theorem 3.7.5 [FL92] *If X is projective and irreducible, the graphing map*

$$\Gamma: \quad \mathcal{C}^s(X; Y) \longrightarrow \mathcal{C}^s(Y/X)^{\text{eq}}$$

is a homeomorphism. Here, $\mathcal{C}^s(Y/X)^{\text{eq}}$ denotes the submonoid of the Chow monoid $\mathcal{C}^s(Y \times X)$ consisting of cycles that are equidimensional over X.

This theorem allows one to have a good grasp of the topology of $\mathcal{C}^s(X; Y)$.

Definition 3.7.6 Define the morphic cohomology of X with values in Y by

$$L^q H^n(X; Y) = \pi_{2q-n}(\mathcal{Z}^q(X; Y)),$$

where $\mathcal{Z}^q(X, Y)$ denotes the group completion $\mathcal{Z}^q(X; Y) := \Omega B(\mathcal{C}^q(X; Y))$ of the monoid $\mathcal{C}^q(X; Y)$.

Returning to the motivation given at the beginning of this section, let T be a compact CW-complex, and let $K(\mathbb{Z}(q), 2q)$ be an Eilenberg–MacLane space. Then one has canonical isomorphisms:

$$\begin{aligned}
\pi_{2q-n}(\text{Map}_\bullet(T, K(\mathbb{Z}(q), 2q))) &\cong [S^{2q-n}; \text{Map}(T, K(\mathbb{Z}(q); 2q))] \\
&= [S^{2q-n} \wedge T; K(\mathbb{Z}(q), 2q)] \\
&\cong [T; \Omega^{2q-n} K(\mathbb{Z}(q); 2q)] \\
&\cong [T; K(\mathbb{Z}(q), n)] = H^n(X; \mathbb{Z}(q)). \quad (19)
\end{aligned}$$

In morphic cohomology, we replace Map by Mor, and use $\mathcal{Z}^q(\mathbb{A}^n)$ as an 'algebraic geometric' model for $K(\mathbb{Z}(q), 2q)$.

Definition 3.7.7 For $n \geq q$, denote

$$\mathcal{Z}^q(X; \mathbb{A}^n) := \mathcal{Z}^q(X; \mathbb{P}^n)/\mathcal{Z}^{q-1}(X; \mathbb{P}^{n-1})$$

(homotopy quotient) and define the *morphic cohomology groups* of X by

$$L^q H^n(X) := \pi_{2q-n}(\mathcal{Z}^q(X; \mathbb{A}^n)).$$

An application of Lawson's suspension theorem (modified arguments for families), gives canonical homotopy equivalences

$$\mathcal{Z}^q(X; \mathbb{A}^n) \simeq \mathcal{Z}^q(X; \mathbb{A}^{n+1}) \simeq \cdots,$$

hence, the definition of morphic cohomology does not depend on n.

Properties 3.7.8

1. Morphic cohomology is contravariantly functorial for arbitrary morphisms.
2. If $f : X \to X'$ is *proper* and *flat* of relative dimension e, there are Gysin maps:

$$f_! : L^q H^n(X) \longrightarrow L^{q+e} H^{n+2e}(X')$$

   ($f_! \circ g_! = (f \circ g)_!$ for two such maps).
3. The *join pairing* induces a ring structure on $L^\bullet H^\bullet$, i.e. the pointwise multiplication

$$(\varphi \# \varphi')(x) = \varphi(x) \# \varphi(x')$$

   for elements $\varphi \in \mathcal{C}^q(X; \mathbb{P}^n)$, $\varphi' \in \mathcal{C}^{q'}(X; \mathbb{P}^{n'})$, induces a pairing

$$L^q H^r(X) \otimes L^{q'} H^{r'}(X) \to L^{q+q'} H^{r+r'}(X)$$

   once one descends to homotopy quotients and takes homotopy groups. This makes $L^\bullet H^\bullet(X)$ a bigraded commutative ring.
4. *Cycle maps*: the forgetful functor from Mor to Map induces a map

$$\mathcal{Z}^q(X; \mathbb{A}^n) \to \mathrm{Map}(X(\mathbb{C}); \mathcal{Z}^q(\mathbb{A}^n)).$$

In the level of homotopy groups, this gives natural homomorphisms

$$L^q H^n(X) := \pi_{2q-n}(\mathcal{Z}^q(X; \mathbb{A}^n)) \to \pi_{2q-n}(\mathrm{Map}(X(\mathbb{C}); \mathcal{Z}^q(\mathbb{A}^n)))$$
$$\cong [X(\mathbb{C}); \Omega^{2q-n} \mathcal{Z}^q(\mathbb{A}^n)] \cong [X(\mathbb{C}); K(\mathbb{Z}(q); n)]$$
$$\cong H^n(X(\mathbb{C}); \mathbb{Z}(q)),$$

cf. Remark 3.7.1 and (19). Hence one gets cycle maps:

$$\Phi : L^p H^n(X) \longrightarrow H^n(X; \mathbb{Z}(p)).$$

Example 3.7.9 It is shown in [FL92] that in the case of divisors one has the following.

Theorem *For any projective variety X there is a natural homotopy equivalence:*

$$\mathcal{Z}^1(X) \cong Pic(X) \times \mathbb{P}^\infty.$$

If X is smooth, then

  (i) $L^1 H^0(X) \cong \mathbb{Z}$.
 (ii) $L^1 H^1(X) \cong H^1(X; \mathbb{Z})$.
(iii) $L^1 H^2(X) \cong N S(X)$.
(iv) *The cycle map:*

$$\Phi\colon\ L^1 H^2(X) \cong N S(X) \longrightarrow H^2(X; \mathbb{Z}(1))$$

   is the first Chern class map.
 (vi) $L^1 H^k(X) = 0$ *for* $k > 2$.

3.7.2 Duality

There is a very simple *duality map* between Lawson homology and morphic cohomology. Recall that one has an inclusion $\mathcal{C}^q(X; \mathbb{P}^n) \hookrightarrow \mathcal{C}^q(X \times \mathbb{P}^n)$, hence one can pass to group completions and homotopy quotients to get a map

$$\mathcal{Z}^q(X; \mathbb{A}^n) \longrightarrow \mathcal{Z}^q(X \times \mathbb{A}^n) = \mathcal{Z}_{d-q+n}(X \times \mathbb{A}^n),$$

where $d = \dim X$. Using the homotopy property one obtains a (homotopy class) map:

$$\mathcal{Z}^q(X; \mathbb{A}^n) \longrightarrow \mathcal{Z}_{d-q}(X)$$

which, when one applies the homotopy group functor π_{2q-r}, gives:

$$\mathcal{D}\colon\ L^q H^r(X) \longrightarrow L_{d-q} H_{2d-r}(X).$$

This is the 'duality homomorphism'.

Theorem [FL97] *For any projective variety of dimension d, one gets a commutative diagram:*

$$
\begin{array}{ccc}
L^q H^n(X) & \xrightarrow{\ \mathcal{D}\ } & L_{d-q} H_{2d-n}(X) \\
\Phi \downarrow & & \downarrow S^{d-q} \\
H^n(X; \mathbb{Z}(q)) & \xrightarrow[\ \mathcal{P}\ ]{} & H_{2d-n}(X; \mathbb{Z}(q - d))
\end{array}
$$

where $\mathcal{P}$ is the Poincaré duality map given by cap product with the fundamental class $[X] \in H_{2d}(X, \mathbb{Z}(d))$ of X.

In fact, as a beautiful consequence of their *moving lemma* for families [FL98], Friedlander and Lawson prove that one has duality isomorphisms:

Theorem 3.7.10 [FL97] *For a smooth projective variety X, of dimension d, the duality map:*

$$\mathcal{D}: \; L^q H^n(X) \longrightarrow L_{d-q} H_{2d-n}(X)$$

is an isomorphism.

Acknowledgements

The author would like to thank the hospitality of the Institut Fourier during the Summer School 2001, and to thank the organizers Chris Peters and Stefan Müller-Stach, for the opportunity to participate in such a stimulating event and for their infinite patience with my tardiness in preparing these notes.

References

[Alm62]   Almgren, F. J. Jr. Homotopy groups of the integral cycle groups, *Topology* **1** (1962) 257–299.

[Blo86]   Bloch, S. Algebraic cycles and higher *K*-theory, *Adv. Math.* **61** (1986) 267–304.

[BO84]   Bloch, S. and Ogus, A. Gersten's conjecture and the homology of schemes, *Ann. Sci. Ec. Norm. Super.* **7** (4) (1984) 181–202.

[Del71]   Deligne, P. Théorie de Hodge II, *Inst. Hautes Etudes Sci. Publ. Math.* **40** (1971) 5–58.

[Del74]   Théorie de Hodge III, *Inst. Hautes Études Sci. Publ. Math.* **44** (1974) 5–77.

[dS]   dos Santos, P. Algebraic cycles on real varieties and $\mathbb{Z}/2$-homotopy theory, Preprint, 2000 (http://www.dosSantos.com/paper.pdf).

[dSLF]   dos Santos, P. and Lima-Filho, P. Quaternionic algebraic cycles and reality, Preprint, 2001.

[DT56]   Dold, A. and Thom, R. Quasifaserungen und unendlich symmetrische Produkte, *Ann. Math.* **67** (2) (1956) 230–281.

[Fed69]   Federer, H. *Geometric Measure Theory*, 1st edn., Springer-Verlag, New York, 1969.

[FG83]   Fulton, W. and Gillet, H. Riemann–Roch for general algebraic varieties, *Bull. Soc. Math. Fr.* **111** (1983) 287–300.

[FG93]   Friedlander, E. M. and Gabber, O. Cycle spaces and intersection theory. *Topological Methods in Modern Mathematics, Conference in Honor of J. Milnor, Austin, TX*, pp. 325–370, Publish or Perish, 1993.

[FL92]   Friedlander, E. M. and Lawson, H. B., Jr. A theory of algebraic cocycles, *Ann. Math.* **136** (2) (1992) 361–428.

[FL97]   Duality relating spaces of algebraic cocycles and cycles, *Topology* **36** (2) (1997) 533–565.

[FL98]   Moving algebraic cycles of bounded degree, *Invent. Math.* **132** (1998) 91–119.

[FM94]   Friedlander, E. M. and Mazur, B. *Filtrations on the Homology of Algebraic Varieties, Mem. Am. Math. Soc.* **110** no. 529, AMS, Providence, RI, 1994.

[Fri91]   Friedlander, E. M. Algebraic cycles, Chow varieties and Lawson homology, *Compositio Math.* **77** (1991) 55–93.

[Fri95]   Filtrations on algebraic cycles and homology, *Ann. Sci. Ec. Norm. Super.* **28** (4) (1995) 317–343.

[Ful84]   Fulton, W. *Intersection Theory*, 1st edn., Springer-Verlag, Heidelberg, 1984.

[GH78]   Griffiths, P. and Harris, J. *Principles of Algebraic Geometry*, John Wiley, New York, 1978.

[GD66]   Grothendieck, A. and Dieudonné, J. Elements de geometrie algebrique IV – part 3, *Inst. Hautes Etudes Sci. Publ. Math.* **28** (1966) 5–255.

[Gri68]   Griffiths, P. Periods of integrals on algebraic manifolds, *Am. J. Math.* **90** (1978) 568–626.

[Hir75]   Hironaka, H. Triangulation of algebraic sets, *Proc. Symp. Pure Math.* **29** (1975) 165–185.

[Hod41]   Hodge, W. V. D. *The Theory and Applications of Harmonic Integrals*, Cambridge University Press, 1941.

[Kin83]   King, J. R. Log complexes of currents and functorial properties of the Abel–Jacobi map, *Duke Math. J.* **50** (1983) 1–53.

[Kol96]   Kollár, J. Rational curves on algebraic varieties, *Ergebnisse der Mathematik und ihrer Grenzgebiete* **32**, Springer-Verlag, Berlin, 1996.

[Law89]   Lawson, H. B., Jr. Algebraic cycles and homotopy theory, *Ann. Math.* **129** (2) (1989) 253–291.

[Law95]   Spaces of algebraic cycles, *Surveys in Differential Geometry* **2**, pp. 137–213, International Press, 1995.

[LF92]   Lima-Filho, P. Lawson homology for quasiprojective varieties, *Compositio Math.* **84** (1992) 1–23.

[LF93a]   Completions and fibrations for topological monoids, *Trans. Am. Math. Soc.* **340** (1993) 127–147.

[LF93b]   On the generalized.cycle map, *J. Diff. Geom.* **38** (1) (1993) 105–129.

[LF94]   The topological group structure of algebraic cycles, *Duke Math. J.* **75** (2) (1994) 467–491.

[LLFM]   Lawson, H. B. Jr., Lima-Filho, P. and Michelsohn, M.-L. Algebraic cycles and the classical groups, I; real cycles, *Topology*, in press.

[LLFM96]   Algebraic cycles and equivariant cohomology theories, *Proc. London Math. Soc.* **73** (3) (1996), no. 3, 679–720.

[LLFM98]   On equivariant algebraic suspension, *J. Algebraic Geometry* **7** (4) (1998), 627–650.

[LM88]   Lawson, H. B. Jr. and Michelsohn, M.-L. Algebraic cycles, Bott periodicity, and the Chern characteristic map, *Proc. Symp. Pure Math.* **48** (1988) 241–264.

[May75]   May, J. P. Classifying spaces and fibrations, *Mem. Am. Math. Soc.* **155**, AMS, Providence, RI, 1975.

[May82]   *Simplicial Objects in Algebraic Topology*, Midway Reprint, University of Chicago Press, Chicago, IL, 1982.

[MV99]   Morel, F. and Voevodsky, V. $\mathbb{A}^1$-homotopy theory of schemes, *Inst. Hautes Etudes Sci. Publ. Math.* **90** (1999) 45–143.

[Nor93]   Nori, M. V. Algebraic cycles and Hodge theoretic connectivity, *Invent. Math.* **111** (1993), no. 2, 349–373.

[Roj71]   Rojtman, A. A. On γ-equivalence of zero-dimensional cycles, *Math. USSR Sbornik* **15** (1971) 555–567.

[Roj72]   Rational equivalence of zero-cycles, *Math. USSR Sbornik* **18** (1972) 571–588.

[Seg74]   Segal, G. Categories and cohomology theories, *Topology* **13** (1974) 193–312.

[Ste67]   Steenrod, N. A convenient category of topological spaces, *Michigan Math. J.* **14** (1967) 133–152.

Part III

Motives and motivic cohomology

4

Lectures on motives

Jacob P. Murre

Department of Mathematics, University of Leiden, Niels Bohrweg 1, P.O. Box 9512,
2300 RA Leiden, The Netherlands

Introduction

These are the notes of my lectures at the Summer School at the Institute Fourier, Grenoble, 2001. I have tried to keep the written version as close as possible to the informal style of the oral lectures.

I would like to stress that by 'motives' we mean in these lectures always 'pure motives', i.e. motives constructed from smooth, projective varieties.

4.1 Lecture 1: Grothendieck's construction of motives

4.1.0 Motivation

Let X be a smooth, projective variety defined over an algebraically closed field k. In the early 1960s Grothendieck, together with Artin and Verdier, constructed for every prime $l \neq p = \mathrm{char}(k)$ the l-adic cohomology groups $H^i_{et}(X, \mathbb{Q}_l)$. Moreover, for the missing $l = p$, Grothendieck outlined the construction of the crystalline cohomology if $p \neq 0$. If $k = \mathbb{C}$ then there are also the classical Betti cohomology $H^i_B(X(\mathbb{C}), \mathbb{Q})$ and the De Rham cohomology groups $H^i_{DR}(X, \mathbb{C})$. This gives infinitely many cohomology theories! However, all these theories have similar properties. In fact if $k = \mathbb{C}$ then there is firstly the famous De Rham isomorphism theorem $H^i_{DR}(X, \mathbb{C}) \xrightarrow{\sim} H^i_B(X, \mathbb{C})$ and next there are also the comparison isomorphisms

$$H^i_B(X, \mathbb{Q}) \otimes \mathbb{Q}_l \xrightarrow{\sim} H^i_B(X, \mathbb{Q}_l) \xrightarrow{\sim} H^i_{et}(X, \mathbb{Q}_l)$$

Transcendental Aspects of Algebraic Cycles ed. S. Müller-Stach and C. Peters.
© Cambridge University Press 2004.

between Betti- and étale-cohomology groups. 'Clearly' there must be an underlying, deeper reason for this. In order to 'understand' and 'explain' this, Grothendieck created, in the mid-1960s, the *theory of motives*. These are 'objects' which 'carry' all these different cohomology groups; in fact these latter are only the 'realizations' or 'incarnations' of the motives in the different cohomology theories.

In fact, it is best to quote Grothendieck himself. In Section 16 (Les motifs – ou la coeur dans la coeur) of the 'En Guise d'Avant-Propos' of his 'Récoltes et Semailles' [G85] Grothendieck writes the following:

> ... Contrairement à ce qui se passait en topologie ordinaire, on se trouve donc placé là devant une abondance déconcertante de théories cohomologiques différentes. On avait l'impression très nette qu'en un sens, qui restait d'abord très flou, toutes ces théories devaient 'revenir au même', qu'elles 'donnaient les mêmes résultats'. C'est pour parvenir à exprimer cette intuition de 'paranté' entre théories cohomologiques différentes, que j'ai dégagé la notion de 'motif' associé à une variété algébrique. Par ce terme j'entends suggérer qu'il s'agit du 'motif commun' (ou de la raison commune) sous-jacent à cette multitude d'invariants cohomologiques différents associés à la variété, à l'aide de la multitude de toutes les théories cohomologiques possibles à priori. [...]. Ainsi, le motif associé à une variété algébrique constituerait l'invariante cohomologique 'ultime', 'par excellence', dont tous les autres (associés aux différentes théories cohomologiques possibles) se déduiraient, comme autant d' 'incarnations' musicales, ou de 'réalisations' différentes [...].

Finally, to try to see what he has in mind, we also quote from the end of Grothendieck's paper given at the Bombay Conference in 1968 [G68, p. 198]:

> ... the 'theory of motives' is a systematic theory of 'arithmetical properties' of algebraic varieties as embodied in their groups of classes of cycles for numerical equivalence.

4.1.1 Notation: algebraic cycle groups

4.1.1.1

In this section we fix the notation and the assumptions and we recall briefly some facts on algebraic cycle groups. For more details of the latter we refer to the lectures of Javier Elizondo (Chapter 1), and to [F84].

Let k be a field and $V = V(k)$ the category of *smooth, projective* varieties defined over k. If $X \in V$ is moreover irreducible (and usually we can restrict ourselves to such X) then X has a dimension d and we write X_d to indicate this.

Let $X_d \in \mathcal{V}$ and $0 \leq i \leq d$; consider

$$\mathcal{Z}^i(X) = \{Z = \sum_\alpha n_\alpha Z_\alpha; Z_\alpha \subset X \text{ irreducible subvariety}$$
$$\text{of codimension } i \text{ in } X\},$$

the group of *algebraic cycles* on X of codimension i (the n_α are integers). Let $\sim$ be a 'good' equivalence relation [F84], [H75] defined via a subgroup $\mathcal{Z}^i_\sim(X) \subseteq \mathcal{Z}^i(X)$. Put

$$C^i_\sim(X) = \mathcal{Z}^i(X)/\mathcal{Z}^i_\sim(X)$$

and

$$C_\sim(X) = \oplus^d_{i=0} C^i_\sim(X),$$

then $C_\sim(X)$ is the *ring of equivalence classes* of algebraic cycles on X modulo the equivalence $\sim$; also put $C_\sim(X; \mathbb{Q}) = C_\sim(X) \otimes \mathbb{Q}$.

4.1.1.2
The most important $\sim$ for us are the following.

1. Rational equivalence $Z \in \mathcal{Z}^i(X)$ is rationally equivalent to zero iff there exist a finite collection $\{Y_\alpha, f_\alpha\}$ with $Y_\alpha \subset X$ irreducible subvarieties of codimension $(i - 1)$ on X and f_α rational functions on Y_α such that $\sum_\alpha \text{div}(f_\alpha) = Z$. (For $i = 0$ we take $Z^0_{\text{rat}}(X) = (0)$). For $i = 1$ (the case of divisors) we get precisely linear equivalence. So rational equivalence generalizes, for higher codimension, the notion of linear equivalence of divisors. If $\sim$ is rational equivalence then instead of $C^i_{\text{rat}}(X)$ we write

$$CH^i(X) = \mathcal{Z}^i(X)/\mathcal{Z}^i_{\text{rat}}(X).$$

This is the ith *Chow group* of X and $CH(X) = \oplus^d_{i=D} CH^i(X)$ is the *Chow ring* of X; we write $CH^i_{\mathbb{Q}}(X) = CH^i(X; \mathbb{Q}) = CH^i(X) \otimes \mathbb{Q}$. (Remark: Fulton [F84] uses the notation $A^i(X)$ but we prefer $CH^i(X)$.)

2. Homological equivalence Fix a *Weil cohomology theory* $H(X)$ (see [Kl68], [Kl94] or [Ja94]). Let $\gamma_X : \mathcal{Z}^i(X) \to H^{2i}(X)$ be the cycle map. Then $Z \in \mathcal{Z}^i(X)$ is by definition homologically equivalent to zero iff $\gamma_X(Z) = 0$. For the Weil cohomology theory we could take, if k is an arbitrary field, the $H^*_{et}(\bar{X}, \mathbb{Q}_l)$, i.e. the étale cohomology groups *over the algebraic closure* (!), i.e. $\bar{X} = X \times_k \bar{k}$

and $l \neq \mathrm{char}(k) = p$; if $k = \mathbb{C}$ we can also take the usual Betti cohomology $H_B^*(X, \mathbb{Q})$.

Remarks

1. If k is not algebraically closed then there is also the cohomology theory $H_{et}^*(X, \mathbb{Q}_l)$, however this is not a Weil cohomology, therefore we have to pass to the algebraic closure!
2. Homological equivalence depends, at least a priori, on the choice of Weil cohomology, but one could avoid this by requiring (somewhat unpleasant!) that the cycle maps to zero in 'all' Weil cohomology theories.

3. Numerical equivalence By definition $Z \in \mathcal{Z}^i(X)$ is numerically equivalent to zero if the intersection number $\#(Z \cup Z') = 0$ for all $Z' \in \mathcal{Z}^{d-i}(X)$ (strictly speaking for all $Z' \in \mathcal{Z}^{d-i}(X)$ for which this intersection number is defined).

4.1.1.3

There is the following relation between the above equivalence relations

$$\mathcal{Z}_{rat}^i(X) \subseteq \mathcal{Z}_{hom}^i(X) \subseteq \mathcal{Z}_{num}^i(X) \subset \mathcal{Z}^i(X), \tag{1}$$

and dividing out by $\mathcal{Z}_{rat}^i(X)$ we therefore have the following subgroups of the Chow group

$$CH_{hom}^i(X) \subseteq CH_{num}^i(X) \subset CH^i(X). \tag{1'}$$

There is the following *fundamental conjecture*

$$\text{Conjecture } D(X) : \mathcal{Z}_{hom}^i(X) = \mathcal{Z}_{num}^i(X).$$

As we shall see later (Section 4.1.6.2) this conjecture plays a crucial role in Grothendieck's theory of motives.

Remarks

1. Recall that for *divisors* ($i = 1$) conjecture $D(X)$ is true (see [H75, Proposition 3.1], [F84, p. 385]).
2. If $k = \mathbb{C}$, $D(X)$ would follow from the Hodge conjecture.

4.1.2 Correspondences and projectors

4.1.2.1 Correspondences

Let X_d and Y_e be objects of $\mathcal{V}(k)$. Fix an equivalence relation $\sim$. The *group of correspondences* between X and Y (with respect to $\sim$) is defined as

$$\text{Corr}_\sim(X, Y) := C_\sim(X \times Y; \mathbb{Q}).$$

If $\sim$ is rational equivalence we write simply

$$\text{Corr}(X, Y) := CH(X \times Y; \mathbb{Q}).$$

If $f \in \text{Corr}_\sim(X, Y)$ then $^t f \in \text{Corr}_\sim(Y, X)$ denotes the *transpose* of f. One defines the *degree* of correspondences by

$$\text{Corr}_\sim^r(X, Y) := C_\sim^{d+r}(X_d \times Y; \mathbb{Q}).$$

Example Let $\varphi : X_d \to Y_e$ be a usual morphism of varieties and let Γ_φ be the graph, then

$$\varphi_* = \Gamma_\varphi \in \text{Corr}_\sim^{e-d}(X, Y),$$
$$\varphi^* = {}^t\Gamma_\varphi \in \text{Corr}_\sim^0(Y, X).$$

Composition of correspondences Let $f \in \text{Corr}_\sim(X, Y)$ and $g \in \text{Corr}_\sim(Y, Z)$ then $g \bullet f \in \text{Corr}_\sim(X, Z)$, where

$$g \bullet f := pr_{XZ}\{(f \times Z) \cap (X \times g)\},$$

where $\cap$ is the intersection product of algebraic cycle classes on $X \times Y \times Z$.

Operation on the algebraic cycle classes A correspondence $f \in \text{Corr}_\sim^r(X, Y)$ operates on $C_\sim(X; \mathbb{Q})$

$$f : C_\sim^i(X, \mathbb{Q}) \to C_\sim^{i+r}(X; \mathbb{Q})$$

by the formula $Z \mapsto f(Z) := (pr_Y)_*\{f \cap (pr_X)^*(Z)\}$ when $Z \in C_\sim^i(X, \mathbb{Q})$. Similarly f *operates on cohomology* $f : H^i(X) \to H^{i+2r}(Y)$ provided $\sim$ is finer than, or equal to, homological equivalence.

Remarks

1. $\text{Corr}_\sim^0(-, -)$ respects degrees.
2. Correspondences with respect to rational equivalence operate both on the Chow groups and on cohomology. The correspondences with respect to homological equivalence operate on cohomology but not on Chow groups.

Finally we mention that correspondences with respect to *numerical equivalence* operate only on cohomology groups provided the conjecture $D(X)$ is
true! This explains already the fundamental importance of this conjecture.
3. Under composition of correspondences $\mathrm{Corr}_\sim(X, X)$ becomes a *ring* and
   $\mathrm{Corr}_\sim^0(X, X)$ is a subring.

4.1.2.2 Projectors

Definition A correspondence $p \in \mathrm{Corr}_\sim^0(X, X)$ is called a *projector* of X (with
respect to $\sim$) if $p \bullet p = p$. Two projectors $p, q \in \mathrm{Corr}_\sim^0(X, X)$ are called *orthogonal* if $p \bullet q = q \bullet p = 0$.

Examples

1. $p = \Delta(X) = id_X$ is a projector ($\Delta(X)$ denotes the *diagonal* of X).
2. If p is a projector then $1 - p$ is also a projector and is orthogonal to p.
3. If $\varphi : X_d \to Y_d$ is a morphism of algebraic varieties which is generically
   of finite degree m, then $p = \frac{1}{m}{}^t\Gamma_\varphi \bullet \Gamma_\varphi$ is a projector of X (by abuse of
   language we write $p = \frac{1}{m}\varphi^* \bullet \varphi_*$).

4.1.3 Motives (pure motives to be precise!)

4.1.3.1

Fix a good equivalence relation $\sim$. Then Grothendieck defined the category
$\mathcal{M}_\sim(k)$ of (pure) motives, with respect to $\sim$, as follows.

(a) First effective (pure) motives This is a category $\mathcal{M}_\sim^+(k)$ which has as *objects*
pairs (X, p) with $X \in \mathcal{V}(k)$ and p a projector of X and which has as *morphisms*
if $M = (X, p), N = (Y, q)$

$$\mathrm{Hom}_{\mathcal{M}_\sim^+}(M, N) := q \bullet \mathrm{Corr}_\sim^0(X, Y) \bullet p$$

$$X \xrightarrow{p} X \xrightarrow{f} Y \xrightarrow{q} Y$$

and composition of morphisms in $\mathcal{M}_\sim^+$ is defined via composition of correspondences.

Remark As an exercise, prove:

$$\mathrm{Hom}_{\mathcal{M}_\sim^+}(M, N) = \{f \in \mathrm{Corr}_\sim^0(X, Y); f \bullet p = q \bullet f\}/\{f; f \bullet p = 0\}.$$

This was the original definition of Grothendieck. The above formulation, which by the exercise is equivalent to the original formulation, is due to Jannsen and is a little easier to work with.

(b) Virtual (pure) motives As in cohomology, it is important to allow 'twisting'. Therefore one introduces the category $\mathcal{M}_\sim(k)$ ($= \mathcal{M}_\sim$ for short), the *objects* of which are triples $M = (X, p, m)$ with $X \in \mathcal{V}(k)$, p a projector of X and $m \in \mathbb{Z}$ and with *morphisms*, if $M = (X, p, m)$, $N = (Y, q, n)$,

$$\text{Hom}_{\mathcal{M}_\sim}(M, N) := q \bullet \text{Corr}_\sim^{n-m}(X, Y) \bullet p$$

and composition of morphisms is as composition of correspondences before. The objects $M = (X, p, m)$ are now called motives with respect to $\sim$. Clearly $\mathcal{M}_\sim^+(k) \subset \mathcal{M}_\sim(k)$ is a full subcategory via $(X, p) \to (X, p, 0)$.

Remark Contrary to the historically spread misunderstanding the *construction* of the category of motives is *not conjectural* at all (and is in fact very easy!), what is conjectural are the 'good' *properties* of the motives (see below, end of Section 4.1.6 and Section 4.2)!

4.1.3.2 Examples

For the moment only trivial examples, for the interesting ones see later!

1. There exists a functor $h_\sim : \mathcal{V}^{opp}(k) \to \mathcal{M}_\sim(k)$ by taking $h_\sim(X) = (X, \Delta(X), 0)$ ($= (X, id_X, 0)$) and if $\varphi : X \to Y$ is a morphism of algebraic varieties then $h_\sim(\varphi) = {}^t\Gamma_\varphi : h_\sim(Y) \to h_\sim(X)$.
2. $\mathbb{1} := (\text{Spec } k, id, 0)$, the 'trivial' motive (i.e. the motive of a point).
3. Fix a point $e \in X(k)$, take $\pi_0 = e \times X$ and $\pi_{2d} = X \times e$ ($d = \dim X$, X irreducible). These are both projectors and are orthogonal to each other. Put $h_\sim^0(X) := (X, \pi_0, 0)$ and $h_\sim^{2d}(X) = (X, \pi_{2d}, 0)$.
   Exercise Show $h_\sim^0(X) \simeq h_\sim^0(Y) = \mathbb{1}$ for any $X_d, Y_e \in \mathcal{V}(k)$. Show $h_\sim^{2d}(X) \simeq$ (Spec $k, id, -d$) and (hence) if $e = d$ we have also $h_\sim^{2d}(X) \simeq h_\sim^{2d}(Y)$.
4. If $M = (X, p, m)$ write $M(i) = (X, p, m + i)$; hence $h_\sim^{2d}(X_d) \simeq \mathbb{1}(-d)$.
5. **Lefschetz motive and Tate motive** Define $\mathbb{L} := (\text{Spec } k, id, -1)$ and $\mathbb{T} := (\text{Spec } k, id, 1)$, $\mathbb{L}$ is called the Lefschetz motive and $\mathbb{T}$ the Tate motive. By the above exercise we have $\mathbb{L} \simeq h_\sim^2(\mathbb{P}_1) = (\mathbb{P}_1, \mathbb{P}_1 \times e, 0)$, where $\mathbb{P}_1$ is the projective line and e some point on it.
6. Let $\varphi : X_d \to Y_d$ be a morphism of algebraic varieties, generically of degree m, then $(X, \frac{1}{m}\varphi^* \bullet \varphi_*, 0)$ is a motive and it is isomorphic to $h_\sim(Y) = (Y, id_Y, 0)$ (prove!)

4.1.4 Properties of $\mathcal{M}_\sim(k)$

1. $\mathcal{M}_\sim$ is an *additive category*, i.e. the $\mathrm{Hom}_{\mathcal{M}_\sim}(M, N)$ are *abelian groups* and there exist *direct sums* $M \oplus N$. For the definition of the latter, if $M = (X, p, m)$ and $N = (Y, q, n)$ with $n = m$ take $M \oplus N = (X \amalg Y, p \amalg q, m)$ with $\amalg$ *disjoint union*. For the general case $n \neq m$, which is a little bit technical, we refer to Scholl [Sch94, 1.14].

   Exercises Show: (a) $h_\sim(\mathbb{P}_1) = 1 \oplus \mathbb{L}$; (b) in example 6 of Section 4.1.3.2 we have $h_\sim(X) = h_\sim(Y) \oplus (X, id_X - p, 0)$ with $p = \frac{1}{m}\varphi^* \bullet \varphi_*$.

2. $\mathcal{M}_\sim$ is a *pseudo-abelian category*, i.e. it is an additive category and projectors have images. In fact $\mathcal{M}_\sim$ is the so-called 'pseudo-abelian completion' of the category $C_\sim(\mathcal{V})$ where $C_\sim(\mathcal{V})$ has the same objects as $\mathcal{V}$ but $\mathrm{Hom}_{C_\sim(\mathcal{V})}(X, Y) = \mathrm{Corr}^0_\sim(X, Y)$ (we refer to [Ma68] and [Sch94, Section 1.6] for more details). In particular $(X, p, 0) = \mathrm{Im}(p)$.

   Remark For use later (Section 4.1.8.3) we remark that if $\mathcal{C}$ is an additive category then, by adding formally images and kernels of projectors in a method similar to that above, it is always possible to construct a 'pseudo-abelian completion' $\mathcal{PC}$ and a functor $\psi : \mathcal{C} \to \mathcal{PC}$ which is universal (in an obvious sense), with respect to functors $\phi : \mathcal{C} \to \mathcal{D}$ with $\mathcal{D}$ a pseudo-abelian category (see [Ma68, Section 5]).

3. $\mathcal{M}$ has a *tensor product* defined as follows: $M \otimes N := (X \times Y, p \times q, m + n)$.

   Exercises Prove (a) $h^{2d}_\sim(X_d) \simeq \mathbb{L}^{\otimes d}$; (b) $M = (X, p, m) = (X, p, 0) \otimes \mathbb{L}^{\otimes -m} = (X, p, 0) \otimes \mathbb{T}^{\otimes m}$ (i.e. the virtual motives are obtained from the effective ones by tensoring with powers of the Tate motive which is itself the inverse, in the sense of the tensor structure, of the Lefschetz motive).

4. $\mathcal{M}_\sim$ has a *multiplication structure*: from $\Delta : X \to X \times X$ we get

$$m_X : h_\sim(X) \otimes h_\sim(X) \simeq h_\sim(X \times X) \xrightarrow{h_\sim(\Delta)} h_\sim(X).$$

5. There is an involution $\hat{} : \mathcal{M}_\sim^{opp} \to \mathcal{M}_\sim$ given by $M = (X, p, m) \mapsto \hat{M} := (X, {}^t p, d - m)$ where (restricting to irreducible X) $d = \dim X$.

4.1.5 Relations between the various $\mathcal{M}_\sim$

4.1.5.1

As mentioned in Section 4.1.1.2 there are three equivalence relations of particular interest for us. They give three different (or at least a priori different) types of motives.

Chow motives Take for $\sim$ the rational equivalence. Write in this case $CH\mathcal{M}(k) := \mathcal{M}_{rat}(k)$ and $ch(X) := h_{rat}(X)$. The objects of $CH\mathcal{M}(k)$ are called *Chow motives*.

Homological motives Fix a Weil cohomology theory (say $H_{et}(X_{\bar{k}}, \mathbb{Q}_l)$ or $H_B(X(\mathbb{C}), \mathbb{Q})$ if $k = \mathbb{C}$). Take for $\sim$ the homological equivalence, write $\mathcal{M}_{hom}(k)$ and $h_{hom}(X)$ in this case.

Numerical motives or Grothendieck motives Take numerical equivalence and write $\mathcal{M}_{num}(k)$ and $h_{num}(X)$ in this case.

Remarks

1. Grothendieck's construction works for every 'good' equivalence on algebraic cycles, however for the purpose he has in mind he used (or wanted to use) numerical equivalence (see [G68, p. 198]) for reasons to be explained below (Section 4.1.6.2) and for that purpose he formulated the standard conjectures (see Section 4.2.1).
2. In the case $k = \mathbb{C}$ Deligne [Detal82] constructed a category of motives $\mathcal{M}_{AH}$ using for the correspondences *absolute Hodge cycles* instead of algebraic cycles (of course, if the Hodge conjecture holds then such cycles are algebraic!)

4.1.5.2 Relations between the various theories (for an arbitrary field k)
Because of relation (1) in Section 4.1.1.3 we have natural functors indicated in the following diagram

$$\mathcal{V}^{opp}(k) \xrightarrow{\;ch\;} CH\mathcal{M}(k) \longrightarrow \mathcal{M}_{hom}(k) \xrightarrow{\;\cong?\;} \mathcal{M}_{num}(k)$$

$$\mathcal{M}_{AH} \quad (\text{if } k = \mathbb{C}).$$

As remarked in Section 4.1.1.3 there is the fundamental *conjecture* $D(X)$ and this would clearly imply that $\mathcal{M}_{hom}(k)$ and $\mathcal{M}_{num}(k)$ would coincide.

Remark If $k = \mathbb{C}$ then there is the *Hodge conjecture* (see Chapters 2 and 6).

$$\text{Hodge conjecture } HC(X, i) : \gamma_X : CH_{\mathbb{Q}}^i(X) \to$$
$$H_B^{2i}(X, \mathbb{Q}) \cap H^{i,i}(X) \text{ is onto.}$$

This would imply not only $\mathcal{M}_{hom}(\mathbb{C}) = \mathcal{M}_{AH}$, but also the Hodge conjecture implies, as is well known, the conjecture $D(X)$ and hence then also $\mathcal{M}_{\text{hom}}(\mathbb{C})$ would coincide with $\mathcal{M}_{\text{num}}(\mathbb{C})$.

4.1.6 Cycle groups and cohomology for motives

4.1.6.1

Let $M = (X, p, m) \in \mathcal{M}_\sim(k)$. As mentioned already in Section 4.1.2, the correspondence p operates on the vector spaces $C^i_\sim(X, \mathbb{Q})$ and in particular

$$p : C^{i+m}_\sim(X; \mathbb{Q}) \to C^{i+m}_\sim(X; \mathbb{Q}).$$

Definition $C^i_\sim(M) := \text{Im}(p) \subset C^{i+m}_\sim(X; \mathbb{Q})$ (note the twist!). In particular if $M \in CH\mathcal{M}(k)$ then M has Chow groups $CH^i(M)$ (or more precisely, Chow vector spaces).

Exercise Prove $C^i_\sim(M) = \text{Hom}_{\mathcal{M}_\sim}(\mathbb{L}^{\otimes i}, M)$.

If $\sim$ is *finer than, or equal to,* homological equivalence, i.e. $\mathcal{Z}_\sim \subseteq \mathcal{Z}_{\text{hom}}$, then p also operates on the cohomology groups

$$p : H^{i+2m}(X_{\bar{k}}) \to H^{i+2m}(X_{\bar{k}})$$

and passing to the algebraic closure $\bar{k}$ and using the obvious functor $\mathcal{M}_\sim(k) \to \mathcal{M}_\sim(\bar{k})$, sending $M = (X, p, m)$ to $\bar{M} = M_{\bar{k}} = (X_{\bar{k}}, p, m)$, one puts:

Definition $H^i(M_{\bar{k}}) := \text{Im}(p) \subset H^{i+2m}(X_{\bar{k}})$.

So we have a *realization functor*:

$$\text{real} : \mathcal{M}_\sim(k) \to (\text{vector spaces over } F)$$
$$M \to H^*(M_{\bar{k}})$$

where F is the field (of char. 0) used in the Weil cohomology (so $F = \mathbb{Q}_l$ if we use $H_{et}(-, \mathbb{Q}_l)$ and $F = \mathbb{Q}$ or $\mathbb{C}$ if we use Betti cohomology if $k = \mathbb{C}$). Moreover, since the cohomology is defined via the action of a correspondence, *extra structures*, like the $\text{Gal}(\bar{k}/k)$-action or the Hodge structure, carry over to the cohomology of the motive.

The diagram of functors from Section 4.1.5.2 can now be extended to the (commutative) diagram

$$\mathcal{V}^{opp}(k) \xrightarrow{\ ch\ } CH\mathcal{M}(k) \longrightarrow \mathcal{M}_{\text{hom}}(k) \xrightarrow{\ \cong?\ } \mathcal{M}_{\text{num}}(k)$$
$$\text{real} \downarrow$$
$$(\text{vector spaces})/F.$$

4.1.6.2 Remark

From the point of view of *Grothendieck's objectives* the above diagram makes the importance of the conjecture $D(X)$ (or of the standard conjectures to be discussed in Section 4.2.1) very clear. Namely, the category $\mathcal{M}_{num}$ is defined intrinsically independent of any cohomology theory and the objects of $\mathcal{M}_{num}$ are the 'true' motives (now called Grothendieck motives). But the problem is that, at least a priori, there is no realization functor! The role of the standard conjectures is that they imply equality between $\mathcal{M}_{num}$ and $\mathcal{M}_{hom}$ so that, a posteriori, the objects of $\mathcal{M}_{num}$ have realizations as vector spaces in various cohomology theories. Moreover it follows also from the standard conjectures that the category $\mathcal{M}_{num}$ is an *abelian, semi-simple category*. This latter part (but not the former!) has now been proved unconditionally by Jannsen (see Section 4.2.2)!

4.1.7 Manin's identity principle

4.1.7.1

In order to avoid a too technical statement we shall give this only in the most simple form; we refer to Scholl's Seattle paper [Sch94, Section 2.3] for a more refined version.

The idea is the following. If we have two correspondences $f, g \in \text{Corr}_\sim$ (X, Y) then we *cannot* test equality (or inequality) of the cycle classes by the action of f and g on the Chow group $CH(X_K)$ for $K \supset k$ a field, i.e. if we are only allowed to base change by a field, we can do it by considering 'T-points'. Namely, let $X, Y \in \mathcal{V}(k)$ and also $T \in \mathcal{V}(k)$, put $X(T) := \text{Corr}_\sim(T, X)$ (with $\sim$ some good equivalence relation). For $f \in \text{Corr}_\sim(X, Y)$ write $f_T :$ $X(T) \to Y(T)$ with $f_T(t) = f \bullet t$ for $t \in X(T)$.

Manin's identity principle *Let $f, g \in \text{Corr}_\sim(X, Y)$ for $X, Y \in \mathcal{V}(k)$. Then the following are equivalent*:

(i) $f = g$,
(ii) $f_T = g_T$ *for all* $T \in \mathcal{V}(k)$,
(iii) $f_X = g_X$.

The proof is straightforward: (i) $\Rightarrow$ (ii) $\Rightarrow$ (iii) $\Rightarrow$ (i); for the latter implication compose with $\Delta(X)$. However, Manin has shown that the principle is very useful! For instance it can be used to compute the motive of a projective bundle and the motive of a blow-up.

4.1.7.2

We give just one simple application to demonstrate how it is used. Consider in $\mathcal{V}(k)$ the inclusion $i : X \to X_1$, where X is a smooth subvariety of X_1 of codimension r. Then we have the formula $i^* i_*(\alpha) = c_r(N) \cap \alpha$ for all $\alpha \in C_\sim(X)$, where $c_r(N)$ is the rth Chern class of the normal bundle of X in X_1 [F84, p. 103]. Since this formula holds 'universally', i.e. after base extension by $T \in \mathcal{V}(k)$, we also have, by the above principle, equality on cycle class level, hence in $C_\sim(X \times X)$ the relation $'\Gamma_i \bullet \Gamma_i = \Delta(X)(c_r(N))$ holds.

4.1.8 Motives of curves

4.1.8.1

Finally we come now to a non-trivial example! For simplicity, let k be algebraically closed and let us work with Chow motives $CH\mathcal{M}(k)$.

Let C be a smooth, projective *curve* defined over k and fix a point $e \in C(k)$. As before, put $\pi_0 = e \times C$ and $\pi_2 = C \times e$ (Section 4.1.3.2, Example 3); now take $\pi_1 = \Delta(C) - \pi_0 - \pi_2$. From the orthogonality of π_0 and π_2 we get that π_1 is also a projector. Put $ch^i(C) = (C, \pi_i, 0)$ $(i = 0, 1, 2)$ so

$$ch(C) = ch^0(C) \oplus ch^1(X) \oplus ch^2(C).$$

Since $ch^0(C)$ and $ch^2(C)$ are essentially trivial parts of $ch(C)$, the $ch^1(C)$ must contain the essential properties of C and, hence, should be interesting and non-trivial. In fact it is, as shown by the following theorem.

4.1.8.2 Theorem (Grothendieck) [De69]

(i) *We have*

$$CH^i(ch^1(C)) = \begin{cases} 0 & i \neq 1, \\ J(C)(k) & i = 1, \end{cases}$$

where $J(C)$ is the Jacobian of C and

$$H^i(ch^1(C)) = \begin{cases} 0 & i \neq 1, \\ H^1(C) & i = 1. \end{cases}$$

(ii) *Let C and C' be two smooth, projective curves. Then*

$$\mathrm{Hom}_{CH\mathcal{M}}(ch^1(C), ch^1(C')) = \mathrm{Hom}_{AV, isog}(J(C), J(C')),$$

where $\mathrm{Hom}_{AV, isog}(-, -)$ means homomorphisms of abelian varieties up to isogeny.

(iii) *Let* $\mathcal{M}' \subset CH\mathcal{M}(k)$ *be the following full subcategory:*

$$\mathcal{M}' = \{M; \exists C \text{ smooth, projective curve such that}$$
$$ch^1(C) = M \oplus M' \text{ for some } M'\}.$$

Then $\mathcal{M}'$ is equivalent to the category of abelian varieties up to isogeny.

Remarks

1. This is also true if we work with homological or with numerical equivalence (of course we must drop then in (i) the assertion about the Chow group).
2. Part (iii) of the theorem shows that the theory of motives is, in some sense, a generalization of the theory of abelian varieties, namely it contains, up to isogeny, that category as a full subcategory! Undoubtedly this fact (iii) must also have been one of the motivations for Grothendieck for his theory of motives (again see his remark on p. 198 of [G68]).

4.1.8.3 Proof of the theorem

In fact for parts (i) and (ii) we can work with motives with $\mathbb{Z}$-coefficients. Write $CH\mathcal{M}_{\mathbb{Z}}$.

Proof of (i) For $Z \in CH^1(C)$ we get $\pi_1(Z) = Z - \deg(Z) \cdot e$. This gives us immediately the result for the Chow group. For the cohomology, since π_0 (respectively π_2) is the identity on $H^0(C)$ (respectively $H^2(C)$) and zero on the other groups, we must have that π_1 is the identity on $H^1(C)$ and zero on the other groups.

Proof of (ii) This statement is a reinterpretation in terms of motives of a well known theorem of Weil on the relation between Jacobians of curves and correspondences between curves ([W48, Theorem 22 p. 79]; see also [La59, p. 153]), and in fact – as Weil remarks himself – it goes back to classical Italian geometry:

Theorem (Weil)

$$CH^1(C \times C')/(\text{horizontal} + \text{vertical components}) \xrightarrow{\sim}$$
$$\operatorname{Hom}_{AV}(J(C), J(C')),$$

where (horizontal + vertical components) *stands for the subgroup generated by the cycles of the form* $C \times Z'$ *and* $Z \times C'$ *with* $Z \in CH^1(C)$ *and* $Z' \in CH^1(C')$ *and* $\operatorname{Hom}_{AV}(-, -)$ *are now the homomorphisms themselves.*

In order to translate this into the statement (ii) (in its more precise form, i.e. in $CH\mathcal{M}_{\mathbb{Z}}$) we note that $(C \times Z') \bullet \pi_1(C) = 0$ and also $\pi_1(C') \bullet (Z \times C') = 0$ and hence we get, from the definition of the morphisms in $CH\mathcal{M}_{\mathbb{Z}}$, that the left-hand side in Weil's theorem is indeed $\operatorname{Hom}_{CH\mathcal{M}_{\mathbb{Z}}}(ch^1(C), ch^1(C'))$.

Proof of (iii) Again this is a reinterpretation by Grothendieck of known theorems for abelian varieties! Namely, still first working with $CH\mathcal{M}_{\mathbb{Z}}$, consider the full subcategory $\mathcal{M}''_{\mathbb{Z}} = \{M; M \simeq ch^1(C), C \text{ curve}\}$. Part (ii) is saying that $\mathcal{M}''_{\mathbb{Z}}$ is equivalent to the category of Jacobians of curves (with the 'true' homomorphisms). Next, if A is an abelian variety it is well known that A is the quotient of a Jacobian of a curve $J(C)$. Now by the *complete reducibility theorem of Poincaré* we have that $J(C)$ is isogenous with $A \oplus A'$ where A' is another abelian variety. However, this means that the *category of abelian varieties up to isogeny* is the *pseudo-abelian completion* of the category of Jacobians up to isogeny. On the other hand applying the natural functor $\varphi : CH\mathcal{M}_{\mathbb{Z}} \to CH\mathcal{M}_{\mathbb{Q}}$ we have that the category $\mathcal{M}'$ of (iii) in the theorem is, by construction, the pseudo-abelian completion of the image $\varphi(\mathcal{M}''_{\mathbb{Z}})$ in $CH\mathcal{M}_{\mathbb{Q}}$. By the universal property of a pseudo-abelian completion, the above two pseudo-abelian completions at issue must therefore also be equivalent, which completes the proof. $\square$

4.1.9 References for Lecture 1

Grothendieck himself has, unfortunately, not published anything on his theory of motives. However, in the spring of 1967 he gave a series of lectures on his theory at the IHES. These lectures are at the origin of the paper by Manin [Ma68] (see the remarks on p. 444) and the paper by Demazure [De69] and, I think, also of Kleiman's paper [Kl70].

The comments, of a historical and philosophical nature, which Grothendieck made much later in his 'Récoltes et Semailles' [G85] are also very interesting.

For more recent papers relevant to Lecture 1 we mention the nice introductory paper [Se91] by Serre, and the beautiful paper [Sch94] of Scholl; this last paper gives a thorough treatment of the fundamentals of the subject and we strongly recommend it. For a simple introduction one can also look at [Mu96]. For the many further developments and the enormous impact of motives, the reader can consult the proceedings of the 1991 Seattle Conference on Motives [JKS94].

Finally, as mentioned in the beginning, and stressed again in Section 4.1.3, these lectures are dealing with *pure motives* only, i.e. motives 'coming' from smooth, projective varieties. However, similar to the case of Hodge structures and mixed Hodge structures, there should also exist a theory of *mixed motives*, i.e. motives 'coming' from quasi-projective and possibly singular varieties. For approaches towards such a theory see, in particular, the works of Hanamura [Ha95], [Ha99], Levine [Le98], Voevodsky [V00] and Nori [No]; for the latter we refer to the lectures of Mouroukos in this Summer School (see also the remark at the end of Section 4.3.2.2).

4.2 Lecture 2: conjectures and results on motives

4.2.1 Grothendieck's standard conjectures (main points)

4.2.1.1 Preparation

Let k now be an algebraically closed field, and $\mathcal{V} = \mathcal{V}(k)$ is as before the category of smooth, projective varieties defined over k. Fix a Weil cohomology. Let $X_d \in \mathcal{V}(k)$ and consider the cycle map (or better its factorization through the Chow group)

$$\gamma_X : CH_{\mathbb{Q}}^i(X) \to H^{2i}(X) \qquad (0 \le i \le d).$$

Let

$$A^i(X) := \mathrm{Im}(\gamma_X) \subset H^{2i}(X)$$

be the $\mathbb{Q}$ vector space of 'algebraic classes' in $H^{2i}(X)$ (note that, depending on the choice of the Weil cohomology, $H^{2i}(X)$ is an F vector space with $F = \mathbb{Q}_l$, or $\mathbb{Q}$, the latter only in the case $k = \mathbb{C}$). Note that by definition (see Lecture 1) $\mathrm{Ker}(\gamma) = CH_{\mathrm{hom}}^i(X; \mathbb{Q})$.

In *cohomology* there is the *Künneth decomposition* of the diagonal $\gamma_{X \times X}(\Delta) \in H^{2d}(X \times X) = \sum_{i=0}^{2d} H^{2d-i}(X) \otimes H^i(X)$

$$\gamma_{X \times X}(\Delta) = \sum_{i=0}^{2d} \Delta_{2d-i,i},$$

where the $\Delta_{2d-i,i}$ are topological classes and there is the old conjecture (usually called the *Künneth conjecture*):

$C(X)$: the Künneth components $\Delta_{2d-i,i}$ are algebraic classes

$$(0 \le i \le 2d).$$

Remarks

1. For cases where $C(X)$ is known see below, Section 4.2.1.2.
2. For $k = \mathbb{C}$ the $C(X)$ would follow from the Hodge conjecture.

4.2.1.2 Grothendieck's standard conjecture of Lefschetz type

Let $X = X_d$ be embedded in $\mathbb{P}_N$ (say), let H be a hyperplane section and $Y = X \cap H$

$$\gamma_X(Y) \in A^1(X) \subset H^2(X).$$

One has the *Lefschetz operator*

$$L : H^i(X) \to H^{i+2}(X)$$
$$\alpha \mapsto L(\alpha) = \alpha \cup \gamma(Y)$$

and by iteration $L^r = L \cdot L \cdots L$ (r-times).
There is the *hard Lefschetz theorem*:

$$L^{d-i} : H^i(X) \xrightarrow{\sim} H^{2d-i}(X)$$

is an *isomorphism*, or equivalently putting $j = d - i$,

$$L^j : H^{d-j}(X) \xrightarrow{\sim} H^{d+j}(X)$$

is an *isomorphism*.

Note that L (and hence every L^j) is given by an *algebraic* cycle, in fact $L = \Delta(Y) \in CH^{d+1}(X \times X)$.

The Λ operator Using the hard Lefschetz theorem one can define a *unique linear map* Λ making the following diagrams commutative:

(i) for $0 \le j \le d - 2$

$$
\begin{array}{ccc}
H^{d-j}(X) & \xrightarrow[\cong]{L^j} & H^{d+j}(X) \\
{\scriptstyle \Lambda} \downarrow & & \downarrow {\scriptstyle L} \\
H^{d-j-2}(X) & \xrightarrow[L^{j+2}]{\cong} & H^{d+j+2}(X),
\end{array}
$$

(ii) for $2 \le j \le d$

$$
\begin{array}{ccc}
H^{d-j+2}(X) & \xrightarrow[\cong]{L^{j-2}} & H^{d+j-2}(X) \\
{\scriptstyle L} \uparrow & & \uparrow {\scriptstyle \Lambda} \\
H^{d-j}(X) & \xrightarrow[L^j]{\cong} & H^{d+j}(X),
\end{array}
$$

(iii) as inverse of L

$$H^{d-1}(X) \xrightarrow[\cong]{L} H^{d+1}(X).$$

Of course, again one can iterate $\Lambda^r := \Lambda \circ \Lambda \circ \cdots \circ \Lambda$ (r-times).

Remarks

1. So Λ is 'almost' the inverse of L, for instance in case (i) it is the inverse of L on $\text{Im}(L) \subset H^{d-j}(X)$.
2. Λ being a linear map on cohomology, using Poincaré duality and the Künneth decomposition it can be interpreted as a *topological correspondence* for X via

$$\text{Hom}(H(X), H(X)) \simeq H(X)^{\vee} \otimes H(X) \subset H(X \times X).$$

Now there is the following conjecture:

> standard conjecture of Lefschetz type $B(X)$: Λ is algebraic.

That is $\Lambda = \gamma_{X \times X}(Z)$ for $Z \in CH_{\mathbb{Q}}^{d-1}(X \times X)$. Also then, clearly, by composition the iterates Λ^r should be algebraic.

Remarks

1. If $k = \mathbb{C}$, then Λ is a Hodge cycle, hence $B(X)$ would follow from the Hodge conjecture. As such $B(X)$ is an 'old conjecture', however Grothendieck put it in the forefront and emphasized its importance.
2. (Again for arbitrary $k = \bar{k}$.) $B(X)$ has a number of consequences and also a number of equivalent statements. We refer to the papers of Kleiman [Kl68], [Kl94] for the details but we mention some of them.
   (a) If $B(X)$ holds for one embedding, i.e. for one L, then it holds for any ample L' (which justifies the notation $B(X)$ instead of $B(X, L)$).
   (b) $B(X) \Rightarrow A(X, L)$ where $A(X, L)$ is the conjecture that the restriction of $L^{d-2i} : A^i(X) \to A^{d-i}(X)$ is an isomorphism in the diagram

$$H^{2i}(X) \xrightarrow[L^{d-2i}]{\cong} H^{2d-2i}(X)$$
$$\uparrow \qquad\qquad \uparrow$$
$$A^i(X) \longrightarrow A^{d-i}(X)$$

   where the vertical arrows are the inclusions.
   (c) $B(X) \Rightarrow C(X)$ (the Künneth conjecture, see Section 4.2.1.1).

Current status of $B(X)$ Except for 'trivial' varieties like projective spaces, Grassmannians, etc., $B(X)$, and hence $C(X)$, is known to be true for curves (trivial), surfaces (Grothendieck, see [Kl68]) and abelian varieties (Lieberman,

Kleiman [Kl68]). $C(X)$, moreover, is true for varieties over finite fields as was proved by Katz and Messing, as a consequence of Deligne's theorem on the Weil conjectures [KM74].

4.2.1.3 Grothendieck's standard conjecture of Hodge type

Primitive elements By the hard Lefschetz theorem, $L^{d-i} : H^i(X) \xrightarrow{\sim} H^{2d-i}(X)$; put $P^i(X) := \text{Ker}\{L^{d-i+1} : H^i(X) \to H^{2d-i+2}(X)\}$. The elements of $P^i(X)$ are called the *primitive elements* of $H^i(X)$. Now consider the intersection

$$A^i(X) \cap P^{2i}(X) \subset H^{2i}(X).$$

Let $x, y \in A^i(X) \cap P^{2i}(X)$ for $i \le d/2$ and consider the pairing

$$x, y \mapsto (-1)^i \langle L^{d-2i}(x), y \rangle,$$

where $\langle -, - \rangle$ stands for the cup-product.

> Standard conjecture of Hodge type $\text{Hdg}(X)$: this pairing is positive definite.

Note that this pairing is $\mathbb{Q}$-valued, so the statement makes sense!

Current status of $\text{Hdg}(X)$ (see [Kl68] and [Kl94])

1. If $\text{char}(k) = 0$ then $\text{Hdg}(X)$ is true. First of all by the so-called Lefschetz principle we can reduce to the case $k = \mathbb{C}$, next by the comparison theorem we can reduce to Betti cohomology and then $\text{Hdg}(X)$ follows from Hodge theory (the Hodge–Riemann bilinear relations, see [GH78, p. 123] and the lectures of Peters and Kosarew (Chapter 2)).
2. In arbitrary characteristic $\text{Hdg}(X)$ is true for surfaces (Segre 1937, Grothendieck 1958, see [Kl68]).

4.2.1.4 Consequences of the standard conjectures

Again we refer to Kleiman ([Kl68] and [Kl94]) for the details, but we mention here two of the most important points.

1. $B(X) + \text{Hdg}(X) \Rightarrow D(X)$, i.e. homological equivalence $=$ numerical equivalence. We have seen in Section 4.1.6.2 the crucial importance of this. This must have been one of the principal goals of the standard conjectures.
2. Moreover, and this must have been one of the other objectives, the $B(X) + \text{Hdg}(X)$ imply that the category $\mathcal{M}_{\text{num}}$ is an abelian, semi-simple category.

However, in 1991 this was proved by Jannsen unconditionally (see Section 4.2.2).

4.2.1.5 Further remarks

1. If char$(k) = 0$ it follows now by Hdg(X) and the results of Lieberman mentioned in Section 4.2.1.2 that on abelian varieties homological equivalence = numerical equivalence. Recently Clozel also proved this for abelian varieties over finite fields [Cl99].
2. Finally we mention that in the case $k = \mathbb{C}$ Lieberman has proved that numerical equivalence = homological equivalence holds, not only for divisors, but also for cycles of codimension 2 and $(d - 1)$ [Li68].

References for 4.2.1

Grothendieck published his standard conjectures in a short paper in the proceedings of the Bombay Conference in 1968 [G68]. The two papers of Kleiman [Kl68] and [Kl94] give a very detailed discussion of these conjectures and their consequences.

4.2.2 Jannsen's theorem

The most important progress in the direction of the standard conjectures was made in 1991 by Jannsen [Ja92].

4.2.2.1

Let us assume (for simplicity) that here in Section 4.2.2 the field k is *algebraically closed*. Let $\sim$ be a 'good' (= 'adequate') equivalence relation; usually we shall write for short $\mathcal{M} = \mathcal{M}_\sim$ for the corresponding category of motives. Furthermore if $F \supset \mathbb{Q}$ is a field then we write $CH(X; F) = CH(X) \otimes_{\mathbb{Z}} F$.

4.2.2.2 Theorem of Jannsen [Ja92]

Let the assumptions and notation be as above. Then the following properties are equivalent:

(i) *$\mathcal{M} = \mathcal{M}_\sim$ is an abelian, semi-simple category,*
(ii) *the equivalence relation $\sim$ is numerical equivalence,*
(iii) *for all $X_d \in \mathcal{V}(k)$ the F-algebra $\mathrm{Corr}^0_\sim(X \times X; F) = CH^d(X \times X; F)/\sim$ is a finite dimensional, semi-simple F-algebra.*

4.2.2.3

We shall outline the main parts of the proof.

(i) $\Rightarrow$ (ii) The proof proceeds by contradiction. So assume that $\mathcal{M}_\sim$ is abelian, semi-simple but $\mathcal{Z}_\sim \neq \mathcal{Z}_{\mathrm{num}}$. First note that it is well known that for a good equivalence relation always $\mathcal{Z}_\sim \subseteq \mathcal{Z}_{\mathrm{num}}$, so assume strict inclusion. Let $Z \in \mathcal{Z}_{\mathrm{num}}^i(X)$ but $Z \notin \mathcal{Z}_\sim^i(X)$. Then we have in $\mathcal{M}$ a morphism f from the identity object $f : 1 = (pt, id, 0) \to h_\sim(X)(i) = (X, id, i)$ given by $f = Z \in \mathrm{Corr}_\sim^i(pt, X) = CH_{\mathbb{Q}}^i(X)/ \sim = C_\sim^i(X)$, and since $Z \notin \mathcal{Z}_\sim^i(X)$ this $f \neq 0$. Since $\mathcal{M}_\sim$ is abelian and semi-simple there exists $g : h_\sim(X)(i) \to 1$ such that $g \bullet f = id_{pt}$. Such a g is given by $W \in CH_\varphi^{d-i}(X)/ \sim$. Then $g \bullet f = pr_{13}\{(pt \times Z) \cap (W \times pt)\} = \#(Z \cdot W) \cdot pt$, where the intersection is on $pt \times X \times pt = X$ and $\#(Z \cdot W)$ stands for the intersection number on X. Since $g \bullet f = id_{pt}$ we have $\#(Z \cdot W) = 1$, but this means that Z is *not* numerically equivalent to zero, a contradiction. Hence $\mathcal{Z}_\sim = \mathcal{Z}_{\mathrm{num}}$. $\qquad\square$

4.2.2.4

(ii) $\Rightarrow$ (iii) (the main step!) Let $X_d \in \mathcal{V}(k)$. In 4.2.1.1 we introduced the notation $A^i(X) = CH_{\mathbb{Q}}^i(X)/(\text{homological equivalence})$, let us also write $B^i(X) = CH_{\mathbb{Q}}^i(X)/(\text{numerical equivalence})$; we always have *surjections* $CH_{\mathbb{Q}}^i(X) \to A^i(X) \to B^i(X)$.

The following was known in the 1960s [Kl68, Theorem 3.5]:

Lemma 1 $\dim_{\mathbb{Q}} B^i(X)$ *is finite and in fact* $\dim_{\mathbb{Q}} B^i(X) \leqq b_{2i}(X) := \dim_{\mathbb{Q}_l} H_{et}^{2i}(X, \mathbb{Q}_l)$.

Proof Consider the cycle map $\gamma_X : \mathcal{Z}^{d-i}(X) \to H_{et}^{2d-2i}(X, \mathbb{Q}_l)$. Let $\alpha_1, \dots, \alpha_m \in \mathcal{Z}^{d-i}(X)$ be such that the $\gamma(\alpha_j)$ give a $\mathbb{Q}_l$-basis for $\mathrm{Im}(\gamma) \subset H_{et}^{2d-2i}(X, \mathbb{Q}_l)$; clearly $m \leq b_{2d-2i}(X) = b_{2i}(X)$ (by Poincaré duality). Consider now the homomorphism

$$\lambda : \mathcal{Z}^i(X) \to \mathbb{Z}^m$$

given, if $\beta \in Z^i(X)$, by $\lambda(\beta) = (\#(\beta \cdot \alpha_1), \dots, \#(\beta \cdot \alpha_m))$ where, as before, $\#(-)$ stands for the intersection number.

Claim: $\mathrm{Ker}(\lambda) = \mathcal{Z}_{\mathrm{num}}^i(X)$. Clearly $\mathcal{Z}_{\mathrm{num}}^i(X) \subseteq \mathrm{Ker}(\lambda)$. For the converse, note that if $\alpha \in \mathcal{Z}^{d-i}(X)$, then $\gamma(\alpha) = \sum_{j=1}^m q_j \gamma(\alpha_j)$ with $q_j \in \mathbb{Q}_l$. Using the compatibility, by the cycle map, of the intersection number with the cup-product we get $\#(\beta \cdot \alpha) = \sum_j q_j \gamma(\beta) \cup \gamma(\alpha_j)$, hence if $\beta \in \mathrm{Ker}(\lambda)$ then $\beta \in \mathcal{Z}_{\mathrm{num}}^i(X)$. From the claim we get $B^i(X) \subseteq \mathbb{Q}^m$, which completes the proof. $\qquad\square$

Now we come to the main point:

Lemma 2 (Key lemma)*The* $\mathbb{Q}$-*algebra* $B^d(X \times X) = CH^d_{\mathbb{Q}}(X \times X)/(\text{numer-})$ ical equivalence) *is semi-simple.*

Proof Let us first recall some facts from algebra (see for instance [P82], [Bo58]). If $F \supset \mathbb{Q}$ and R is a finite dimensional F-algebra then R is artinian and the (Jacobson) radical $J(R)$ is the largest two-sided nilpotent ideal of R [Bo58, Theorem 3, p. 69]. Furthermore, R is semi-simple if and only if $J(R) = (0)$ [Bo58, Corollary 2, p. 70]. Moreover if $F_1 \supset F$ is an extension of fields then $J(R \otimes_F F_1) = J(R) \otimes_F F_1$ [Bo58, Corollary 2, p. 85]. Therefore it suffices to prove the lemma (and in fact the entire statement (iii)) after making the base extension $F = \mathbb{Q}_l \supset \mathbb{Q}$.

Put $A = A^d(X \times X) \otimes_{\mathbb{Q}} \mathbb{Q}_l$ and $B = B^d(X \times X) \otimes_{\mathbb{Q}} \mathbb{Q}_l$, there is clearly a surjective map $\phi : A \to B$, and both are finite dimensional $\mathbb{Q}_l$-algebras. Put $J = J(B)$ and $J' = J(A)$.

We must prove $J = 0$. So take $f \in J$, we must show $f = 0$. In view of the above mentioned behaviour of the radical by base extension it suffices to start with $f \in (B^d(X \times X) \cap J)$.

Claim: $\phi(J') = J$.

Proof Since ϕ is surjective, $\phi(J')$ is two-sided, nilpotent ideal in B, hence $\phi(J') \subset J$. On the other hand we have

$$A/J' \to B/\phi(J') \to B/J.$$

Since A/J' is semi-simple we have $B/\phi(J')$ semi-simple (see [P82, p. 42]), but J is the smallest ideal in B making the quotient semi-simple [Bo58, p. 66, Corollary) hence $\phi(J') = J$.

Returning now to $f \in J$ we can lift f to $f' \in J'$, and f' is nilpotent in A. Let $g \in A^d(X \times X)$ be arbitrary. Now comes the crucial point! Apply the *Lefschetz trace formula* to f' and g in A. By this formula

$$f' \cup {}^t g = \sum_{i=0}^{2d} (-1)^i Tr_i(f' \bullet g)$$

(see [Kl68, Proposition 1.3.6]) where ${}^t g$ is the transpose and the trace is the trace of $f' \bullet g$ operating on $H^i_{et}(X, \mathbb{Q}_l)$. However $f' \in J'$ and hence $f' \bullet g \in J'$ since J' is a two-sided ideal, hence $f' \bullet g$ is nilpotent, hence all these traces are zero. But $f' \cup {}^t g = \#(f \cdot {}^t g)$ on $X \times X$, and since g was completely arbitrary in $A^d(X \times X)$ this implies that $f = 0$ which proves Lemma 2 and hence, as we have remarked above, the inclusion (ii) $\Rightarrow$ (iii). $\qquad \square$

4.2.2.5

(iii) $\Rightarrow$ (i) This part is rather formal, so we do not repeat it here. It depends on the following two lemmas.

Lemma 3 *Condition* (iii) *implies that for every* $M \in \mathcal{M}_\sim$ *we have*

$\dim_F \mathrm{End}_{\mathcal{M}_\sim}(M)$ *is finite,*
$\mathrm{End}_{\mathcal{M}_\sim}(M)$ *is a semi-simple* F-*algebra.*

Lemma 4 *If* $\mathcal{M}$ *is an* F-*linear, pseudo-abelian category such that* $\mathrm{End}_{\mathcal{M}}(M)$ *is a finite-dimensional, semi-simple* F-*algebra for every* $M \in \mathcal{M}$, *then* $\mathcal{M}$ *is an abelian, semi-simple category.*

For the details, see [Ja92].

4.2.2.6 Final remark

As we have seen, the main ingredient in Jannsen's ingenious proof is the Lefschetz trace formula, a formula well known to Grothendieck in the 1960s! On the other hand, in spite of this beautiful result the standard conjectures, and in particular the relation between homological and numerical equivalence (conjecture $D(X)$), remain open.

Reference for 4.2.2

[Ja92]

4.2.3 Chow–Künneth decomposition

4.2.3.1

We shall introduce now a refined version of the concept of Künneth decomposition and a stronger conjecture than the algebraicity of the Künneth components (conjecture $C(X)$, see Section 4.2.1.1). The motivation for this comes from the results in Section 4.3.1.

4.2.3.2

Let k be a field and $X_d \in \mathcal{V}(k)$, i.e. X is a smooth, projective variety defined over k.

Definition We say that X has a *Chow–Künneth decomposition* over k if $\exists \pi_i \in \mathrm{Corr}^0(X, X) = CH^d(X \times X, \mathbb{Q})$, for $0 \leq i \leq 2d$, such that

(1) $\pi_j \bullet \pi_i = \begin{cases} \pi_i & \text{if } i = j \\ 0 & \text{if } i \neq j \end{cases}$ i.e., the π_i are *projectors, mutually orthogonal,*

(2) $\sum_i \pi_i = \Delta(X)$,

(3) π_i (modulo homological equivalence over $\bar{k}$) $= \Delta_{2d-i,i}$, i.e. the usual ith Künneth component.

If we have such a Chow–Künneth decomposition, put $ch^i(X) := (X, \pi_i, 0) \in CH\mathcal{M}(k)$ and then

$$(*) \qquad ch(X) = \sum_{i=0}^{2d} ch^i(X).$$

Remarks

1. The π_i themselves are *not unique* (even for curves $X = C$ the π_0 and π_2 depend already on the choice of the point $e \in C(k)$!). However, we expect that the Chow *motive* $ch^i(X) = (X, \pi_i, o)$ itself will be *unique up to an isomorphism* (as is trivially the case for $ch^0(X)$ and $ch^{2d}(X)$ as was seen in Lecture 1, Section 4.1.3.2, Example 3).
2. We expect moreover that we can take the π_i such that $\pi_{2d-i} = {}^t\pi_i$ $(0 \le i \le d)$.
3. From the decomposition $(*)$ it will follow of course that the $ch^i(X)$ do not only 'carry' the cohomology of X but each of the $ch^i(X)$ must also take care of part of the Chow groups. However, for the Chow groups this will not give a direct sum decomposition but *only a filtration* on the $CH^j(X, \mathbb{Q})$. We shall return to this in the next lecture (Section 4.3.2).

4.2.3.3 Example of curves

Taking as before $\pi_0 = e \times C, \pi_2 = C \times e, \pi_1 = \Delta(C) - \pi_0 - \pi_2$ we have (see Section 4.1.8)

$$ch(C) = ch^0(C) \oplus ch^1(C) \oplus ch^2(C).$$

4.2.3.4 Chow–Künneth conjecture

We conjecture [Mu93]:

$CK(X)$: Every $X_d \in \mathcal{V}(k)$ has a Chow–Künneth decomposition over $\bar{k}$.

Remarks

1. Of course $CK(X) \Rightarrow C(X)$ (see Section 4.2.1.1). In fact, $CK(X)$ is saying that the – conjectural – algebraic Künneth components $\Delta_{2d-i,i}$ in $\mathrm{Corr}^0_{\mathrm{hom}}(X, X)$ can be lifted to $\mathrm{Corr}^0(X, X) = CH^d(X \times X; \mathbb{Q})$ itself.
2. $CK(X)$ is part of a more complete set of conjectures to be discussed in Section 4.3.2.

4.2.3.5 Evidence and current status of $CK\,(X)$

$CK(X)$ is true for:

  (i) curves (trivial),
 (ii) surfaces ([Mu90], see Section 4.3.1),
(iii) if X has CK and Y has CK then $X \times Y$ has CK, in fact $\pi_i(X \times Y) = \sum_{p+q=i} \pi_p(X) \times \pi_q(Y)$ (immediate); hence CK is true for products of curves and surfaces [Mu93],
 (iv) abelian varieties ([Sh74], [DM91], see Section 4.3.3),
  (v) uniruled threefolds [AMS98], (recall: a threefold X is uniruled if there exists a dominant rational map $S \times \mathbb{P}_1 - - \rightarrow X$, with S a surface),
 (vi) certain classes of threefolds X with a special condition on $H^2_{\mathrm{trans}}(X)$ [AMS00],
(vii) elliptic modular varieties ([GHM], see Section 4.4).

Remark As we have seen above in Section 4.2.1.2 for (most of) the above cases the $C(X)$ itself was known much earlier. There is, at least, one important case for which the $C(X)$ is known but for which the $CK(X)$ is not yet known, namely the case of varieties over a finite field [KM74].

References for 4.2.3

[Mu93]

4.3 Lecture 3: conjectures and results on motives (continued)

4.3.1 Picard and Albanese motive: Chow–Künneth decomposition of a surface

Let k be a field. For every $X_d \in \mathcal{V}(k)$ we can construct in Chow theory, besides the two trivial projectors π_0 and π_{2d}, two more projectors π_1 and π_{2d-1} which are (candidates for) part of the (conjectural) Chow–Künneth projectors. The corresponding motives are closely related to the Picard and Albanese variety. In the case of a surface $X_2 = S$ this indeed gives a Chow–Künneth decomposition.

4.3.1.1 Theorem [Mu90]
Let $X_d \in \mathcal{V}(k), d \geq 1$. There exist two classes π_1 and π_{2d-1} in $CH^d(X \times X; \mathbb{Q})$ which are projectors, mutually orthogonal, orthogonal to π_0 and π_{2d}, with $\pi_{2d-1} = {}^t\pi_1$ and which have furthermore the following properties:

(i) *put* $ch^1(X) = (X, \pi_1, 0)$ *then*

(a) $CH^i(ch^1(X)) = \begin{cases} 0 & i \neq 1 \\ \mathrm{Pic}^0_{\mathbb{Q}}(X)(k) & i = 1 \end{cases}$

(b) $\pi_1 \equiv \Delta_{2d-1,1}$ *(modulo homological equivalence over $\bar{k}$), i.e.*

$$H^i(ch^1(\bar{X})) = \begin{cases} 0 & i \neq 1 \\ H^1(\bar{X}) & i = 1 \end{cases}$$

(ii) *put* $ch^{2d-1}(X) = (X, \pi_{2d-1}, 0)$, *then*

(a) $CH^i(ch^{2d-1}(X)) = \begin{cases} 0 & i \neq d \\ \mathrm{Alb}_{\mathbb{Q}}(X)(k) & i = d \end{cases}$

(b) $\pi_{2d-1} \equiv \Delta_{1,2d-1}$ *(modulo homological equivalence over $\bar{k}$), i.e.*

$$H^i(ch^{2d-1}(\bar{X})) = \begin{cases} 0 & i \neq 2d-1 \\ H^{2d-1}(\bar{X}) & i = 2d-1 \end{cases}$$

(*where $\bar{X} = X \times_k \bar{k}$, i.e. base extension to the algebraic closure of k*).

Remark In view of the above properties we call $ch^1(X)$ the *Picard motive* and $ch^{2d-1}(X)$ the *Albanese motive* of X (these motives should not be confused with the Picard and Albanese 1-motives of Ramachandran and Barbieri Viale-Srinivas!).

4.3.1.2 Some remarks on the proof

Since the proof is rather involved we do not reproduce it here, but we refer to the original paper [Mu90] or to the excellent exposition of it by Scholl in the Seattle proceedings [Sch94]. However we shall make some remarks here.

The result is closely related to the fact that the Λ^{d-1} operator is *algebraic* (this was well known to Grothendieck, see [Kl94]). However, since we want here not (only) the Künneth, but the Chow–Künneth, components we need a refinement of this based upon the theory of divisorial correspondences and abelian varieties of Weil–Severi (see [La59, Chapter VI.2]). The construction, first for π_1, is done by fibering X via repeated hyperplane sections into a family of curves $\{C_t\}$ and π_1 is obtained as a *divisor* on $C_{t_0} \times X$ (with C_{t_0} a fixed smooth curve of the family). Next π_{2d-1} is constructed as the transpose of π_1, hence supported on $X \times C_{t_0}$ (in the case of a surface we need some slight modifications).

Finally as to the proof of their properties the most delicate is (ii) (a). We remark that from the construction it follows rather easily that π_{2d-1} acts as the identity on the Albanese variety and this gives easily that the kernel of the action of π_{2d-1} on $CH^d_{\deg(0)}(X, \mathbb{Q})$ is *contained* in the 'Albanese kernel', i.e. in the kernel of alb: $CH^d_{\deg(0)}(X; \mathbb{Q}) \to \mathrm{Alb}_{\mathbb{Q}}(X)$, however the *crucial fact* is that these two kernels are *equal*. This requires a more careful analysis depending on the precise construction of π_{2d-1}.

4.3.1.3 About uniqueness

The construction of the *cycle classes* π_1 and π_{2d-1} depends upon choices and is not unique, however as remarked already in Section 4.2.3.1 we expect that the corresponding Chow motives are unique up to isomorphisms. In this direction we have for $ch^1(X)$ and $ch^{2d-1}(X)$ a partial result: up to isomorphism they do not depend upon the choice of the polarization [Mu90, Proposition 5.2].

4.3.1.4 The case of a surface

Now let $d = 2$, so $X_2 = S$ is a surface. Put $\pi_2 = \Delta(S) - \pi_0 - \pi_1 - \pi_3 - \pi_4$, then due to the orthogonality of the π_0, π_1, π_3 and π_4, the π_2 is again a projector (and also $^t\pi_2 = \pi_2$).

Theorem *Let* $X_2 = S \in \mathcal{V}(k)$ *be a smooth, projective surface. Then:*

(i) *the above projectors* $\pi_i (0 \leq i \leq 4)$ *give a Chow–Künneth decomposition of* S *with* $ch^i(S) = (S, \pi_i, 0)$, *hence we have*

$$ch(S) = \oplus_{i=0}^4 ch^i(S),$$

(ii) *the distribution of the Chow groups and cohomology over the* $ch^i(S)$ *is summarized in the following diagram:*

	$ch^0(S)$	$ch^1(S)$	$ch^2(S)$	$ch^3(S)$	$ch^4(S)$
$H^*(S)$	H^0	H^1	H^2	H^3	H^4
$CH^0_\mathbb{Q}(S)$	$CH^0_\mathbb{Q}(S)$	0	0	0	0
$CH^1_\mathbb{Q}(S)$	0	$Pic^0_\mathbb{Q}(S)$	$NS_\mathbb{Q}(S)$	0	0
$CH^2_\mathbb{Q}(S)$	0	0	$CH^2_{alb}(S;\mathbb{Q})$	$Alb_\mathbb{Q}(S)$	$\mathbb{Q}$

Remarks

1. In the above $NS(S)$ is the *Néron–Severi group* of S and $CH^2_{alb}(S)$ is the *Albanese kernel* of S, i.e.

$$CH^2_{alb}(S) := \text{Ker}\{CH^2_{\deg(0)}(S) \xrightarrow{\text{alb}} \text{Alb}(S)\}.$$

2. In the above constructions in Section 4.3.1.3 for π_1, and hence for π_{2d-1}, we do not need to go all the way to $\mathbb{Q}$-coefficients in the Chow groups, for a given X a sufficiently large – but difficult to keep track of – denominator suffices.

References for Section 4.3.1
[Mu90], [Sch94]

4.3.2 Conjectural filtration on the Chow groups based upon the Chow motives: relation with the Bloch–Beilinson filtration

In the 1970s Bloch started to conjecture that there should be a *'natural'* *filtration* on the Chow groups of a smooth projective variety (see his famous Duke lectures, [Bl80, pp. 1.12–1.15]). Independently this was also conjectured by Beilinson [Be87]. Beilinson based his conjectures on his, also conjectural, *theory of mixed motives*. The conjectures of Beilinson have been discussed thoroughly, and made more explicit, by Jannsen first in [Ja90] and later in his beautiful lecture [Ja94] at the Seattle conference.

By the properties of the Picard and Albanese motive and the distribution of the Chow groups over the Chow motives of a surface, as discussed in Section 4.3.1, the present author was led to believe that a *filtration on the Chow groups could also be based on the theory of Chow motives*. Therefore, he formulated a set of four conjectures [Mu93], to be discussed below. These four conjectures were communicated in 1990 in a letter to Jannsen and discussed by Jannsen in his Seattle lecture. Shortly after the Seattle conference Jannsen proved that these conjectures are equivalent to (the first version of) Beilinson's conjectures (see [Ja94, Section 5]) and that the filtrations, if they exist, coincide. Finally, to avoid misunderstanding, irrespective of which point of view one adopts, these conjectures still remain far open at present!

4.3.2.1 Conjectures on a filtration on Chow groups

Let $X_d \in \mathcal{V}(k)$; i.e. X_d is a smooth, projective variety defined over k of dimension d (and, say, irreducible for simplicity).

Conjecture I, CK conjecture (see 4.2.3.3) *There exists a Chow–Künneth decomposition for every $X_d \in \mathcal{V}(k)$.*

Now the Chow–Künneth projectors π_i ($0 \le i \le 2d$) operate on the Chow groups $CH^j(X; \mathbb{Q})$. Fix j ($0 \le j \le d$).

Conjecture II *The $\pi_{2d}, \pi_{2d-1}, \ldots, \pi_{2j+1}$ and the $\pi_0, \ldots, \pi_{j-1}$ operate as zero on $CH^j(X; \mathbb{Q})$.*

Definition of the filtration on CH^j ($X; \mathbb{Q}$) Assuming I and II define:
$$F^0 := CH^j(X, \mathbb{Q}),$$

$F^1 := \mathrm{Ker}(\pi_{2j})$,
$F^2 := \mathrm{Ker}(\pi_{2j-1}|F^1) = \mathrm{Ker}\,\pi_{2j} \cap \mathrm{Ker}\,\pi_{2j-1}$,

etc., inductively, i.e.

$F^\nu := \mathrm{Ker}(\pi_{2j-\nu+1}|F^{\nu-1}) = \mathrm{Ker}\,\pi_{2j} \cap \mathrm{Ker}\,\pi_{2j-1} \cdots \cap \mathrm{Ker}\,\pi_{2j-\nu+1}$.

Lemma 1 *This is a decreasing filtration on $CH^j(X;\mathbb{Q})$ and $F^{j+1} = 0$.*

Proof Decreasing is clear from the definition. Next

$$F^{j+1} = \mathrm{Ker}\,\pi_{2j} \cap \mathrm{Ker}\,\pi_{2j-1} \cap \cdots \cap \mathrm{Ker}\,\pi_j = 0$$

because of Conjecture II and $\sum_{i=0}^{2d} \pi_i = id$. □

Lemma 2 *For the graded pieces we have*

$$Gr_{F^\bullet}^\nu(CH^j(X;\mathbb{Q})) = CH^j(ch^{2j-\nu}(X)).$$

Proof Exercise. □

Conjecture III *The above filtration is independent of the ambiguity in the choices of the π_i.*

(As remarked before, the π_i are not unique themselves but we expect the corresponding Chow motives $ch^i(X)$ to be unique up to isomorphism.)

Lemma 3 $F^1 \subsetneqq CH_{\mathrm{hom}}^j(X;\mathbb{Q})$ *(for a Weil cohomology).*

Proof This follows immediately from the following commutative diagram

$$
\begin{array}{ccc}
CH^j(X;\mathbb{Q}) & \xrightarrow{\ \pi_{2j}\ } & CH^j(X;\mathbb{Q}) \\
\gamma \downarrow & & \downarrow \gamma \\
H^{2j}(\bar{X}) & \xrightarrow[\Delta_{2d-2j,2j}]{\cong} & H^{2j}(\bar{X}).
\end{array}
$$

□

Conjecture IV $F^1 = CH_{\mathrm{hom}}^j(X;\mathbb{Q})$.

Remark Similarly (if $k = \mathbb{C}$ say) $F^2 \subseteq CH_{AJ}^j(X;\mathbb{Q})$ where CH_{AJ}^j is the 'Abel–Jacobi' kernel, and it seems plausible to expect equality.

Some evidence for the conjectures The above conjectures are true for the following:

- curves (trivial);
- surfaces [Mu93], starting from the results of Section 4.3.1 this is easy;
- $S \times C$ (surface × curve [Mu93]);
- uniruled threefolds (del Angel–Müller-Stach [AMS98]);
- elliptic modular threefolds (Gordon–Murre [GM99]);
- parts of the conjectures are true for abelian varieties as follows from results of Shermenev and Beauville (see Section 4.3.3).

4.3.2.2 Bloch–Beilinson filtration: Beilinson's conjectures

Next we give Beilinson's list of conjectures (as formulated by Jannsen [Ja94]).

Conjectures of Beilinson *For $X_d \in \mathcal{V}(k)$ there exists on $CH^j(X, \mathbb{Q})$ a decreasing filtration $F^\nu (\nu \geq 0)$ with the following properties.*

(i) $F^0 = CH^j(X, \mathbb{Q})$, $F^1 = CH^j_{\text{hom}}(X, \mathbb{Q})$ *(for some fixed Weil theory).*

(ii) $F^r \cdot F^s \subset F^{r+s}$ *under the intersection product.*

(iii) $F^\bullet$ *is functorial for morphisms $f : X \to Y$.*

(iv) *Assuming the algebraicity of the Künneth components, $Gr^\nu_F. CH^j(X, \mathbb{Q})$ depends only on the motive $h^{2j-\nu}(X) = (X, \Delta_{2d-2j+\nu, 2j-\nu}, 0)$ in $\mathcal{M}_{\text{hom}}$ (for the precise meaning see Jannsen's paper [Ja94]).*

(v) $F^{j+1} = 0$ *(or, equivalently, $F^m = 0$ for $m \gg 0$).*

In the above version mixed motives do not appear explicitly, but in his Seattle lecture Jannsen discussed in detail several versions of Beilinson's conjectures all based, as said before, on the philosophy of mixed motives (and this philosophy is itself based on Deligne's theory of mixed Hodge structures!). In the more refined versions there appears the beautiful formula (of course conjecturally!)

$$ Gr^\nu_F CH^j(X; \mathbb{Q}) = \text{Ext}^\nu_{\mathcal{M}\mathcal{M}(k)}(1, h^{2j-\nu}(X)(j)), $$

Beilinson's formula, where $\mathcal{M}\mathcal{M}(k)$ is the conjectural category of mixed motives containing $\mathcal{M}_{\text{hom}}(k)$ as a full subcategory and $h^{2j-\nu}(X) \in \mathcal{M}_{\text{hom}}(k)$ (see Jannsen [Ja94, Conjecture 2.3]).

Remark Recently Nori has constructed an interesting theory of mixed motives – see the beautiful lectures of Mouroukos in this Summer School. As far as I understand, it is not yet clear whether this theory will throw new light upon the above conjectures.

4.3.2.3 Relation between the two sets of conjectures

Theorem [Ja94, Theorem 5.2] *The following are equivalent.*

(i) *The set of conjectures I–IV of Section* 4.3.2.1.
(ii) *The set of conjectures (i)–(v) of Section* 4.3.2.2.

Moreover, if the conjectures are true, the filtrations agree.

Jannsen proved moreover

Theorem [Ja94, Corollary 5.7] *A filtration satisfying the above properties is unique.*

4.3.2.4 Consequences

Note the following consequence.

Corollary *If the above conjectures are true, then in the exact sequence of rings (with respect to composition of correspondences)*

$$0 \to I \to \mathrm{Corr}^0(X, X) \to \mathrm{Corr}^0_{\mathrm{hom}}(X, X) \to 0$$

the ideal I is two-sided nilpotent.

Proof If $T \in I$ then $T \in F^1 C H^d_{\mathbb{Q}}(X \times X)$ where $d = \dim X$, but then $T^2 = T \bullet T \in F^2$, etc., hence $T^N = 0$ for large N (in fact for $N = d + 1$). $\square$

The above conjectures also imply the famous *Bloch conjecture*: if $S \in \mathcal{V}(k)$ is a surface such that $H^2(\bar{S})_{\mathrm{trans}} = 0$ then the Albanese map

$$\mathrm{alb} : C H^2_{\deg(0)}(\bar{S}) \to \mathrm{Alb}(\bar{S})$$

is an *isomorphism* (see [Ja94, Lemma 3.2]).

4.3.2.5 Candidates for the Bloch–Beilinson filtration via cohomology: Saito's filtration

Based upon a Weil cohomology, S. Saito has constructed an *unconditional* filtration $F^\nu_S C H^j(X; \mathbb{Q})$ satisfying most of the conditions for a Bloch–Beilinson filtration, *except* maybe condition (v) (see Section 4.3.2.2) [SSa96], [SSa00]. This is a beautiful construction and moreover, building upon infinitesimal methods, started originally by Griffiths and continued and extended by Green and Voisin, S. Saito and his student Asakura have been able to detect interesting non-trivial elements in the higher parts of this filtration (in the case of complex varieties $k = \mathbb{C}$).

There are several other interesting approaches to this still conjectural Bloch–Beilinson filtration. We mention for arbitrary fields the paper of Raskind [Ra95], the paper of Jannsen [Ja00] (based upon a method of H. Saito) and over the complex field a paper of Lewis [Le01], and (added in January 2002) a paper of Green and Griffiths [GrGr02]. All these constructions are based upon cohomology, all of them are unconditional, but by all of them the property (v) of Section 4.3.2.2 (or, equivalently, Lemma 1 of Section 4.3.2.1) remains open.

There is no doubt that each of these constructions is highly interesting and has its own merits, however to keep a correct perspective on our present state of knowledge, the condition (v) is, in the opinion of the author, the 'crux' of a Bloch–Beilinson filtration because the vector space $\cap_{m \geq 0} F^m C H^j(X; \mathbb{Q})$ measures precisely whether we do, or do not, capture rational equivalence using cohomological methods.

References for 4.3.2

[Mu93], [Ja94]

4.3.3 The case of abelian varieties

There is some evidence for the conjectures on abelian varieties. The results in this case are based upon the so-called 'abstract Fourier–Mukai' theory. This Fourier theory was first employed by Lieberman and Kleiman on the level of cohomology in order to prove conjecture $B(X)$ for abelian varieties [Kl68], next it was developed by Mukai in great generality to study coherent sheaves on such varieties, and finally it was extended and employed by Beauville for the study of algebraic cycles on abelian varieties ([B83], [B86], see also [Mu00]).

4.3.3.1 Fourier transform on abelian varieties

Let $k = \bar{k}$ be algebraically closed, and let A_d be an abelian variety defined over k, of dimension d. Let $\hat{A}$ be its dual and $P \in \mathrm{Pic}(A \times \hat{A})$ the normalized *Poincaré* divisor. The *Fourier correspondence* $F_A \ (= F$ for short) between A and $\hat{A}$ is defined as the *Chern character* of P

$$F_A := 1 + \frac{P}{1!} + \frac{P^2}{2!} + \cdots + \frac{P^{2d}}{(2d)!} = e^P = ch(P).$$

So $F_A \in CH_{\mathbb{Q}}(A \times \hat{A})$, and it operates in an obvious way on the Chow groups and cohomology

$$F_A : CH_{\mathbb{Q}}(A) \to CH_{\mathbb{Q}}(\hat{A}) \text{ and } F_A : H(A) \to H(\hat{A}),$$

if $\alpha \in CH_\mathbb{Q}(A)$ then $F(\alpha) \in CH_\mathbb{Q}(\hat{A})$ is called the *Fourier transform* of α. Note that $^t F_A = F_{\hat{A}}$.

4.3.3.2 Main theorem (Mukai, Beauville)

(i) $F_{\hat{A}} \bullet F_A = (-1)^d \Gamma_\sigma$ *(inversion formula) where* $\sigma : A \to A$ *is the involution* $x \to \sigma(x) = -x$ *and* Γ_σ *is its graph.*

(ii) *For* $\alpha, \beta \in CH_\mathbb{Q}(A)$ *we have*

$$F(\alpha \cap \beta) = (-1)^d F(\alpha) * F(\beta)$$
$$F(\alpha * \beta) = F(\alpha) \cap F(\beta)$$

where $\alpha \cap \beta$ *is the intersection product and* $\alpha * \beta$ *is the Pontryagin product.*

(iii) F *is functorial with respect to isogenies.*

Remarks

1. Hence the Fourier transform gives an isomorphism between $CH_\mathbb{Q}(A)$ and $CH_\mathbb{Q}(\hat{A})$ with (essential) inverse $F_{\hat{A}}$, transforming intersection product into Pontryagin product and visa versa.

2. The main ingredients of the proof are the Grothendieck–Riemann–Roch theorem and Mumford's computation of the cohomology of the Poincaré sheaf $\mathcal{P}$ on $A \times \hat{A}$.

4.3.3.3

Next we state the *decomposition theorem of Beauville*. Let k and A_d be as before, let $n \in \mathbb{Z}$. Consider on the abelian variety A the homomorphism

$$n : A \to A$$
$$x \to n \cdot x$$

and let n^* be the corresponding action on $CH_\mathbb{Q}(A)$, $H(A)$, etc. For $n \in \mathbb{Z}$ put

$$CH_s^i(A) = \{\alpha \in CH_\mathbb{Q}^i(A); n^*\alpha = n^{2i-s}\alpha, \text{ for all } n\}.$$

Decomposition theorem (Beauville) *Let A be an abelian variety defined over* $k = \bar{k}, d = \dim A$. *Then*

$$CH_\mathbb{Q}^i(A) = \oplus_{s=i-d}^{i} CH_s^i(A).$$

Exercise Work this out for divisors. Prove that only $s = 0, 1$ occur in this case.

Remarks

1. The main tool in the proof of the theorem is the inversion formula, see [B86].
2. *Beauville conjecture*: $CH_s^i(A) = 0$ for $s < 0$. This is true for divisors ($i = 1$).
3. The Fourier theory and Beauville's theorem are true more generally for an abelian scheme A over a base S where S is a smooth, quasi-projective variety over a field [DM91] (in this case $s'' \leq s \leq s'$ with $s'' = \text{Max}(i - d, 2i - 2d)$ and $s' = \text{Min}(2i, i + \dim S)$ where d is the fiber dimension). This is essential for the following.

4.3.3.4 Chow–Künneth decomposition for abelian varieties
Theorem (Shermenev [Sh74], Deninger–Murre [DM91]) *Let A be an abelian variety defined over $k = \bar{k}$ of dimension d. Then*

(i) *A has a Chow–Künneth decomposition.*
(ii) *There exist Chow–Künneth components π_i ($i = 0, \ldots, 2d$) such that*

$$n^* \bullet \pi_i = \pi_i \bullet n^* = n^i \pi_i$$

for all n; moreover such π_i are unique.

Remarks

1. The existence of *algebraic Künneth* components was proved by Lieberman and Kleiman [Kl68].
2. The decomposition of A into *Chow motives* was proved by Shermenev already in 1974 (using in an ingenious way Jacobian varieties) [Sh74]; this proves, in particular, point (i) of the above theorem (Shermenev did not introduce the concept of Chow–Künneth decomposition, so he used different terminology). Moreover, Shermenev proved $ch^i(A) = \Lambda^i ch^1(A)$ (i.e. the ith exterior power of $ch^1(A)$; compare with cohomology!).
3. The above theorem was proved (i.e. (i) re-proved!) by Deninger and Murre using the Fourier theory for the scheme $A \times A \to A$.
4. Building further on this Fourier method, Künnemann reproved $ch^i(A) = \Lambda^i ch^1(A)$ in 1991 and next in a beautiful paper in 1993 he developed a complete Lefschetz theory for Chow motives [Kü93], [Kü94].

4.3.3.5 On Conjectures II and III for abelian varieties
So by the theorem in Section 4.3.3.4, Conjecture I of Section 4.3.2.1 (i.e. the Chow–Künneth conjecture) is *true for abelian varieties*, and in fact we have

'canonical' projectors π_i acting in such a way that $n^* \bullet \pi_i = \pi_i \bullet n^* = n^i \pi_i$ if n is multiplication by n on A.

Now using the decomposition theorem of Beauville from Section 4.3.3.3 we get [Mu93]:

Lemma

$$\pi_i \mid CH_s^j(A) = \begin{cases} 0 & \text{if } s \neq 2j - i \\ id & \text{if } s = 2j - i. \end{cases}$$

Proof Using $n^* \bullet \pi_i = \pi_i \bullet n^* = n^i \pi_i$, we get a commutative diagram

$$
\begin{array}{ccc}
CH^j(A; \mathbb{Q}) & \xrightarrow{\pi_i} & CH^j(A; \mathbb{Q}) \\
n^* \downarrow & \searrow{n^i \pi_i} & \downarrow n^* \\
CH^j(A; \mathbb{Q}) & \xrightarrow[\pi_i]{} & CH^j(A; \mathbb{Q}).
\end{array}
$$

Let $\alpha \in CH^j(A; \mathbb{Q})$, by Beauville $\alpha = \sum_s \alpha_s$ with $\alpha_s \in CH_s^j(A)$. This gives on the one hand $\pi_i \bullet n^*(\alpha) = \pi_i \bullet n^*(\sum_s \alpha_s) = \pi_i(\sum_s n^{2j-s}\alpha_s) = \sum_s n^{2j-s}\pi_i(\alpha_s)$, and on the other hand $\pi_i \bullet n^*(\alpha) = n^i \pi_i(\alpha) = n^i(\sum_s \pi_i(\alpha_s))$. Hence (if $d = \dim A$)

$$\sum_{s=j-d}^{j} (n^i - n^{2j-s})\pi_i(\alpha_s) = 0.$$

Since this is true for all $n \in \mathbb{Z}$ we get (cf. [Kl68, p. 377]) $\pi_i(\alpha_s) = 0$ if $i \neq 2j - s$; but then π_{2j-s} must act as the identity on α_s. $\quad\square$

Corollary 1

(i) *Part of Conjecture* II *is true, namely* π_i *operates as zero on* $CH^j(A; \mathbb{Q})$ *for* $i < j$ *and for* $i > j + d$.

(ii) *For the 'remaining' part of Conjecture* II, *i.e. for* $2j + 1 \leq i \leq j + d$ *we have that this remaining part* $\Leftrightarrow$ *the Beauville conjecture (see Section 4.3.3.3).*

Proof

(i) π_i acts different from zero only if there is an s such that $i = 2j - s$, i.e. if $s = 2j - i$. But $i < j \Leftrightarrow s = 2j - i > j$ and this does not occur by the theorem of Beauville, similarly $i > j + d \Leftrightarrow s = 2j - i < j - d$ and this does not occur by Beauville.

(ii) Same argument: $2j + 1 \leq i \Leftrightarrow s = 2j - i \leq -1$. $\quad\square$

Corollary 2 *Assuming Conjecture* II *to be true we get* $F^\nu C H^j(A; \mathbb{Q}) = \oplus_{s \geq \nu} C H_s^j(A)$ *for* $0 \leq \nu \leq j$, *and* $F^{j+1} = 0$; *i.e. the filtration is 'natural' and hence Conjecture* III *is true.*

4.3.3.6 Résumé of the results for abelian varieties

1. Conjecture I (conjecture CK) is true
   Conjecture II: part of it is true, the remaining part is equivalent to a conjecture of Beauville.
   Conjecture III: if Conjecture II is true then Conjecture III is true for the natural projectors π_i.
2. Finally, Beauville has proved his conjecture for cycles of codimension $j = 0, 1, d - 2, d - 1$ and d [B86, Proposition 3]; hence Conjecture II and III are true in that case. In particular, they are true for abelian varieties for dimension at most 4.

References for 4.3.3
[B83], [B86], [Mu93], [Mu00]

4.4 Lecture 4: Chow–Künneth decomposition for elliptic modular varieties

In this lecture we report on joint work with Gordon and Hanamura. This work is still in progress and we shall discuss here only the main lines, referring for the details to forthcoming papers. Also our results are more general than the case of elliptic modular varieties (they apply for instance also to families of abelian varieties over Hilbert modular surfaces), but we restrict the discussion here to this particular case.

4.4.1 Elliptic modular varieties

4.4.1.1

In this lecture we work only over the complex numbers. Fix an integer $N \geq 3$; since N is fixed, we shall drop it further in the notation. Let $M = M_N$ be the *modular curve* belonging to the *principal congruence subgroup* $\Gamma(N)$ of the modular group $SL_2(\mathbb{Z})$, i.e. to a point $t \in M$ corresponds an elliptic curve E_t defined over $\mathbb{C}(t)$ together with a fixed isomorphism $E_t[N] \simeq \mathbb{Z}/N \oplus \mathbb{Z}/N$, where as usual $E_t[N]$ denotes the subgroup of the N-torsion points on E_t (see [Si86, Appendix C, Sections 12 and 13]). Varying $t \in M$ we get a curve $\varphi_0 : E \to M$ over M, the 'universal elliptic curve' of level N. Next, compactifying over the *cusps* $c \in M_\infty$ we get a curve $\varphi : \bar{E} \to \bar{M}$ over $\bar{M}$ and a diagram

$$
\begin{array}{ccccc}
E & \longrightarrow & \bar{E} & \longleftarrow & E_\infty \\
\varphi_0 \downarrow & & \varphi \downarrow & & \downarrow \varphi_\infty \\
M & \longrightarrow & \bar{M} & \longleftarrow & M_\infty.
\end{array}
$$

$\bar{E}$ is a *smooth* surface, the so-called *elliptic modular surface (Shioda)* of level N (of course φ is not smooth but φ_0 is smooth). Over the cusps $c \in M_\infty$ we get a so-called N-gon consisting of N copies of $\mathbb{P}_1$, intersecting as indicated, for $N = 3$, by the picture:

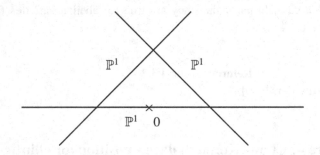

4.4.1.2

Next fix an integer $n \geq 1$, put $m = n + 1$. Consider the n-fold products $E^n = E \times_M E \times \cdots \times_M E$ and $\bar{E}^n = \bar{E} \times_{\bar{M}} \bar{E} \times \cdots \times_{\bar{M}} \bar{E}$ (n times). For $n > 1$ the $\bar{E}^n$ has (simple) singularities. There is a 'natural' desingularization, constructed by Deligne in 1968 and later again by Scholl [Sch90], which we shall denote by $\tilde{E}^n$ and, for simplicity of notation, also by $X_m := \tilde{E}^n$; this is called the *elliptic modular variety* of dimension m and of level N. The fibers $Y_c := X_{m,c}$ over the cusps $c \in M_\infty$ are *toric varieties*; write

$$
Y = \cup_{c \in M_\infty} Y_c.
$$

Finally let $\tilde{Y}$ be the normalization of Y (in this case it is the disjoint union of the irreducible components of Y). Since the Y is a toric variety, we have

$$
CH_*(\tilde{Y}) \cong H_*(\tilde{Y})
$$

(i.e. the Chow groups in the sense of Fulton coincide with the homology groups). We summarize the situation (and notation) in the following diagram

$$E^n \hookrightarrow \overline{E}^n \leftarrow \widetilde{E}^n = X_m \xleftarrow{\alpha} Y \xleftarrow{\beta} \widetilde{Y}$$

$$\begin{array}{ccccc} \downarrow & \downarrow & \downarrow \Psi & \downarrow & \downarrow \\ M \hookrightarrow \overline{M} & = & \overline{M} & \hookleftarrow M_\infty & = & M_\infty \end{array}$$

and denote $\lambda := \alpha \bullet \beta : \widetilde{Y} \to X_m$.

4.4.1.3 Theorem (Gordon–Hanamura–Murre)
The elliptic modular variety X_m has a Chow–Künneth decomposition.

The proof goes in two steps.

Step 1 Construction of a *relative Chow–Künneth decomposition* (to be explained below) [GHM01].

Step 2 Construction of the (absolute) Chow–Künneth decomposition (this part has as yet not been written down in all details).

We shall now give some indications of both parts.

4.4.2 Construction of a relative Chow–Künneth decomposition for elliptic modular varieties (Step 1)

4.4.2.1 Theory of Chow motives over a base (Corti–Hanamura)
Corti and Hanamura have developed a general theory of Chow motives over a base [CH00]. We recall here the *main points*; for the details we refer to their original paper. The general setting is summarized in the diagram

$$\begin{array}{ccc} X^0 & \longrightarrow & X \\ f^0 \downarrow & & f \downarrow \\ S^0 & \longrightarrow & S \\ & & \downarrow \\ & & \text{Spec } k. \end{array}$$

with the following assumptions: S is a quasi-projective variety over a field k, $f : X \to S$ is *projective*, X is *smooth* and $S^0 \subset S$ is the open set over which f is smooth, i.e. $f^0 : X^0 \to S^0$ is *smooth* ($X^0 = f^{-1}(S^0)$).

One works with *relative correspondences*:

$$\text{Corr}(X/S, Y/S) = CH_*(X \times_S Y; \mathbb{Q})$$

where $CH_*(\ )$ are the *Chow homology* groups as defined (for possible singular varieties) in Fulton's book [F84]. The *composition of correspondences* is defined via the refined Gysin map of Fulton [F84, Chapter 6] as follows. Let $\Gamma_1 \in \text{Corr}(X/S, Y/S)$, $\Gamma_2 \in \text{Corr}(Y/S, Z/S)$, then $\Gamma_2 \bullet \Gamma_1 \in \text{Corr}(X/S, Z/S)$ with

$$\Gamma_2 \bullet \Gamma_1 := \text{pr}_{XZ}\{\delta!(\Gamma_1 \times_k \Gamma_2)\}$$

defined via the commutative diagram (with Cartesian square)

$$
\begin{array}{ccccc}
(X \times_S Y) \times_k (Y \times_S Z) & \longleftarrow & X \times_S Y \times_S Z & \xrightarrow[pr_{XZ}]{} & X \times_S Z \\
\downarrow & & \downarrow & & \\
Y \times_k Y & \xleftarrow[\delta = \Delta(Y)]{} & Y & &
\end{array}
$$

and note that this makes sense, since δ is *regular* because Y is smooth.

Finally, the category of *Chow motives over S* (or 'relative' Chow motives) $CHM(S)$ is defined as follows: *objects* are triples $(X/S, p(X/S), r)$ with $X \to S \to \text{Spec}(k)$ as above, $p = p(X/S)$ is a relative correspondence $p \in CH_{\dim X}(X \times_S X)$ which is a *projector*, i.e. $p \bullet p = p$ and $r \in \mathbb{Z}$. Morphisms in $CHM(S)$ are defined as before (see Lecture 1, Section 4.1.3) with, of course, composition as explained above.

4.4.2.2 Relative Chow–Künneth decomposition (Corti–Hanamura)

Let now $K = \mathbb{C}$ and let $f : X \to S \to \text{Spec}(\mathbb{C})$ be as in Section 4.4.2.1. Corti and Hanamura have extended the notion of Chow–Künneth decomposition to this case [CH00, Section 5]. Let n be the fiber dimension of f and put $m = \dim S + n$ (for simplicity, we now assume S irreducible).

Definition $f : X \to S$ has a *Chow–Künneth decomposition over S*, also called a *relative Chow–Künneth decomposition*, if there exists a finite set $p_{i,\alpha} = p_{i,\alpha}(X/S) \in CH_m(X \times_S X, \mathbb{Q})$ with $0 \le i \le 2n, \alpha \in A$, with A a finite set of integers, such that the following conditions hold:

1. $p_{i,\alpha}$ are orthogonal projectors,
2. $\sum_{i,\alpha} p_{i,\alpha} = \Delta(X/S)$,
3. $p_{i,\alpha}$ satisfy a certain 'Künneth-type' of condition for intersection cohomology (to be explained below in Section 4.2.3).

If we have such $p_{i,\alpha}$, put $M_{i,\alpha} = (M/S, p_{i,\alpha}(X/S), 0)$, then $M_{i,\alpha} \in CH\mathcal{M}(S)$ and $ch(X/S) = \sum M_{i,\alpha}$, where $ch(X/S) = (X/S, \Delta(X/S), 0)$.

4.4.2.3 The topological decomposition theorem and Condition 3

Let $f : X \to S$ over $\mathrm{Spec}(\mathbb{C})$ be as above. Then there is $D^b_{cc}(S)$, the triangulated category of bounded complexes of cohomologically constructible $\mathbb{Q}$-sheaves on S. Let $\mathbb{Q}_X$ be the constant sheaf $\mathbb{Q}$ on X, then

$$Rf_*\mathbb{Q}_X \in D^b_{cc}(S)$$

Now there is the *fundamental (topological) decomposition theorem* of Beilinson–Bernstein–Deligne:

Theorem [BBD82] *Let* $f : X \to S$ *over the complex numbers with* X *smooth and* f *projective. Then in* D^b_{cc} *there is a direct sum decomposition*

$$(*) \qquad Rf_*\mathbb{Q}_X = \sum_i {}^p R^i f_*\mathbb{Q}_X[-i]$$

where ${}^p R^i f_*\mathbb{Q}_X$ *denotes the* i*th perverse cohomology of* $Rf_*\mathbb{Q}_X$.

Moreover there exist a 'good (Whitney) stratification' S_α *of* S *by locally closed subsets* S_α *of* S *and local systems* $\mathcal{V}^i_\alpha$ *on* S_α *such that*

$$(**) \qquad {}^p R^i f_*\mathbb{Q}_X \simeq \sum_\alpha IC_{S_\alpha} \mathcal{V}^i_\alpha[-i + \dim S_\alpha]$$

where $IC_{S_\alpha}(\mathcal{V}^i_\alpha)$ *is the intersection complex associated to* $\mathcal{V}^i_\alpha$.

Condition 3 The above Condition 3 now means the following. The relative projectors $p_{i,\alpha}$ operate on $Rf_*\mathbb{Q}_X$ and give a direct sum decomposition of $Rf_*\mathbb{Q}_X$ in $D^b_{cc}(S)$; the condition is that this decomposition should coincide with the decomposition $(*)$ and $(**)$ above.

4.4.2.4

Return to the case of the *elliptic modular variety* of level N $(N \geq 3)$ of Section 4.1.2:

$$\psi : X_m = \tilde{E}^n \to \bar{M} =: S.$$

Theorem [GHM01] *This elliptic modular variety has a relative Chow–Künneth decomposition.*

The following two facts are the essential ingredients for the proof.

(i) Over the open part over which ψ is smooth, i.e. for $\psi_0 : E^n \to M$, there is a relative Chow–Künneth decomposition [DM91], but this is easy in this case.

(ii) Over the cusps, i.e. over M_∞, the fibers are *toric* varieties.

4.4.2.5

Very rough indication of the proof.

Lemma *With the notation of the diagram in Section 4.4.1.2 there exists*

$$\tilde{p}_\infty \in CH_m(\tilde{Y} \times_{\tilde{M}} \tilde{Y})$$

such that

(i)
$$\lambda^* \lambda_* \tilde{p}_\infty \lambda^* \lambda_* = \lambda^* \lambda_*$$

(ii)
$$\tilde{p}_\infty \lambda^* \lambda_* \tilde{p}_\infty = \tilde{p}_\infty.$$

Proof Since we have toric varieties

$$CH_*(\tilde{Y} \times_{\tilde{M}} \tilde{Y}) = H_*(\tilde{Y} \times_{\tilde{M}} \tilde{Y}),$$

this reduces over every cusp $c \in M_\infty$ to simple linear algebra

$$H^q(\tilde{Y}_c) \xrightarrow{\lambda_*} H^{q+2}(\tilde{Y}_c) \xrightarrow{\lambda^*} H^{q+2}(\tilde{Y}_c) \xrightarrow{\tilde{p}_\infty} H^q(\tilde{Y}_c),$$

i.e. construct linear maps $\tilde{p}_\infty$ satisfying the requirements of the lemma.

Remark Note that in fact

$$\tilde{p}_\infty = \sum_q \tilde{p}_{q,\infty} \quad (q = 0, \ldots, 2n - 2).$$

After the lemma we proceed as follows: 'push' $\tilde{p}_\infty$ to X itself, i.e. take

$$p_\infty := \lambda_* \tilde{p}_\infty \lambda^*$$

$$
\begin{array}{ccc}
\tilde{Y} & \xrightarrow{\ \tilde{p}_\infty\ } & \tilde{Y} \\
\lambda \downarrow & & \downarrow \lambda \\
X & \xrightarrow{\ p_\infty\ } & X.
\end{array}
$$

From (ii) of the lemma it follows that

$$p_\infty \bullet p_\infty = p_\infty$$

i.e. p_∞ is a relative *projector* in $CH_m(X \times_{\bar{M}} X)$. Now writing for simplicity $S = \bar{M}$ and $S^0 = M$, we have an exact sequence of rings

$$0 \to I \to CH_m(X \times_S X) \to CH_m(X^0 \times_{S^0} X^0) \to 0,$$

where the ring structure comes from the composition of correspondences, and where $X^0 = \psi^{-1}(S^0)$ and I is a two-sided ideal defined as kernel. Consider now the *orthogonal complement* R of p_∞ in $CH_m(X \times_S X)$, i.e. $R = \{\alpha \in CH_m(X \times_S X); \alpha \bullet p_\infty = p_\infty \bullet \alpha = 0\}$. Next consider the sequence

$$0 \to I \cap R \to R \to CH_m(X^0 \times_{S^0} X^0).$$

Facts:

(a) $R \to CH_m(X^0 \times_{S^0} X^0)$ is surjective (easy),
(b) $I \cap R$ is a nilpotent ideal! (not so easy, but follows from (i) of the lemma!),
(c) $\Delta(X/S) - p_\infty = 1_R$ (easy).

4.4.2.6 Construction of the relative projectors and end of the proof for Step 1

On the open set $S^0 = M \subset S = \bar{M}$ we have $\psi_0 : X^0 \to S^0$, with $X^0 = E^n$ an abelian scheme over $S^0 = M$; in fact the fibers are products of elliptic curves. By [DM91] there exists a set of projectors, pairwise orthogonal $p_0^0, p_1^0, \ldots, p_{2n}^0 \in CH_m(X^0 \times_{S^0} X^0)$, which indeed give a relative Chow–Künneth decomposition for $\psi_0 : X^0 \to S^0$.

Since the ideal $I \cap R$ is nilpotent, by a lemma of Jannsen [Ja94, Lemma 5.4] these projectors can be lifted to idempotents $p_0, p_1, \ldots, p_{2n}$, in R, pairwise orthogonal and summing up to 1_R.

Claim The $p_0, p_1, \ldots, p_{2n}$, together with the p_∞ from above (or better the $p_{q,\infty}$ $(q = 0, \ldots, 2n - 2)$), give a *relative Chow–Künneth decomposition* for the elliptic modular variety $\psi : X_m \to S$. (Remark: by abuse of language we write here in Section 4.4.2 p instead of $p(X/S)$.)

As to the proof: the conditions 1 and 2 of Section 4.4.2.2 are immediate from the above. For Condition 3 one has the natural stratification ($S^0 = M$, $S_\infty = S - S^0 = M_\infty$) of $S = \bar{M}$. On this one has local systems

$$\mathcal{V}_{S^0}^i = R^i(\psi_0)_* \mathbb{Q}_X \text{ on } S^0$$

and local systems $\mathcal{V}^q_{S_\infty}$ on S_∞ coming from the 'extra' algebraic components in the fibers over the cusps. To prove that the above projectors indeed give the right decomposition of [BBD82] one has to use, besides general properties of the category $D^b_{cc}(S)$, again the properties of $\tilde{p}_\infty$ as stated in the lemma in Section 4.4.2.5.

References for 4.4.2

[CH00], [GHM01]

4.4.3 The absolute Chow–Künneth decomposition for elliptic modular varieties (Step 2)

4.4.3.1 Remark

In general to go from a relative Chow–Künneth decomposition to an absolute Chow–Künneth decomposition is an *unsolved problem* which seems very difficult (see below). However, in our present case of elliptic modular varieties it can be solved since we have rather precise information on the the the cohomology [Go93].

4.4.3.2 General fact

Let $f : X \to S$ over $\mathrm{Spec}(k)$ be as in Section 4.4.2.1, i.e. X *smooth* and f projective. Consider the natural embedding $i : X \times_S X \to X \times_k X$. Let $m = \dim X$, then i gives a homomorphism $i_* : CH_m(X \times_S X) \to CH_m(X \times_k X)$ of Chow groups.

Lemma 1 i_* *is a ring homomorphism for the composition of correspondences. In particular, a relative projector $p(X/S)$ gives an absolute projector $p := i_* p(X/S)$, and orthogonal ones give orthogonal ones.*

Proof This follows from the properties of the refined Gysin homomorphism [F84, Theorem 6.2].

Corollary *There exists a functor $\Phi : CH\mathcal{M}(S) \to CH\mathcal{M}(k)$ defined as*

$$M(X/S) = (X/S, p(X/S), r) \to M = (X, p, r)$$

$$(\text{where } p = i_* p(X/S)).$$

4.4.3.3 Remaining problem!

So suppose that we have a relative Chow–Künneth decomposition with relative motives $M_{i,\alpha}(X/S) = (X/S, p_{i\alpha}(X/S), 0)$, then we get, by the corollary, absolute motives

$$M_{i,\alpha} = \Phi(M_{i,\alpha}(X/S)) = (X, p_{i\alpha}, 0).$$

However, the problem is that in general the cohomology $H^*(M_{i,\alpha})$ is 'mixed', i.e. is *not concentrated in one dimension*. Therefore it will be necessary to decompose $M_{i,\alpha}$ further and, with our present knowledge, we are only able to do this under additional assumptions. In the case of elliptic modular varieties this can be done.

4.4.3.4

Let $\psi : X_m \to \bar{M} = S$ be the elliptic modular variety of dimension $m = n + 1$ ($X_m = \tilde{E}^n$) and level $N(N \geq 3)$. In Section 4.4.2.6, we have seen that we have a relative Chow–Künneth decomposition via relative projectors $p_i(X/S)$ ($i = 0, \ldots, 2n$) and $p_{q,\infty}(X/S)$ ($q = 0, \ldots, 2n - 2$) with the latter being supported over the cusps. So we get, as indicated above, absolute Chow motives $M_i = (X, p_i, 0)$ ($i = 0, 1, \ldots, 2n$) and $M_{q,\infty} = (X, p_{q,\infty}, 0)$ ($q = 0, \ldots, 2n - 2$). The latter only have cohomology in one degree and do not give difficulty. For the former, we have the following lemma. Let $\mathbb{L}_S = (S, id, -1)$ be the *relative Lefschetz motive* over S.

Lemma 2 *For the elliptic modular variety $\psi : X \to S$ and with the notation as explained above, consider the motives*

$$M_i = (X, p_i, 0) \ (0 \leq i \leq 2n).$$

Then we have the following.

(i) $H^\bullet(M_{2j-1}) \subset H^{2j}(X)$ ($j = 1, \ldots, n$).
(ii) *There exist projectors λ_{2j} and integers $m(j)$ ($j = 0, \ldots, n$) such that:*
   (a) $\lambda_{2j} \bullet p_{2j} = p_{2j} \bullet \lambda_{2j} = \lambda_{2j}$ *(i.e. λ_{2j} is a so-called 'constituent' of p_{2j}, note that now λ_{2j} and $p_{2j} - \lambda_{2j}$ are orthogonal projectors);*
   (b) $(X, \lambda_{2j}, 0) \simeq m(j)(S, id, -j) = m(j)\mathbb{L}_S^{\otimes j}$;
   (c) $H^\bullet(X, p_{2j} - \lambda_{2j}, 0) \subset H^{2j+1}(X)$.

Remark We get the λ_{2j} via algebraic cycles, obtained by 'spreading out' algebraic cycles in the generic fiber $X_t(t \in S)$.

Corollary *Using the decomposition $ch(S) = \oplus_{i=0}^2 ch^i(S)$ of the base and the isomorphisms in* (ii) (b) *of the above lemma, we can split the $(X, \lambda_{2j}, 0)$*

further, obtaining finally a splitting of $(X, id, 0)$ *into pieces each of which has cohomology in only one degree. Collecting the right pieces together, we obtain – finally! – the desired Chow–Künneth decomposition.*

4.4.3.5 Example

Rather than considering the general case, we shall demonstrate the above description in the special case $n = 2$. So we have the elliptic modular threefold $X = X_3$ (belonging to some fixed level N). In this case we have projectors p_0, p_1, p_3, p_4 coming via the open part $\psi_0 : X^0 \to S^0$ and (as one can check) two types of projectors 'at infinity' $p_{2,\infty}$, $p_{4,\infty}$ (i.e. over the cusps) giving motives $M_{2,\infty}$ and $M_{4,\infty}$. These $p_{2,\infty}$ and $p_{4,\infty}$ are supported over the cusps and the corresponding motives $M_{2,\infty}$ and $M_{4,\infty}$ 'take care' of the 'extra' cohomology coming from the 'extra' algebraic cycles in the fibers over the cusps; they only have cohomology in H^2, respectively H^4. Next, turning to the $M_i = (X, p_i, 0)$ for $i = 0, 1, 2, 3, 4$; they come (via closure) from the open part $\psi_0 : X^0 \to S^0$ and their contribution to the cohomology of X comes from the locally constant sheaves $R^j(\psi_0)_* \mathbb{Q}_X$ and this can be studied via the action of the fundamental group $\pi_1(S^0) \cong \Gamma(N)$ of the modular curve S^0 (of level N) on the cohomology $H^j(E_{\bar{\eta}}^2, \mathbb{Q})$ of the geometric generic fiber $E_{\bar{\eta}}^2$. This reduces the problem to representation theory of the modular group. From this one concludes that M_1 and M_3 contribute only to cohomology in one degree, namely H^2, respectively H^4. On the other hand, M_0, M_2 and M_4 have 'mixed' cohomology. However, using the zero section $\epsilon : S \to X$, one sees easily that $M_0 \simeq ch(S)$ and $M_4 \simeq ch(S)(-2)$ (this is entirely similar to the absolute case, say $Y_d \to \mathrm{Spec}(k)$, where $ch^0(Y) \simeq 1 = ch(\text{point})$ and $ch^{2d}(Y) \simeq 1(-d)$; see Examples 3 and 4 of Section 4.1.3.2). It remains to study M_2. For this, we take an orthogonal base of the Néron–Severi group $NS(E_\eta^2)$ of the generic fiber $\psi^{-1}(\eta)$ and we 'spread' these classes out over X; this gives a projector λ_2 which is a constituent of p_2, i.e. $p_2 \bullet \lambda_2 = \lambda_2 \bullet p_2 = \lambda_2$. Since $rk(NS(E_\eta^2)) = 3$ (the generic fiber has no complex multiplication!), λ_2 consists of three mutually orthogonal projectors and we get $(X, \lambda_2, 0) \simeq 3ch(S)(-1)$ (again think of the absolute case $Y_2 \to \mathrm{Spec}(k)$, where we split off the algebraic part of $ch^2(Y_2)$). From representation theory we get that $M_2' := (X, p_2 - \lambda_2, 0)$ has only cohomology in H^3. Finally, using the splitting of the base $ch(S) = ch^0(S) \oplus ch^1(S) \oplus ch^2(S) = A \oplus B \oplus C$ (for simplicity of notation), and the above isomorphism, we get $M_0 \simeq ch(S)$, $M_2 = (X, \lambda_2, 0) \oplus (X, p_2 - \lambda_2, 0) \simeq 3ch(S)(-1) \oplus M_2'$ and $M_4 \simeq ch(S)(-2) \simeq \mathbb{L}_S^{\otimes 2}$. We can summarize this now in the following table:

	Splitting		
$M_4 \simeq \mathbb{L}_S^{\otimes 2}$	$A(-2)$	$B(-2)$	$C(-2)$
M_3		M_3	
$M_2 \simeq 3\mathbb{L}_S \oplus M_2'$	$3A(-1)$	$3B(-1) \oplus M_2'$	$3C(-1)$
M_1		M_1	
$M_0 \simeq ch(S)$	A	B	C

All pieces in columns two, three and four of the table only have cohomology in one degree and they contribute to the total cohomology $H^*(X)$ in a way similar to the pattern of the Leray spectral sequence. So for instance

$$H^0(X) = H(A),\ H^1(X) = H(B),\ H^2(X) = H(C) + H(M_1) +$$
$$3H(A(-1)) + H(M_{2,\infty}),\ H^3(X) = 3H(B(-1)) + H(M_2'),\ \text{etc.}$$

Taking the right pieces together, we get the Chow–Künneth components $ch^i(X)$ for $i = 0, 1, \ldots, 6$. For instance $ch^2(X) \cong C \oplus M_1 \oplus 3A(-1) \oplus M_{2,\infty}$ and $ch^3(X) \cong 3B(-1) \oplus M_2'$.

Reference for 4.4.3
[GHM01, Part 2]

Acknowledgements

I wish to thank the organizers Chris Peters and Stefan Müller-Stach for inviting me to this very interesting Summer School. In October 2000 I gave a similar set of lectures at the Universidad Nacional Aotonoma de Mexico at the invitation of Javier Elizondo; this was very useful for me and I take the opportunity to thank him also. I thank both Institutions for the warm hospitality and stimulating atmosphere and the audience for their patience and the – for me – very valuable discussions.

Many thanks go also to Mrs. Kahl and Mrs. Combes from the University of Essen, for the typesetting of the manuscript.

References

[AMS98]   Angel, P. Luis del and Müller-Stach, S. Motives of uniruled 3-folds, *Compositio Math.* **112** (1998) 1–16.

[AMS00]   On Chow motives of 3-folds, *Trans. Am. Math. Soc.* **352** (2000) 1623–1633.

[B83]   Beauville, A. *Quelques Remarques sur la Transformation de Fourier dans l'Anneau de Chow d'une Variété Abélienne, Springer Lecture Notes in Mathematics* **1016**, pp. 238–260, Springer, 1983.

[B86]   Sur l'anneau de Chow d'une variété abélienne, *Math. Annal.* **273** (1986)
                647–651.

[BBD82]   Beilinson, A., Bernstein, J. and Deligne, P. Faisceaux pervers, *Astérisque*
                100 (1982) 5–171.

[Be87]   Beilinson, A. *Height Pairing Between Algebraic Cycles, Springer Lecture
                Notes in Mathematics* **1289**, pp. 1–26, Springer, 1987; *Contemp. Math.*
                AMS **67** (1987) 1–24.

[Bl80]   Bloch, S. Lectures on algebraic cycles, *Duke Univ. Math. Series* **IV**, 1980.

[Bo58]   Bourbaki, N. *Algèbre*, Book 8, Chapter 8, Hermann, Paris, 1958.

[CH00]   Corti, A. and Hanamura, M. Motivic decomposition and intersection Chow
                groups I, *Duke Math. J.* **103** (2000) 459–522.

[Cl99]   Clozel, L. Equivalence numérique et équivalence cohomologique pour les
                variétés abéliennes sur les corps finis, *Ann. Math.* **150** (1999) 151–163.

[De69]   Demazure, M. *Motifs des Variétés Algébriques, Sém. Bourbaki 365,
                Springer Lecture Notes in Mathematics* **180**, Springer, 1971.

[Detal82]   Deligne, P., Milne, J.S., Ogus, A. and Shih, K. *Hodge Cycles, Motives and
                Shimura Varieties, Springer Lecture Notes in Mathematics* **900**, Springer,
                1982.

[DM91]   Deninger, C. and Murre, J. P. Motivic decomposition of abelian schemes
                and the Fourier transform, *J. reine angew. Math.* **422** (1991) 201–219.

[F84]   Fulton, W. *Intersection Theory, Ergeb. Math,* 3 Folge, Bd 2., Springer
                Verlag, 1984.

[G68]   Grothendieck, A. Standard conjectures on algebraic cycles. In: *Algebraic
                Geometry Bombay Colloquium 1968,* pp. 193–199, Oxford University
                Press, 1969.

[G85]   *Récoltes et Semailles: Réflexions et Témoignages sur un Passé de
                Mathématicien,* Montpellier, 1985 (unpublished).

[GH78]   Griffiths, Ph. and Harris, J. *Principles of Algebraic Geometry,* Wiley, 1978.

[GHM01]   Gordon, B. B., Hanamura, M. and Murre, J. P. Relative Chow–Künneth
                projection for modular varieties, preprint 2001.

[GM99]   Gordon, B. B. and Murre, J. P. Chow motives of elliptic modular surfaces
                and threefolds, *J. reine angew. Math.* **514** (1999).

[Go93]   Gordon, B. B. Algebraic cycles and the Hodge structure of a Kuga fiber
                variety, *Trans. Am. Math. Soc.* **336** (1993) 933–947.

[GrGr02]   Green, M. and Griffiths, Ph. Hodge-theoretic invariants for algebraic cycles,
                preprint 2002.

[H75]   Hartshorne, R. Equivalence relations on algebraic cycles and subvarieties
                of small codimension. In *Algebraic Geometry, Arcata 1974, AMS Proc.
                Symp. Pure Math.* **29**, pp. 129–164, AMS Providence, RI, 1975.

[Ha95]   Hanamura, M. Mixed motives and algebraic cycles I, *Math. Res. Lett.* **2**
                (1995) 811–821.

[Ha99]   Mixed motives and algebraic cycles III, *Math. Res. Lett.* **6** (1999) 61–82.

[Ja90]   Jannsen, U. *Mixed Motives and Algebraic K-Theory, Springer Lecture
                Notes in Mathematics* **1400**, Springer, 1990.

[Ja92]   Motives, numerical equivalence and semi-simplicity, *Invent. Math.* **107**
                (1992) 447–452.

[Ja94]   Motivic sheaves and filtration on Chow groups. In *Seattle Conf. on Motives 1991, AMS Proc. Symp. Pure Math.* **55**, pp. 245–302, AMS Providence, RI, 1994.

[Ja00]   Equivalence relations on algebraic cycles. In: *Arithmetic and Geometry of Algebraic cycles, Banff Conf. 1998*, pp. 225–260 Kluwer, Dordrecht, 2000.

[JKS94]   Jannsen, U., Kleiman, S. L. and Serre J. P. *Motives, Seattle Conf. on Motives 1991, AMS Proc. Symp. Pure Math.* **55**, Parts I and II, AMS Providence, RI, 1994.

[Kl68]   Kleiman, S. L. Algebraic cycles and the Weil conjectures. In: *Dix Exposés sur la Cohomologie des Schémas*, pp. 359–386, North-Holland, 1968.

[Kl70]   Kleiman, S. L. *Motives in Algebraic Geometry Conf. Oslo*, pp. 53–82, 1970.

[Kl94]   The standard conjectures. In: *Seattle Conf. on Motives 1991, AMS Proc. Symp. Pure Math.* **55**, pp. 3–20, AMS Providence, RI, 1994.

[KM74]   Katz, N. and Messing, W. Some consequences of the Riemann hypothesis for varieties over finite fields, *Invent. Math.* **23** (1974) 73–77.

[Kü93]   Künnemann, K. A Lefschetz decomposition for Chow motives of abelian schemes, *Invent. Math.* **113** (1993) 85–102.

[Kü94]   On the Chow motive of an abelian scheme. In *Seattle Conf. on Motives 1991, AMS Proc. Symp. Pure Math.* **55**, pp. 189–205, AMS Providence, RI, 1994.

[La59]   Lang, S. *Abelian Varieties*, Interscience, 1959 and Springer Verlag 1983.

[Le98]   Levine, M. *Mixed Motives, Math. Surveys and Monographs* **57**, AMS, Providence, RI, 1998.

[Le01]   Lewis, J. D. A filtration on the Chow groups of a complex projective variety, *Compositio Math.* **128** (2001) 299–322.

[Li68]   Lieberman, D. Numerical and homological equivalence of algebraic cycles on Hodge manifolds, *Am. J. Math.* **90**, (1968) 366–374.

[Ma68]   Manin, Y. I. Correspondences, motives and monoidal transformations, *Math. USSR Sbornik* **6** (1968) 439–470.

[Mu90]   Murre, J. P. On the motive of an algebraic surface, *J. reine angew. Math.* **409** (1990) 190–204.

[Mu93]   On a conjectural filtration on the Chow groups of an algebraic variety, Parts I and II, *Indag. Math.* **4** (1993) 177–201.

[Mu96]   Introduction to the theory of motives, *Boll. UMI* **7**(10-A) (1996) 477–489.

[Mu00]   Algebraic cycles on abelian varieties and applications of abstract Fourier theory. In: *Arithmetic and Geometry of Algebraic Cycles, Banff Conf. 1998*, pp. 307–320, Kluwer, Dordrecht, 2000.

[No]   Nori, M. V. Mixed motives" (unpublished manuscript).

[P82]   Pierce, R. S. *Associative Algebras, Graduate Texts in Mathematics* **88**, Springer Verlag, 1982.

[Ra95]   Raskind, W. Higher *l*-adic Abel–Jacobi mappings and filtrations on Chow groups, *Duke Math. J.* **78** (1995) 33–57.

[Sch90]   Scholl, A. J. Motives for modular forms, *Invent. Math.* **100**, (1990) 419–430.

[Sch94]　　　Classical motives. In: *Seattle Conf. on Motives 1991, AMS Proc. Symp. Pure Math.* **55**, pp. 163–187, AMS Providence, RI, 1994.

[Se91]　　　Serre, J.-P. Motifs, *Astérisque* **198–200** (1991) 333–349.

[Sh74]　　　Shermenev, A. M. The motive of an abelian variety, *Funkt. Anal.* **8** (1974) 47–53.

[Si86]　　　Silverman, J. H. *The Arithmetic of Elliptic Curves, Graduate Texts in Mathematics* **106**, Springer Verlag, 1986.

[SSa96]　　　Saito, S. Motives and a filtration on Chow groups, *Invent. Math.* **125** (1996) 149–196.

[SSa00]　　　Motives and a filtration on Chow groups II. In: *Arithmetic and Geometry of Algebraic Cycles, Banff Conf. 1998*, pp. 321–346, Kluwer, Dordrecht, 2000.

[V00]　　　Voevodsky, V. Triangulated categories of motives over a field. In: *Cycles, Transfers and Motivic Cohomology Theories* (V. Voevodsky, A. Suslin and E. M. Friedlander, eds.), Princeton University Press, Princeton, NJ, 2000.

[W48]　　　Weil, A. *Variétés Abéliennes et Courbes Algébriques*, Hermann, Paris, 1948.

5

A short introduction to higher Chow groups

Philippe Elbaz-Vincent

Institut de Mathématiques et Modélisation de Montpellier, UMR CNRS 5030, CC 51,
Université Montpellier II, 34095 Montpellier Cedex 5, France

Introduction

The goal of this text is to give a short (and fast) introduction to the theory of higher Chow groups and related concepts. It is designed as a 'quick user's guide' and not at all as a 'reference manual'. Few proofs are given, instead we try to explain why (and how) the hypotheses of the statements are used, and how to use the theory. In the same way, no historical background and no motivations are given. A complete treatment of the theory of higher Chow groups is beyond the scope of the present text. It is designed primarily for people who just need to know the basics of the constructions and the theory, and some main results. The prerequisites are standard: graduate commutative algebra, algebraic geometry, homological algebra, category theory and (classical) intersection theory. Some knowledge of (algebraic) number theory and algebraic topology could help to understand the theory and several examples. A final word: *if you need to quote a result from this paper, please use the original reference instead, in order to avoid misleading ownership.*

Conventions, notation and basic definitions Unless specified, in this paper, ring means *commutative domain*, and mainly *noetherian*. If R is a ring, 0 and 1 always denote, respectively, its neutral for the laws $+$ and $\times$. If R is a ring, we will denote by $R^\times$ its units and if $\{r_0, \ldots, r_m\}$ (with m a natural number) is a collection of elements of R, we will denote by $(r_0, \ldots, r_m)$ the ideal of R spanned by those elements. The set of natural numbers will be denoted by $\mathbb{N}$, the integers by $\mathbb{Z}$, the rational numbers by $\mathbb{Q}$, the real numbers by $\mathbb{R}$, and the complex numbers by $\mathbb{C}$. If K is a field, $\overline{K}$ will be its algebraic closure (unique up to isomorphism) and $char(K)$ will denote its characteristic. If p is a prime

Transcendental Aspects of Algebraic Cycles ed. S. Müller-Stach and C. Peters.
© Cambridge University Press 2004.

number and n a positive integer, we will denote by $\mathbb{F}_{p^n}$ a field with p^n elements (again unique up to isomorphism). If $n = 1$, we just set $\mathbb{F}_p = \mathbb{Z}/(p)$. If X is a scheme (or a ring), $dim(X)$ will denote its Krull dimension. Closed subschemes A and B of an algebraic k-scheme X (i.e. scheme of finite type over a field) will be said to intersect *properly* if $codim_X(A \cap B) \geqslant codim_X(A) + codim_X(B)$. The symbol $\cong$ will denote an isomorphism (taken in the underlying category, if no further specifications are given). A (semi)local ring R is said to be of geometric type if it is a localization of an algebra of finite type over a field. We recall also the following abbreviations: UFD unique factorization domain, PID principal ideal domain, DVR discrete valuation domain. If $B_{*,*}$ is a bicomplex of modules over a ring R, we will denote by $Tot(B_{*,*})$ its total (homological) complex. Unless specified, tensor products are over $\mathbb{Z}$ (as abelian groups).

5.1 Definitions and basic constructions

5.1.1 Higher Chow groups

We fix a ground field k (or equivalently a base scheme $Spec(k)$). This means that all products (or pull-back) will be in the category of schemes over $Spec(k)$. For $n \in \mathbb{N}$, define the 'simplex' Δ^n by

$$\Delta^n = Spec\left(k[T_0, \ldots, T_n]\Big/\left(\sum T_i - 1\right)\right).$$

Notice that Δ^n is isomorphic (but not canonically) to $\mathbb{A}_k^n$, and is identified with the linear subvariety of $\mathbb{A}_k^{n+1}$ given by the equation $T_0 + \cdots + T_n = 1$. By setting the coordinates $t_i = 0$, where t_i is the class of T_i in the quotient, one obtains $(n + 1)$ linear hypersurfaces in Δ^n called *codimension one faces of the simplex* Δ^n. As usual, the points $p_i = (0, \ldots, 1, \ldots, 0)$ (1 placed in the ith coordinate), $0 \leqslant i \leqslant n$, are called the vertices of the 'simplex' Δ^n.

By iterating this, one gets codimension $(n - m)$-faces isomorphic to Δ^m for every $m < n$ inside of Δ^n, these $(n - m)$-faces are given by intersecting the codimension one faces. The faces are parametrized by strictly increasing maps $\rho : \{1, \ldots, m\} \longrightarrow \{1, \ldots, n\}$, which are characterized by the conditions p_i goes to $p_{\rho(i)}$. This makes Δ^n a simplicial set.

The higher Chow groups of an algebraic k-scheme X in codimension p, i.e. scheme of finite type over k (later we will assume X equidimensional) are defined as the homology groups of a chain complex. Let $Z^p(X, n) \subset Z^p(X \times \Delta^n)$ be the subset of cycles Z of $X \times \Delta^n$ of codimension p such that every irreducible component of Z meets all faces $X \times \Delta^m$ again in codimension p for $m < n$. We will say that Z *meets* $X \times \Delta^n$ *properly*.

Observation 5.1.1 As the elements of $Z^p(X, n)$ are a formal sum of cycles of $X \times \Delta^n$ with some transversality conditions, we can use classical constructions from intersection theory as described in [13, 14] (without forgetting to check carefully the transversality condition). Nevertheless, as will be seen in the following, the fundamental setting (from a computational point of view) is when $X = Spec(k)$. In such a setting, the generators of $Z^p(Spec(k), n)$ are subvarieties of Δ^n intersecting its faces in the correct codimension. Among such varieties, the 'simplest' family is the one given by linear subvarieties of Δ^n. In order to construct an admissible such variety, we can use the following recipe: take $n + 1$ (assuming $n \geqslant p$) distinct vectors of k^p which are in general position (i.e. any p of them form a basis of k^p). Such a family of vectors can be represented as a $p \times (n + 1)$ matrix M. Then we can check that the linear subvariety $\ker(M) \cap \Delta^n$, of codimension p in Δ^n, is an element of $Z^p(Spec(k), n)$.

Let $\partial_i : Z^p(X, n) \longrightarrow Z^p(X, n - 1)$ be the restriction map to the ith codimension one face for $i = 0, \ldots, n$ and let $\partial = \sum (-1)^i \partial_i$. We then have:

Proposition and definition 5.1.2 *The map ∂ is a differential of degree -1 (i.e. $\partial_n \circ \partial_{n+1} = 0$), and endows the family $Z^p(X, *)$ with a structure of (homological) complex of abelian groups. The nth homology group of the complex*

$$\cdots \longrightarrow Z^p(X, n + 1) \xrightarrow{\partial} Z^p(X, n) \xrightarrow{\partial} Z^p(X, n - 1) \longrightarrow \cdots$$

is denoted by $CH^p(X, n)$ (see [2]), and is called the nth higher Chow group of X in codimension p. Furthermore, if $m > 0$, we will denote by $CH^p(X, n; \mathbb{Z}/m\mathbb{Z})$ the homology of the previous complex tensored (over $\mathbb{Z}$) by $\mathbb{Z}/m\mathbb{Z}$. It is called the higher Chow groups with coefficients.

Remark 5.1.3 As is often the case with homology groups, it is difficult to give explicit (non-trivial) homological cycles, despite the fact that we can describe (formally) the generators.

If X is an (algebraic) affine k-scheme, say $X = Spec(R)$, we will just denote $CH^p(X, n)$ by $CH^p(R, n)$. We should note that this will change the 'variance' of the functorialities (see below) due to the fact that $Spec()$ is a contravariant functor from the category of (commutative) rings to the category of schemes.

In the remaining text we will assume (unless specified) that all schemes are algebraic over a field k, equidimensional (i.e. all the irreducible components have the same dimension, and in this setting maps are supposed to be equidimensional, too) and quasi-projective (this last hypothesis is sometimes not needed).

As a direct consequence of the definition we have:

Proposition 5.1.4 *Let X be any smooth k-scheme. Then*

$$CH^0(X, n) = \begin{cases} H^0_{Zar}(X, \mathbb{Z}) & n = 0 \\ 0 & n \neq 0. \end{cases}$$

Furthermore, $CH^i(X, n) = 0$ if $i > n + dim(X)$.

Proof If $i > n + dim(X)$, then no subvariety of $X \times \Delta^n$ can have codimension $i > dim(X \times \Delta^n)$. So we already have $Z^i(X, n) = 0$. $\qquad\qquad\qquad \square$

Remark 5.1.5 In fact, we will see later that for any smooth *semilocal* affine k-scheme X, we have $CH^i(X, n) = 0$ as soon as $i > n$.

Let $m \in \mathbb{N}$, with $m \geqslant 2$. The usual sequence $0 \to \mathbb{Z} \xrightarrow{\times m} \mathbb{Z} \to \mathbb{Z}/m\mathbb{Z} \to 0$ gives rise to a universal coefficient sequence:

Proposition 5.1.6 (Universal coefficients) *Let $m \in \mathbb{N}$, with $m \geqslant 2$, and X be an algebraic k-scheme. Then, for all $p, n \in \mathbb{N}$, we have exact sequences*

$$0 \to CH^p(X, n)/m \to CH^p(X, n; \mathbb{Z}/m\mathbb{Z})$$
$$\to {}_mCH^p(X, n-1) \to 0.$$

5.1.2 First properties

We will begin by exploring the functorialities.

Proposition 5.1.7 ([2, 1.3, p. 273], [25, 4.5.12]) *Let X and Y be algebraic k-schemes, and $f : X \to Y$ a map of k-schemes. If f is proper we have, for all p, a map of complexes*

$$f_* : Z^p(X, *) \to Z^{p-d}(Y, *),$$

with $d = dim(X) - dim(Y)$. If f is flat, we have, for all p, a map of complexes

$$f^* : Z^p(Y, *) \to Z^p(X, *).$$

The above proposition is often summarized by saying that higher Chow groups are covariant functorial for proper maps and contravariant functorial for flat maps. Furthermore, they are contravariant for all maps between smooth affine schemes. Let us explain how these maps are defined. If $f : X \to Y$ is a flat map, the pull-back of cycles defines a natural map (of complexes) $f : Z^p(Y, *) \to Z^p(X, *)$. Indeed, we may assume that X has pure dimension $dim(Y) + m$, for some $m \in \mathbb{N}$, and f is flat of relative dimension m (see [14, B.2.5, p. 429]). Thus, $f \times 1 : X \times \Delta^n \to Y \times \Delta^n$ is again flat and of the same relative dimension. The morphism $f \times 1$ will be denoted again by f. If

$Z \subset Y \times \Delta^n$ is a codimension p cycle (we can assume it to be irreducible, without loss of generality), then $f^{-1}(Z)$ is a codimension p cycle of $X \times \Delta^n$; hence a map $f^* : CH^p(Y, n) \to CH^p(X, n)$.

In the same way, if $f : X \to Y$ is proper, then so is $f \times 1 : X \times \Delta^n \to Y \times \Delta^n$, and we get an induced map on cycles $f_* : Z^p(X \times \Delta^n) \to Z^{p-d}(Y \times \Delta^n)$ (see [14, Section 1.4]), which induces a morphism $f_* : CH^p(X, n) \to CH^{p-d}(Y, n)$.

Example 5.1.8 Let L/K be a finite field extension, and denote by π the morphism $Spec(L) \to Spec(K)$. Then we have $\pi_* \pi^* = [L : K]$, and thus, rationally, $CH^p(K, n)_{\mathbb{Q}} \subset CH^p(L, n)_{\mathbb{Q}}$. We can generalize this to any finite flat morphism.

We will now describe the case of a 'general' morphism with X smooth over k. We follow Levine [27, p. 5]. Suppose X is smooth over k, and let $f : Y \to X$ be a morphism of quasi-projective k-schemes. For each n, let $Z^p(X, n)_f$ be the subgroup of $Z^p(X, n)$ generated by the subvarieties $W \subset X \times \Delta^n$, of codimension p, such that W meets all the faces of $X \times \Delta^n$ properly (i.e. $W \in Z^p(X, n)$), and each component of $(f \times 1)^{-1}(W)$ has codimension p on $Y \times \Delta^n$ and meets the faces of $Y \times \Delta^n$ properly. The $Z^p(X, n)_f$ form a subcomplex $Z^p(X, *)_f$ of $Z^p(X, *)$, and if X is affine, the inclusion $Z^p(X, *)_f \hookrightarrow Z^p(X, *)$ is a quasi-isomorphism. This gives a functorial pull-back map

$$f^* : CH^p(X, n) \to CH^p(Y, n).$$

In fact, such a kind of technique involves 'a moving lemma', since a priori an arbitrary pull-back has no reason to give a well defined cycle of $Z^p(Y, n)$.

Remark 5.1.9

(1) The 'quasi-projectivity' assumption is used to embed Y as a locally closed subset of a projective space (of large enough dimension) in order to get some intermediate quasi-isomorphisms, and also to ensure some proper intersections (see for instance Section 4 of [2]).

(2) It is possible to replace the smoothness by some technical assumptions, such as that the graph of f is a local complete intersection.

5.1.3 Homotopy invariance

An often used property is the so-called 'homotopy invariance', which asserts that higher Chow groups of X are the same as higher Chow groups of $X \times \mathbb{A}^1_k$. More precisely, we have the following (see [2, Section 2]).

Proposition 5.1.10 (Homotopy invariance property) *Let X be an algebraic k-scheme, and $pr_1 : X \times \mathbb{A}_k^1 \to X$ be the canonical first projection. Then the map pr_1^* is a quasi-isomorphism.*

Corollary 5.1.11 *Let X be an algebraic k-scheme, and m a natural number. Then for all $p \in \mathbb{N}$, the (graded) group $CH^p(X \times \mathbb{A}_k^m, *)$ is isomorphic to $CH^p(X, *)$.*

If we rewrite the above result in terms of rings, we have, for all $p \in \mathbb{N}$, the isomorphism $CH^p(k, *) \cong CH^p(k[T_1, \ldots, T_m], *)$, induced by the maps $k \hookrightarrow k[T_1, \ldots, T_m]$ and evaluation in 0.

5.1.4 The product structure

Let X and Y be quasi-projective algebraic k-schemes. We want to define a product

$$CH^p(X, n) \otimes CH^q(Y, m) \to CH^{p+q}(X \times Y, n+m).$$

When X is smooth, we can compose with pull-back along the diagonal $X \to X \times X$ to get an internal product structure

$$CH^p(X, n) \otimes CH^q(X, m) \to CH^{p+q}(X, n+m).$$

Remark 5.1.12 Here we use the smoothness hypothesis to get the (contravariant) functoriality at the level of higher Chow groups as seen in Proposition 5.1.7. We can again get a more general construction by using some technical assumptions (e.g. the diagonal $X \to X \times X$ is a local complete intersection map, and X is normal).

Let $Z^{p+q}(X, Y; *)$ be the subcomplex of the total complex of $Z^p(X, *) \otimes Z^q(Y, *)$ generated by products $V \otimes W$ such that V and W are irreducible varieties of $X \times \Delta^n$ and $Y \times \Delta^m$, respectively, and $V \times W \subset X \times Y \times \Delta^n \times \Delta^m$ meets all faces of $\Delta^n \times \Delta^m$ properly. Then, modulo an adequate triangulation of $\Delta^n \times \Delta^m$, we get a map from $Z^{p+q}(X, Y; *)$ to $Z^{p+q}(X \times Y, *)$. We just need another moving lemma to prove that we can always put the cycles in such a position:

Proposition 5.1.13 [2, Theorem 5.1, p. 283] *Let X and Y be quasi-projective algebraic k-schemes. Then*

$$Z^{p+q}(X, Y; *) \hookrightarrow Tot\left(Z^p(X, *) \otimes Z^q(Y, *)\right)$$

is a quasi-isomorphism.

As an application, suppose Y is smooth and we are given an arbitrary map $f : X \to Y$. The graph of f defines a map $\Gamma_f : X \to X \times Y$. This gives an action of $CH^p(Y, *)$ on $CH^q(X, *)$ by

$$CH^q(X, m) \otimes CH^p(Y, n) \to CH^{q+p}(X, m+n).$$

Furthermore, if X is smooth, we have a structure of a (commutative) graded ring on $CH^*(X, \bullet)$. For $x \in CH^*(X, m)$, $y \in CH^*(X, n)$, we have

$$x \cdot y = (-1)^{mn} y \cdot x.$$

This provides the usual 'projection formula'.

Proposition 5.1.14 (Projection formula) *Let Y and X be smooth over k, $f : X \to Y$ be proper, $y \in CH^*(Y, \bullet)$, $x \in CH^*(X, \bullet)$. Then*

$$f_* \left(x \cdot f^*(y) \right) = f_*(x) \cdot y.$$

Remark 5.1.15

(1) As already pointed out, the smoothness hypothesis is mainly used to get the functoriality at the level of higher Chow groups.
(2) The isomorphism $X \times Y \to Y \times X$, induced by the permutation of the factors, induces in turn, as expected, an isomorphism at the level of higher Chow groups.
(3) Almost all the constructions on higher Chow groups are compatible with the product structure (as localization for instance).
(4) We will see in Section 5.5.1, that using the cubical setting instead of the simplicial one will give an easier construction of the product.

Searching for indecomposable elements With the product structure we can consider several interesting situations, and in particular look for some 'decomposability property'. First we will consider some 'general indecomposable elements'. Let X and Y be quasi-projective algebraic k-schemes. Suppose as above that Y is smooth and that we are given a proper map $f : X \to Y$. We have the map $\Pi_{f,r,s}$ defined as follows

$$\bigoplus_{p+q=r,\; m+n=s} CH^p(X, m) \otimes_{\mathbb{Z}} CH^q(Y, n) \xrightarrow{\Pi_{r,s}} CH^r(X, s).$$

We will say that an element of $CH^r(X, s)$ is decomposable relative to f (or to Y) if it is in the image of $\Pi_{f,r,s}$. In the case where $Y = X$ and f is the identity, we can drop the subscript f.

As we always have a proper map, given by the structural morphism pr : $X \to$ $Spec(k)$, we can consider as another particular case the map $\Pi_{\text{pr},r,s}$.

A general problem is to study the behaviour of the map $\Pi_{f,r,s}$. Under which conditions is this map surjective or injective? In general, this map behaves badly (for to be injective in general, and rarely surjective). Nevertheless, the main point is that when we want to study say $CH^r(X, s)$, we do not want to consider the elements that we already know from a lower dimension/codimension. More precisely, the crucial part (relative to Y or f) of $CH^r(X, s)$ is given by the cokernel of this map (i.e. the Y-indecomposable elements).

5.2 Localization and Mayer–Vietoris sequences

The main (and difficult) result is the following.

Theorem 5.2.1 (Localization) [2, 3] *If $W \subset X$ is a closed subvariety of pure codimension r with X quasi-projective over a field, then one has the localization*

$$\cdots \to CH^*(W, n) \to CH^{*+r}(X, n) \to CH^{*+r}(X - W, n)$$
$$\to {*}CH^*(W, n-1) \to \cdots$$

Example 5.2.2

(1) If $X = Spec(A)$ with A a DVR over k, denote by F its fraction field and k_v its residue field. Then the localization sequence looks like

$$\cdots \to CH^*(k_v, n) \to CH^{*+1}(A, n) \to CH^{*+1}(F, n)$$
$$\to CH^*(k_v, n-1) \to \cdots$$

Owing, for instance, to 'Gersten's conjecture' (see Section 5.3), the sequence splits into short exact sequences of the form

$$0 \to CH^{*+1}(A, n) \to CH^{*+1}(F, n) \to CH^*(k_v, n-1) \to 0$$

and thus we have $CH^{n+1}(A, n) = 0$.

(2) Applying the localization sequence to $X \times Spec(k[T, T^{-1}]) \subset X \times Spec(k[T])$ (and thanks to homotopy invariance) yields:

$$CH^p(X \times Spec(k[T, T^{-1}]), n) \cong CH^p(X, n) \oplus CH^{p-1}(X, n-1).$$

(3) Applying the localization sequence to $X \times \mathbb{A}_k^1 \subset X \times \mathbb{P}_k^1$ yields:

$$CH^p(X \times \mathbb{P}_k^1, n) \cong CH^p(X, n) \oplus CH^{p-1}(X, n).$$

As a consequence we have a Mayer–Vietoris sequence:

Corollary 5.2.3 *Let $X = U \cup V$ be a Zariski cover. Then we have a long exact sequence*

$$\cdots \to CH^*(U \cup V, n) \to CH^*(U, n) \oplus CH^*(V, n)$$
$$\to CH^*(U \cap V, n) \to CH^*(U \cup V, n-1) \to \cdots$$

Remark 5.2.4 What is difficult in this setting is that all the cycles are k-cycles. Thus, we are not allowed to have mixed characteristic. This problem was part of the motivation for Levine's work [26, 28]. The main interest of mixed characteristic is the ability to use some tricks 'from characteristic 0 to characteristic p', which is helpful as we sometimes need resolution of singularities and we do not (yet) have resolution of singularities in characteristic p.

5.3 Gersten resolution for higher Chow groups

We will now introduce a new tool, which will reduce conceptually the computation of higher Chow groups of a smooth (quasi-projective) variety to the computation of 'its function fields'. So, in theory, if we can compute the higher Chow groups of fields, we should be able to compute everything. Of course, as we will see, computing these groups for fields is already a challenge, and for the time being, no effective techniques are known.

Let X be a quasi-projective k-scheme. Set

$$X^n = \{x \in X, \text{ Zariski closure of } x \text{ has codim } n \text{ in } X\}.$$

For A an abelian group, $i_x A$ denotes the constant sheaf with stalk A on the Zariski closure $\{\bar{x}\}$. Following Bloch [2, Section 10], define a decreasing filtration $Fil_* Z^p(X, *)$ by

$$Fil_r Z^p(X, *) = \{z \in Z^p(X, *), \text{ projection of the support of}$$
$$Z \text{ on } X \text{ has codim} \geq r\}.$$

This filtration gives rise, for all $p \in \mathbb{N}$, to the following spectral sequence

$$E_1^{r,s} = \bigoplus_{x \in X^{-s}} CH^{p-r}(k(x), -r-s) \Rightarrow CH^p(X, -r-s).$$

Denote by $\mathcal{CH}_X^p(n)$ the sheaf associated to the presheaf $U \mapsto CH^p(U, n)$. Then the complex of E_1 terms (i.e. $E_1^{*,s}$) can be localized for the Zariski topology on X giving

$$\bigoplus_{x \in X^0} i_x CH^p(k(x), n) \to \bigoplus_{x \in X^1} i_x CH^{p-1}(k(x), n-1)$$

$$\to \cdots \to \bigoplus_{x \in X^n} i_x CH^{p-n}(k(x), 0) \to 0$$

We then have the main result:

Theorem 5.3.1 (Gersten's resolution) [2, Section 10] *For X smooth there are flasque (universal) resolutions:*

$$0 \to \mathcal{CH}_X^p(n) \to \bigoplus_{x \in X^0} i_x CH^p(k(x), n) \to \bigoplus_{x \in X^1} i_x CH^{p-1}(k(x), n-1)$$

$$\to \cdots \to \bigoplus_{x \in X^n} i_x CH^{p-n}(k(x), 0) \to 0.$$

As a consequence, for X smooth, we get a *local to global spectral sequence* [2, Theorem 3.2, p. 277] and [3] (for the correction of Theorem 3.3 of [2]):

$$E_2^{r,s} = H^r(X, \mathcal{CH}_X^p(-s)) \Rightarrow CH^p(X, -r-s).$$

In particular we get a Bloch–Ogus–Quillen formula

$$H^p(X, \mathcal{CH}_X^p(p)) \cong CH^p(X).$$

A direct consequence of the Gersten resolution is that if R is a smooth (semi)local k-algebra (of geometric type) then the map $CH^p(R, n) \to CH^p(Frac(R), n)$ is injective. This also shows that $CH^p(R, n) = 0$ if $p > n$ for such R, and the largest codimension for, a priori, a non-trivial group for a smooth (semi)local k-algebra (of geometric type) is $p = n$.

Remark 5.3.2 'On the proof of the Gersten resolution for higher Chow groups' The original proof (without the universal exactness mentioned in Theorem 5.3.1) is due to Bloch [2], but we note that we can get at least three other proofs of this result. We will indicate in the following some hints and will leave the details to the interested reader.

(1) *First strategy of proof (including the universal exactness)*: by using the work of Colliot-Thélène, Hoobler and Kahn [7] (see Sections 5 and 6 and Example 7.3.5, p. 69). Indeed, we can check that higher Chow groups fulfil the axioms developed in their work and as a consequence we get the universal exactness of the Gersten complex (for higher Chow groups).

(2) *Second strategy of proof*: by using the work of Voevodsky [49, Sections 4.4 and 4.6].

(3) *Third strategy of proof*: by using the work of Colliot-Thélène and Ojanguren [6] and checking that higher Chow groups fulfil the conditions P1–P3 of

Theorem 1.1 of [6, p. 100], we can get the injectivity of the morphism $CH^p(R, n) \to CH^p(Frac(R), n)$. The step from Gersten conjecture to Gersten resolution is obtained via standard arguments.

As an illustration of this machinery, we have[1],

Corollary 5.3.3 (**Removing lemma**) *Suppose that R is a regular semilocal ring (of geometric type) over k. Let $Z \in Z^p(R, n)$ be a (algebraic) cycle with no boundary (i.e. $\partial Z = 0$) and which is non-dominant over $Spec(R)$. Then Z is equivalent to zero.*

Proof As Z is non-dominant, we have

$$Supp(pr_*(Z)) \subset \cup_{i \in I} V(\mathfrak{p}_i),$$

for some prime ideals, with I finite as R is noetherian. Thus

$$Z \in Im(\oplus_{i \in I} Z^{p-1}(V(\mathfrak{p}_i), n) \longrightarrow Z^p(R, n)).$$

Hence by the localization exact sequence, the class of Z is in the kernel of the canonical map $CH^p(R, n) \longrightarrow CH^p(Frac(R), n)$. But as higher Chow groups fulfil Gersten's conjecture (for a quasi-projective variety over a field), the class of Z is just zero. □

Applications 5.3.4 Combining the previous machinery we can already get some interesting results. Let $n \in \mathbb{N}$ with $n > 2$. Then, we have the following 'rigidity property'

$$CH^2(k, n) \cong CH^2(k(t), n),$$

where the isomorphism is induced by the structural morphism $k \hookrightarrow k(t)$, and t is an indeterminate over k. Set $A = k[t]_{(t)}$. Then A is smooth and $Frac(A) = k(t)$. For codimension reasons, by the Gersten resolution, we have

$$CH^2(A, n) \cong CH^2(k(t), n).$$

By the localization sequence, we have

$$\cdots \to CH^2(k[t], n) \to CH^2(A, n)$$
$$\to \bigoplus_{\mathfrak{p} \in Spec(k[t]) - (t)} CH^1(k[t]/\mathfrak{p}, n) \to \cdots$$

Again, for codimension reasons, $CH^1(k[t]/\mathfrak{p}, n) = 0$. Furthermore, by

---

[1] This result was in an earlier version of [9] and has thus transformed into the 'removed lemma' ...

homotopy invariance, we have $CH^2(k, n) \cong CH^2(k[t], n)$, induced by the structural morphism $k \hookrightarrow k[t]$. But the structural morphism $k \hookrightarrow A$ is split (by the evaluation in 0 which is well defined on A), and as A is smooth, the map $CH^2(k, n) \to CH^2(A, n)$ is a split monomorphism by functoriality, showing that

$$CH^2(k, n) \cong CH^2(A, n).$$

The claim follows. We can generalize this result to the case of several indeterminates. We will see later that for $n = 2$ the result no longer holds. Furthermore, using the previous ideas we can check that, for all $p \in \mathbb{N}$, the canonical map $CH^p(k, n) \to CH^p(k(t), n)$ is injective and, more generally, if F is a purely transcendental extension of k, we have a canonical embedding $CH^p(k, n) \hookrightarrow CH^p(F, n)$. In some situations, such a result could allow us to restrict our arguments to the case of infinite fields.

5.4 K-theory and the motivic spectral sequence

Let X be a scheme. Denote by $K_n(X)$ its nth K-group (see Chapter 5 of [39] for a definition of Quillen K-theory and [29] for an overview). In the case where $X = Spec(R)$ with R a ring, we can set $K_n(X) = \pi_n(BGL(R)^+), n > 0$, where $BGL(R)^+$ is the Quillen +-construction (see [39]) applied to the classifying space of $GL(R)$ (which can be seen as the union of the $GL_m(R)$ with $m \in \mathbb{N}$). We can also construct a K-theory with coefficients, and they have functorial properties similar to those described for higher Chow groups. In low degrees, we have:

Proposition 5.4.1 *If R is a field or a semilocal ring, then $K_1(R) \cong R^\times$. If, moreover, R is a field or a semilocal ring with infinite residue fields, we also have an isomorphism between $K_2(R)$ and the quotient of $R^\times \otimes R^\times$ by the subgroup generated by elements $a \otimes b$ with $a + b = 0, 1$.*

It is difficult to compute these groups in general, but we have the following results.

Example 5.4.2

(1) Let F be a number field, $\mathcal{O}_F$ its ring of integers, r_1 (respectively r_2) its number of real (respectively non-conjugate complex) embeddings. Then $K_m(\mathcal{O}_F) \otimes \mathbb{Q} \cong K_m(F) \otimes \mathbb{Q}$. Furthermore (see [5]), $K_m(F) \otimes \mathbb{Q}$ is trivial for $m > 0$ even, it is isomorphic to $\mathbb{Q}^{r_1+r_2-1}$ when $m = 1$, to $\mathbb{Q}^{r_2}$ when $m = 3$ mod 4, and to $\mathbb{Q}^{r_2}$ when $m = 1$ mod 4 with $m > 1$. Without neglecting the

torsion, we also have $K_1(\mathbb{Z}) = \mathbb{Z}/(2)$, $K_2(\mathbb{Z}) = \mathbb{Z}/(2)$, $K_3(\mathbb{Z}) = \mathbb{Z}/(48)$, $K_4(\mathbb{Z}) = 0$ [37], $K_5(\mathbb{Z}) = \mathbb{Z}$ [10] and the order of $K_6(\mathbb{Z})$ is a power of 3 [10].[2] The 2-torsion part of the K-theory of $\mathbb{Z}$ is computed in [38].

(2) By [36], we have $K_i(\mathbb{F}_q)$ (respectively $K_i(\overline{\mathbb{F}_q})$) is trivial for $i > 0$ even, it is isomorphic to $\mathbb{Z}$ if $i = 0$, and to $\mathbb{Z}/(q^k - 1)$ (respectively $\mathbb{Q}/\mathbb{Z}[1/p]$ with $p = char(\mathbb{F}_q)$) if $i = 2k - 1$ with $k > 0$.

As for the classical 'Riemann–Roch–Grothendieck–Hirzebruch' theorem, we have a higher analogue.

Theorem 5.4.3 (Riemann–Roch for higher Chow groups) [2] (See also [24, 25]). *Let X/k be a smooth quasi-projective variety. Then there exist Chern maps, $c_{n,p}^{Chow} : K_n(X) \longrightarrow CH^p(X, n)$, which induce an isomorphism (via a Chern character):*

$$ch_n : K_n(X) \otimes \mathbb{Q} \cong \bigoplus_{p \in \mathbb{N}} CH^p(X, n) \otimes \mathbb{Q}.$$

Moreover this Chern character is functorial and compatible with localization exact sequences.

A more precise relationship between K-theory and higher Chow groups is captured by the 'motivic spectral sequence'.

Theorem 5.4.4 (Motivic spectral sequence) *Let X be a quasi-projective smooth k-scheme. There exists a third quadrant spectral sequence ($p, q \leqslant 0$):*

$$E_2^{p,q} = CH^{-q}(X, -p - q) \Rightarrow K_{-p-q}(X).$$

Furthermore, this spectral sequence also holds with coefficients, and is compatible with the product structure.

Remark 5.4.5 Several people have given a construction of the motivic spectral sequence: Bloch and Lichtenbaum [4] in the case of a field, Friedlander and Suslin [12], Levine [27] and Suslin [45] (using Grayson's motivic cohomology) for the general case. In this paper we refer to the construction of Levine [27]. For the product structure we refer to Rognes and Weibel in Appendix B of [38]. In the case of a field, this spectral sequence is often called the 'Bloch–Lichtenbaum spectral sequence'. In the general setting, this spectral sequence is sometime called the 'BLLFS spectral sequence' (for Bloch–Lichtenbaum–Levine–Friedlander–Suslin).

---

[2] In fact the authors have recently shown that $K_6(\mathbb{Z}) = 0$.

In theory, we should be able to compute the K-groups via the higher Chow groups. But, we actually have more computations for the former than for the latter. So we will use these results in order to deduce some computations at the level of higher Chow groups.

5.5 Some explicit computations

Despite what it seems, it is not easy to construct 'true' homological cycles in $Z^p(X, n)$, even if $X = Spec(k)$. Of course we can define a smooth irreducible cycle W by a family of nice equations (or by a parametrized subvariety), but we will first have to show that it is a homological cycle (i.e. $\partial W = 0$, for instance if W has no boundaries), and then, that this cycle is non-trivial. We will see that we can do this in some particular dimensions. In fact, the main difference with classical Chow groups is that higher Chow groups of a field are non-trivial. In the following we will identify some higher Chow groups with already known objects, and will give some cycle construction. Deeper and more extensive results will be shown in Section 5.5.3.

5.5.1 Cubical version of higher Chow groups

Set $\square_k^n = (\mathbb{P}_k^1 - \{1\})^n$ with coordinates t_i and codimension one faces obtained by setting $t_i = 0, \infty$. Let $C^p(X, n)$ be the free abelian group generated by subvarieties of $X \times \square_k^n$ of codimension p and meeting all the faces of the cubes properly. Let ∂_i^∞ (respectively ∂_i^0) be the pull-back to the face $t_i = \infty$ (respectively $t_i = 0$). By setting

$$d_n = \sum_{i=1}^n (-1)^i \left(\partial_i^\infty - \partial_i^0 \right),$$

we get, for each p, a complex (of $\mathbb{Z}$-modules) $(C^p(X, *), d_*)$. Let $D^p(X, n)$ be the subgroup of $C^p(X, n)$ generated by cycles which are the pull-back of some cycle on $X \times \square_k^{n-1}$ via a projection of the form $(t_1, \ldots, t_n) \mapsto (t_1, \ldots, \hat{t_i}, \ldots, t_n)$, $\square_k^n \to \square_k^{n-1}$. We call such cycles 'degenerate' (or 'decomposable' in a certain sense), as we have an isomorphism of k-varieties

$$\square_k^1 \times \square_k^{n-1} \cong \square_k^n.$$

Of course, we do not have such a kind of isomorphism in the simplicial setting. This isomorphism will give a direct product structure on higher Chow groups. The restriction of the differential of $C^p(X, *)$ to $D^p(X, *)$ makes $(D^p(X, *), d_*)$

a subcomplex of $C^p(X, *)$. Then we set

$$Z^p(X, *)_c = C^p(X, *)/D^p(X, *).$$

The crucial result is

Proposition 5.5.1 [23, 4.7, p. 296]. *There is a quasi-isomorphism between* $Z^p(X, *)_c$ *and* $Z^p(X, *)$, *which induces an isomorphism*

$$H_n\left(Z^p(X, *)_c\right) \cong CH^p(X, n).$$

Thus we can use indifferently both settings at our convenience. As we will see, some computations are easier at the cubical level. If X and Y are arbitrary algebraic k-schemes, we have an isomorphism in the category of algebraic k-schemes

$$\left(X \times \square_k^m\right) \times \left(Y \times \square_k^n\right) \cong (X \times Y) \times \square_k^{m+n},$$

which gives the description of the product structure on higher Chow groups in the cubical setting.

Remark 5.5.2

(1) Instead of $\mathbb{P}^1 - \{1\}$, we can use $\mathbb{P}^1 - \{\infty\}$. In the latter, we get an isomorphism $\mathbb{A}_k^1 \hookrightarrow \mathbb{P}^1 - \{\infty\}$ induced by an embedding $\mathbb{A}_k^1 \subset \mathbb{P}^1$. Of course, both versions of the cube are isomorphic, via the automorphism of $\mathbb{P}^1$ given by $x \mapsto 1 - 1/x$. We can see $\square_k^n$ as an affine sheet mapped on the torus $(\mathbb{P}^1)^n$. This will have some consequences when we look at the linear part of the higher Chow complex.

(2) The map which gives the isomorphism of Proposition 5.5.1 is indirect and is via an intermediate bicomplex.

5.5.2 Computing $CH^p(R, q)$ for $p = 1$ with $q \geqslant 0$, and $p \geqslant 0$ with $q = 0$

Let us denote by $CH^i(X)$ the classical Chow group of codimension i cycles on X modulo rational equivalence as defined by Fulton [14]. Then

Proposition 5.5.3 *If X is a quasi-projective variety, we can identify* $CH^p(X, 0)$ *with* $CH^p(X)$.

As usual, if $X = Spec(k)$, then $CH^*(X) = \mathbb{Z}$. If we look at the localization sequence in low dimension, assuming that $W \subset X$ is a closed subvariety of pure codimension r with X quasi-projective over a field, we get

$$CH^*(W, 1) \to CH^{*+r}(X, 1) \to CH^{*+r}(X - W, 1) \to CH^*(W)$$
$$\to CH^{*+r}(X) \to CH^{*+r}(X - W) \to 0.$$

In other words, via the higher Chow groups we are able to describe the kernel of the map $CH^*(W) \to CH^{*+r}(X)$.

Now we want to compute the higher Chow groups in codimension one.

Proposition 5.5.4 [2, 34] *For X smooth, we have*

$$CH^1(X, n) = \begin{cases} Pic(X), & n = 0, \\ H^0_{Zar}(X, \mathcal{O}_X^\times), & n = 1, \\ 0, & n > 1. \end{cases}$$

If, in the previous result, we set $X = Spec(R)$, then $CH^1(R, 1) = R^\times$, and if moreover R is semilocal then $Pic(R) = 0$.

5.5.3 More computations in higher range

In the following we will give further information on $CH^n(R, n)$ for R a smooth semilocal ring or a field, and on some other groups in particular range. The material of this section is mainly extracted from [9]. First we will need some complementary constructions.

Milnor K-theory

As the definition is more self-contained than for Quillen K-theory, we will give the basic facts here.

Definition 5.5.5 [18] *Let R be a ring, we define $K_*^M(R)$ as the graded ring generated by the symbols $\ell(a)$, $a \in R^\times$, in degree one and with the following relations:*

(1) *for all a, b in $R^\times$, we have $\ell(ab) = \ell(a) + \ell(b)$;*
(2) *for all a, b in $R^\times$ such that $a + b = 0$ or 1, we have $\ell(a)\ell(b) = 0$ (also called 'Steinberg relations').*

It follows that the abelian group $K_n^M(R)$ is generated by symbols $\ell(a_1) \cdots \ell(a_n)$, with $a_1, \ldots, a_n \in R^\times$. We will denote the symbol $\ell(a_1) \cdots \ell(a_n)$ by

$\{a_1, \ldots, a_n\}$. We have a canonical surjective morphism,

$$\ell : R^\times \longrightarrow K_1^M(R)$$
$$a \mapsto \ell(a).$$

In particular we have $K_0^M(R) = \mathbb{Z}$ and $K_1^M(R) = R^\times/[R^\times, R^\times]$. Moreover, we know that if R is a commutative ring with 'enough' units, the Milnor K_2 is isomorphic to the Quillen K_2 (see [20, 8.5, p. 512]). The following proposition is straightforward.

Proposition 5.5.6

(1) *For all* $a, b \in R^\times$, *we have* $\ell(a)\ell(b) + \ell(b)\ell(a) = 0$.
(2) *If* $\xi \in K_n^M(R)$ *and* $\eta \in K_m^M(R)$ *then* $\xi\eta = (-1)^{nm}\eta\xi$.
(3) *If* $\sigma \in \mathfrak{S}_n$, $\{a_{\sigma(1)}, \ldots, a_{\sigma(n)}\} = \varepsilon(\sigma)\{a_1, \ldots, a_n\}$, *with* ε *the signature map.*
(4) *For all* $a \in R^\times$, *we have* $\ell(a)^2 = \ell(a)\ell(-1) = \ell(-1)\ell(a)$, *and* $\ell(-a^{-1})\ell(a) = 0$.

Remark 5.5.7 As both Milnor K-theory and Quillen K-theory have multiplicative structures, we deduce a canonical map $K_n^M(R) \xrightarrow{\text{can}} K_n(R)$. Rationally this morphism is injective (see for example [18, 35]), but integrally the morphism 'can' is known to behave badly. In particular, if R is a global field, Weibel[3] (see also Shapiro [40, Proposition 2, p. 96], and Musikhin and Suslin [33] for related questions) has shown that 'can' is zero for $n \geqslant 5$.

The graph map

From now on $Z^p(X, *)$ *will denote exclusively the cubical version of the higher Chow complex.* Let R be a k-algebra of finite type. Any element $f \in R$ gives a map $ev_f : k[T] \longrightarrow R$, $P \mapsto P(f)$. This defines a k-morphism of schemes $f : Spec(R) \longrightarrow \mathbb{P}_k^1$, $x \mapsto f(x)$. The restriction of the graph of f, $graph(f) \cap Spec(R) \times \square_k^1 \subset Spec(R) \times \square_k^1$, is an algebraic cycle in $Z^1(R, 1)$, since it is defined by the algebraic equation $\{y = f(x)\}$ inside $Spec(R) \times \square_k^1$.

In general, we have a map $\varphi_n : (R^\times)^n \longrightarrow CH^n(R, n)$, defined as

$$(f_1, \ldots, f_n) \mapsto graph(f_1, \ldots, f_n) \cap Spec(R) \times \square_k^n.$$

We will call the cycles in the image of φ_n 'graph cycles'. By a result of Bloch [2, Theorem 6.1, p. 287], we know that the regularity of R implies that φ_1 is an isomorphism. We want to define a map at the level of $K_n^M(R)$, thus we need

---

[3] Unpublished work.

to prove bilinearity and the 'Steinberg relation'. We will give the proof, since it could help the reader to gain some understanding about the manipulations of parametrized cycles.

Lemma 5.5.8 [9, Lemma 2.1] *We have the following relations:*

(1) *If* $f, 1 - f, f_i \in R^\times$, *then* $\varphi_n(\{f, 1 - f, f_3, \ldots, f_n\}) = 0$ *in* $CH^n(R, n)$.

(2) *If* $f, g, f_i \in R^\times$, *then*

$$\varphi_n(\{fg, f_2, \ldots, f_n\}) = \varphi_n(\{f, f_2, \ldots, f_n\}) + \varphi_n(\{g, f_2, \ldots, f_n\})$$
$$\in CH^n(R, n).$$

(3) *The cycle* $\varphi_n(\{f, g, f_3, \ldots, f_n\}) + \varphi_n(\{g, f, f_3, \ldots, f_n\})$ *is* 0 *in* $CH^n(R, n)$, *for all* $f, g, f_i \in R^\times$.

(4) *The cycle* $\varphi_n(\{f, -f, f_3, \ldots, f_n\})$ *is* 0 *in* $CH^n(R, n)$, *for all* $f, f_i \in R^\times$.

Proof

(1) As in the following proofs, the goal is to produce a cycle $W \in Z^n(R, n + 1)$ whose boundary is a linear combination of the cycles involved in the statement. The irreducible cycle $W \in Z^n(R, n + 1)$, given as a parametrized subvariety by the image of the morphism

$$Spec(R) \times \square_k^1 \longrightarrow Spec(R) \times \square_k^{n+1},$$

$$(p, x) \mapsto \left(p, x, 1 - x, \frac{f(p) - x}{1 - x}, f_3(p), \ldots, f_n(p) \right),$$

has only one boundary $\varphi_n(\{f, 1 - f, f_3, \ldots, f_n\})$ which proves the assertion.

(2) The boundary of the cycle $V \in Z^n(R, n + 1)$ given by the parametrization

$$Spec(R) \times \square_k^1 \longrightarrow Spec(R) \times \square_k^{n+1},$$

$$(p, x) \mapsto \left(p, x, \frac{f(p)x - f(p)g(p)}{x - f(p)g(p)}, f_2(p), \ldots, f_n(p) \right),$$

is equal to $-\varphi_n(\{f, f_2, \ldots, f_n\}) - \varphi_n(\{g, f_2, \ldots, f_n\}) + \varphi_n(\{fg, f_2, \ldots, f_n\})$.

(3) It is sufficient to establish the result for $n = 2$ since the remaining coordinates do not contribute to the boundaries. We compute the boundary of the parametrized cycle

$$W : (p, t) \mapsto \left(p, t + u(p), t + v(p), \frac{t + s(p)}{t + r(p)} \right) \in Z^2(R, 3)$$

with some given elements $u, v, s, r \in R^\times$ such that the cycle is admissible. For simplicity, such parametrized cycles will be written using the symbolic notation

$$\left[t + u, t + v, \frac{t + s}{t + r}\right].$$

Continuing with a similar notation for cycles in $Z^2(R, 2)$ and substituting $x = u - s, y = v - s, z = u - r$ and $w = v - r$, we get the following boundary of W:

$$[x, y] - [z, w] - \left[x - y, \frac{y}{w}\right] + \left[y - x, \frac{x}{z}\right] = 0,$$

again assuming that all cycles involved are admissible. Then we add this relation to the one that we obtain by permuting the roles of x and y, as well as the roles of w and z. Thus we obtain the relation

$$[x, y] + [y, x] = [w, z] + [z, w].$$

Specializing to $w = 1, x = f$ and $y = g$ (the cycle is still admissible), we get the desired relation.

(4) Again, we give the proof for $n = 2$ only. Given $f \in R^\times$, we compute the sum of the boundaries of the following two cycles in $Z^2(R, 3)$:

$$\partial\left[t + f, -\frac{t}{f}, \frac{t - f}{t + 1}\right] = -[f, -f] + [2f, -1] - \left[f - 1, \frac{1}{f}\right],$$

$$\partial\left[t - 1, \frac{1}{t}, \frac{f - t}{1 - t}\right] = [-1, f] + \left[f - 1, \frac{1}{f}\right].$$

Therefore $[f, -f] = [2f, -1] + [-1, f] \in Z^2(R, 2)$. Now use (2) and (3) with $g = -1$ to obtain $[f, -f] = [2f, -1] - [f, -1] = [2, -1] = 0$ by (1). $\qquad \square$

This lemma provides us with a well defined map (even when R is not smooth or semilocal),

$$\varphi_n : K_n^M(R) \longrightarrow CH^n(R, n), \qquad n \geqslant 1.$$

Moreover we have the following property.

Lemma 5.5.9 *If R is smooth over k, the maps φ_n are compatible with products.*

Remark 5.5.10

(1) As the graph map φ_n is a local complete intersection, its construction is exactly the one that we get with the product structure in higher Chow groups.

(2) The map φ_*^R corresponds to the 'cycle map' cl_X^M of Levine, with $X = Spec(R)$, as defined in [25, 1.1.8.1, p. 298].

(3) *Graph map and Chern map.* The reader should be warned that φ_n is not the composition of the maps

$$K_n^M(R) \xrightarrow{\text{can}} K_n(R) \xrightarrow{c_{n,n}^{Chow}} CH^n(R, n),$$

where $c_{n,n}^{Chow}$ is the Chern map of Theorem 5.4.3 ([2] (see 7.3.2, p. 293 and Lemma 9.2, p. 297)). Indeed, according to Remark 5.5.7, the map 'can' is zero when R is a global field and $n \geq 5$, while on the other side φ_n is an isomorphism (and $K_n^M(R)$ is non-trivial [1]). Nevertheless, up to torsion, both maps coincide.

Proposition 5.5.11 *Let R be a smooth k-algebra. Then we have a well defined homomorphism of graded rings $\varphi_*^R : K_*^M(R) \longrightarrow CH^*(R, *)$. Moreover, if R' is any extension of R of k-algebras, we have, for all $n \geq 0$, the following commutative diagram*

$$
\begin{array}{ccc}
K_n^M(R) & \xrightarrow{\iota_{Milnor}} & K_n^M(R') \\
\varphi_n^R \downarrow & & \downarrow \varphi_n^{R'} \\
CH^n(R, n) & \xrightarrow{\iota_{Chow}} & CH^n(R', n).
\end{array}
$$

Here ι_{Milnor} and ι_{Chow} are the natural maps induced by the embedding $\iota : R \hookrightarrow R'$.

Proof This is a consequence of a diagram chase using Lemma 5.5.8. □

Proposition 5.5.12 *If R is any smooth k-algebra of geometric type, then φ_1 is an isomorphism. If moreover R is semilocal and each residue field contains at least six elements, then φ_2 is an isomorphism.*

We now state the main result.

Theorem 5.5.13 ([9, Theorem 3.4] and [35, 47] for the field case) *If R is a smooth (semi)local k-algebra of geometric type, with k infinite, then the map $\varphi_n : K_n^M(R) \longrightarrow CH^n(R, n)$ is surjective. If moreover $R = k$, then φ_n is an isomorphism, even if k is finite.*

As a direct consequence, we have the following two corollaries.

Corollary 5.5.14 *Let p be a prime number and set $q = p^\nu$. Then, for all $n \in \mathbb{N}$, with $n > 1$, the group $CH^n(\mathbb{F}_q, n)$ is trivial.*

Corollary 5.5.15 *Let R be a smooth (semi)local k-algebra of geometric type, with k infinite. Then, the graded ring $\bigoplus_{n \in \mathbb{N}} CH^n(R, n)$ is generated by its elements of degree one.*

Remark 5.5.16

(1) In the special case of finite fields, Totaro [47] also showed that the map is onto for $n > 1$, and as the Milnor K-theory of a finite field is trivial as long as $n > 1$, he could deduce the triviality of the involved groups.

(2) In the case where R is not smooth one can use the example of Srinivas [42] to show that Theorem 5.5.13 does not necessarily hold, even if R is a two-dimensional normal domain. In addition, examples of smooth, non-semilocal k-algebras of geometric type, hence with *'few units'*, are known where the map φ_n is not surjective even rationally.

(3) We notice that, as R is smooth, it is always UFD. Indeed, if R is local then, by the well known result of Auslander–Buchsbaum [30, Theorem 20.3, p. 163], the smoothness implies that R is a UFD. We deduce the general semilocal case by localization using Corollary 9.4 of [11, p. 40].

(4) *Why Milnor K-theory?* The fact that we have interest in Milnor K-theory is not mysterious and comes in part from Beilinson's conjectures (see [44]) and the fact that higher Chow groups are a candidate for motivic cohomology. We should see the Milnor K-theory as a simple presentation of the graded ring structure of $\bigoplus_{n \in \mathbb{N}} CH^n(R, n)$ and the fact that the product map $CH^1(R, 1)^{\otimes n} \to CH^n(R, n)$ is onto. This phenomenon also appears in other frameworks (e.g. in étale cohomology).

Thus, the above theorem also shows that in general $CH^n(F, n)$ for an infinite field is non-trivial (for instance if F is a number field or more generally a global field [1]).

Some general results for higher Chow groups with coefficients

The higher Chow groups with coefficients have a nice relationship with the étale cohomology groups. The following result was first proved by Suslin in characteristic 0 and later generalized to an arbitrary characteristic by Geisser.

Theorem 5.5.17 (*Higher Chow groups and étale cohomology*) [16, Theorem 3.6] *Let X be an (equidimensional) quasi-projective smooth scheme over an algebraically closed field k, and let $p \geqslant d = dim(X)$. Then for any m prime* *to the characteristic of k, we have,*

$$CH^p(X, n; \mathbb{Z}/m\mathbb{Z}) \cong H_{\acute{e}t}^{2p-n}(X, \mathbb{Z}/m(p)).$$

In the case of fields, there is a more precise result.

Theorem 5.5.18 [17, Theorem 7.1] *Let k be a field of characteristic p, then for $i \neq n$ (and any positive integer r),*

$$CH^n(k, 2n - i; \mathbb{Z}/p^r\mathbb{Z}) = 0.$$

In particular, $CH^n(k, 2n - i)$ is uniquely p-divisible for $i \neq n$.

Remark 5.5.19 Let X be a smooth variety over a field k and ℓ a prime number distinct from $char(k)$. Using some techniques from K-cohomology (see [43, Corollary 23.4]), we can obtain the exact sequence

$$0 \to CH^2(X, 2)/\ell^r \to H_{\acute{e}t}^2(X, \mathbb{Z}/\ell^r(2)) \to {}_{\ell^r}CH^2(X, 1) \to 0.$$

By comparison with the universal coefficients sequence, we get an isomorphism

$$CH^2(X, 2; \mathbb{Z}/\ell^r\mathbb{Z}) \cong H_{\acute{e}t}^2(X, \mathbb{Z}/\ell^r(2)),$$

without assuming that k is algebraically closed. Applying this to $X = Spec(R)$ for R a smooth semilocal k-algebra of finite type, we get

$$CH^2(X, 2)/\ell^n \cong H_{\acute{e}t}^2(X, \mathbb{Z}/\ell^r(2)).$$

In other words, assuming k with 'enough elements', we get a direct proof of the 'generalized Kato conjecture' for $K_2^M(R)$.

Revisiting the motivic spectral sequence

Now using the previous result we can compute some higher Chow groups in 'low degree'. Let R be a smooth semilocal k-algebra of geometric type. The first non-trivial part of the motivic spectral sequence gives the following exact sequence:

$$K_4(R) \to CH^2(R, 4) \to CH^3(R, 3) \to K_3(R) \to CH^2(R, 3) \to 0. \quad (1)$$

On the other hand we have the exact sequence:

$$K_3^M(R) \to K_3(R) \to K_3(R)_{ind} \to 0,$$

where the map $K_3^M(R) \to K_3(R)$ is induced by the product structure in both theories and $K_3(R)_{ind}$ is the cokernel of the previous map. Furthermore, by the 'Gersten resolution', the map $CH^3(R, 3) \to CH^3(Frac(R), 3)$ is injective, and we know that $K_3^M(Frac(R)) \cong CH^3(Frac(R), 3)$ if k is infinite. By the result of Kahn and Levine [19, 27], we know that the map $K_3^M(Frac(R)) \to K_3(Frac(R))$ is injective. By a diagram chase, this shows that $CH^3(R, 3) \to K_3(R)$ is injective. Hence, by the surjectivity of φ_3 we have the isomorphism $K_3(R)_{ind} \cong CH^2(R, 3)$ and the surjectivity of the map $K_4(R) \to CH^2(R, 4)$. If, moreover, R is a PID, by [9 (using [46])], we have an isomorphism $K_3^M(R) \cong CH^3(R, 3)$.

If, on the other hand, we suppose that $R = \mathbb{F}_q$ is an arbitrary finite field, in (1), we then get $CH^2(\mathbb{F}_q, 4) = 0$ (since $K_4(\mathbb{F}_q) = 0$) and $\mathbb{Z}/(q^2 - 1) \cong K_3(\mathbb{F}_q)_{ind} \cong CH^2(\mathbb{F}_q, 3)$. Notice that by using the 'universal coefficients' this shows that $CH^2(\mathbb{F}_q, 4; \mathbb{Z}/m\mathbb{Z}) \neq 0$ if m divides $q^2 - 1$. Furthermore, as $K_4(\mathbb{F}_q) = 0$, the abutment of the spectral sequence gives $CH^3(\mathbb{F}_q, 4) = 0$. Summarizing the results, we have:

Proposition 5.5.20 *Let R be a smooth semilocal k-algebra of geometric type.*

(1) *If k is infinite, the morphism $K_4(R) \to CH^2(R, 4)$ is onto, and we have an isomorphism $K_3(R)_{ind} \cong CH^2(R, 3)$. If R is PID, then $K_3^M(R) \cong CH^3(R, 3)$.*

(2) *If $R = \mathbb{F}_q$ is an arbitrary finite field, we have $CH^p(\mathbb{F}_q, 4) = 0$ for all $p \in \mathbb{N}$, and $CH^2(\mathbb{F}_q, 3) \cong \mathbb{Z}/(q^2 - 1)$.*

As a consequence of Applications 5.3.4 we also have $CH^2(\mathbb{F}_q(t_1, \ldots, t_m), 4) = 0$ for all $m \in \mathbb{N} - \{0\}$.

Remark 5.5.21 *Higher Chow groups of finite fields and values of zeta functions*: another approach to the understanding of higher Chow groups of finite fields is via a conjectural relationship between values of zeta functions of algebraic varieties over finite fields and their higher Chow groups. The conjecture (see Conjecture 1 in [41, p. 259]) states that: *if X is a smooth and proper scheme of finite type over a finite field with $d = dim(X)$, the following identity holds for any integer $i > d$*

$$|\zeta_X(d - i)| = \prod_{n=0}^{2d+1} \#CH^i(X, 2i - n)^{(-1)^n}.$$

In particular, all groups $CH^i(X, 2i - n)$ are finite if $0 \leqslant n \leqslant 2d + 1$.

Further reading

Due to lack of time, several aspects of higher Chow groups were left out. The reader interested in more details on the K-cohomologies side could read the last sections of [9], the survey article [31] and also [21, 43]. For the relationship between higher Chow groups and Deligne cohomology a start is the survey paper [32]. For further details on motivic cohomology, higher Chow groups and étale cohomology, one can read the survey [44]. The identification between motivic cohomology 'à la Voevodsky' and higher Chow groups is given in [48]. You can find different kinds of explicit computations in higher Chow groups, related to polylogarithms, in [15]. For a more 'commutative algebra oriented' point of view (and some motivations) see the expository paper [22].

Acknowledgements

I am grateful to S. Müller-Stach and C. Peters for their invitation to the 'Lecture Summer School, Grenoble 2001: Transcendental Aspects of Algebraic Cycles', and to the participants for their comments on an early version of this document. I am also very grateful to H. Gangl for his suggestions and to the referee who helped me to improve the paper.

References

[1] Bass, H. and Tate, J. The Milnor ring of a global field, *Algebraic K-Theory II, Proc. Conf. Battelle Inst. 1972, Lecture Notes Math.* **342** (1973) 349–446.

[2] Bloch, S. Algebraic cycles and higher K-theory, *Adv. Math.* **61** (1986) 267–304.

[3] — The moving lemma for higher Chow groups, *J. Algebraic Geom.* **3** (1994) 537–568.

[4] Bloch, S. and Lichtenbaum, S. A spectral sequence for motivic cohomology, preprint (K-theory Preprint Archives 62, 1995 March 3).

[5] Borel, A. Stable real cohomology of arithmetic groups, *Ann. Sci. Ec. Norm. Super. Ser. 4* **7** (1974) 235–272.

[6] Colliot-Thélène, J. L. and Ojanguren, M. Espaces homogènes principaux, *Publ. Math. Inst. Hautes Etudes Sci.* **75** (1992) 97–122.

[7] Colliot-Thélène, J. L., Hoobler, R. T. and Kahn, B. The Bloch–Ogus–Gabber theorem. In: *Algebraic K-theory* (V. Snaith ed.), *Fields Institute Communications* **16**, pp. 31–94, AMS, Providence, RI, 1997.

[8] Esnault, H. Some examples of computation of a regulator map on singular varieties. In: *Algebra, Proc. Int. Conf. in Memory of A. I. Mal'cev, Novosibirsk, USSR 1989, Contemp. Math.* **131** (3) (1992) 399–417.

[9] Elbaz-Vincent, Ph. and Müller-Stach, S. Milnor K-theory of rings, higher Chow groups and applications, *Invent. Math.* **148** (1) (2002) 177–206.

[10] Elbaz-Vincent, Ph., Gangl, H. and Soulé, C. Quelques calculs de la cohomologie de $GL_N(\mathbb{Z})$ et de la K-théorie de $\mathbb{Z}$, *C. R. Acad. Sci. Paris, Ser. I* **335** (2002) 321–324.

[11] Fossum, R. M. *The Divisor Class Group of a Krull Domain, Ergebnisse der Mathematik und ihrer Grenzgebiete* **74**, Springer-Verlag, Berlin, 1973.

[12] Friedlander, E. and Suslin, A. The spectral sequence relating algebraic K-theory to motivic cohomology, preprint (K-theory Preprint Archives 360, 1999 August 15).

[13] Fulton, W. *Introduction to Intersection Theory in Algebraic Geometry, CBMS Regional Conf.* 54, AMS, Providence, RI, 1996.

[14] *Intersection Theory*, 2nd edn, Springer-Verlag, 1998.

[15] Gangl, H. and Müller-Stach, S. Polylogarithmic identities in cubical higher Chow groups. In: *Algebraic K-Theory*, *Proc. AMS-IMS-SIAM Summer Research Conf.*, *Seattle, WA, July 13–24, 1997, Proc. Symp. Pure Math.* 67, pp. 25–40 American Mathematical Society, Providence, RI, 1999.

[16] Geisser, Th. Applications of de Jong's theorem on alterations, In: *Resolution of Singularities. A Research Textbook in Tribute to Oscar Zariski* (H. Hauser ed.), *Prog. Math.* 181, pp. 299–314, Birkhauser, Basel, 2000.

[17] Geisser, Th. and Levine, M. The K-theory of fields in characteristic p, *Invent. Math.* 139 (3) (2000) 459–493.

[18] Guin, D. Homologie du groupe linéaire et K-théorie de Milnor des anneaux, *J. Algebra* 123 (1989) 27–59.

[19] Kahn, B. K-theory of semi-local rings with finite coefficients and étale cohomology, preprint (K-theory Preprint Archives 537, 2002 January 10).

[20] van der Kallen, W. The K_2 of rings with many units, *Ann. Sci. Ec. Norm. Super.* 10 (1977) 473–515.

[21] Landsburg, S. Relative Chow groups III, *J. Math.* 35 (4) (1991) 618–641.

[22] Some new K-theoretic invariants for commutative rings. *Methods in Module Theory* (G. Abrams *et al.* eds.), *Lecture Notes Pure Appl. Math.* 140, pp. 175–202, Marcel Dekker, New York, 1993.

[23] Levine, M. Bloch's higher Chow groups revisited, *Proc. Congress of K-Theory*, *Strasbourg 1992*, *Astérisque* 226 (1994) 235–320.

[24] Lambda operations, K-theory and motivic cohomology. In: *Algebraic K-Theory* (V. Snaith ed.), *Fields Institute Communications* 16, pp. 131–184, AMS, Providence, RI, 1997.

[25] *Mixed Motives, Mathematical Surveys and Monographs* 57, AMS, Providence, RI, 1998.

[26] Techniques of localization in the theory of algebraic cycles, preprint (K-theory Preprint Archives 335, 1999 February 26).

[27] K-Theory and motivic cohomology of schemes, preprint (K-theory Preprint Archives 336, 1999 February 26).

[28] Blowing up monomial ideals, preprint (K-theory Preprint Archives 334, 1999 February 26).

[29] Lluis-Puebla, E., Loday, J-L., Gillet, H., Soulé, C. and Snaith, V. *Higher Algebraic K-Theory: an Overview, Lecture Notes in Mathematics* 1491, Springer-Verlag, Berlin, 1992.

[30] Matsumura, H. *Commutative Ring Theory, Cambridge Studies in Advanced Mathematics* 8, Cambridge University Press, 1986.

[31] Müller-Stach, S. Algebraic cycle complexes: basic properties. In: *Proc. 1998 Banff NATO Conf. on the Arithmetic of Algebraic Cycles* (B. B. Gordon, J. D. Lewis, S. Müller-Stach, S. Saito and N. Yui, eds.), *Nato ASI Science Series* 548, pp. 285–305, Kluwer, Dordrecht, 2000.

[32]   Algebraic cycles and Hodge theory, preprint, 2002 (see http://www.uni-essen.de/~mat930).

[33] Musikhin, A. and Suslin, A. A. Triviality of the higher Chern classes in the *K*-theory of global fields, *Algebraic K-Theory, Adv. Sov. Math.* **4** (1991) 145–154.

[34] Nart, E. The Bloch complex in codimension one and arithmetic duality, *Number Theory* **32** (1989) 321–331.

[35] Nesterenko, Y. P. and Suslin, A. A. Homology of the full linear group over a local ring and Milnor's *K*-theory, *Math. USSR Izv.* **34** (1) (1990) 121–145.

[36] Quillen, D. On the cohomology and *K*-theory of the general linear group over a finite field, *Ann. Math.* **96** (1972) 552–586.

[37] Rognes, J. $K_4(\mathbb{Z})$ is the trivial group, *Topology* **39** (2000) 267–281.

[38] Rognes, J. and Weibel, C. Two-primary algebraic *K*-theory of rings of integers in number fields (with an appendix by M. Kolster), *J. Am. Math. Soc.* **13** (1) (2000) 1–54.

[39] Rosenberg, J. *Algebraic K-Theory and its Applications*, GTM **147**, Springer, 1994.

[40] Shapiro, J. M. Relations between the Milnor and Quillen *K*-theory of fields, *J. Pure Appl. Algebra* **20** (1981) 93–102.

[41] Soulé, C. *K*-theory and values of zeta functions. In: *Algebraic K-Theory and its Applications, Proc. Workshop and Symp., ICTP, Trieste, Italy, September 1–19, 1997*, pp. 255–283, World Scientific, Singapore, 1999.

[42] Srinivas, V. K_1 of the cone over a curve, *J. reine angew. Math.* **381** (1987) 37–50.

[43] Suslin, A. A. Algebraic *K*-theory and the norm-residue homomorphism, *J. Sov. Math.* **30** (1985) 2556–2611.

[44]   *Algebraic K-Theory and Motivic Cohomology, Proc. ICM 1994*, Birkhäuser, Basel, 1995.

[45]   On the Grayson spectral sequence, preprint (*K*-theory Preprint Archives 588, 2002 August 19).

[46] Suslin, A.A. and Yarosh, Y. Milnor's K_3 of a discrete valuation ring. In: *Algebraic K-Theory* (A.A. Suslin ed.), *Adv. Sov. Math.* **4**, pp. 155–170, AMS, Providence, RI, 1991.

[47] Totaro, B., Milnor *K*-theory is the simplest part of algebraic *K*-theory, *K-theory* **6** (1992) 177–189.

[48] Voevodsky, V. Motivic cohomology groups are isomorphic to higher Chow groups, preprint (*K*-theory Preprint Archives 378, 1999 December 17).

[49]   Cohomological theory of presheaves with transfer. In: *Cycles, Transfers, and Motivic Homology Theories* (V. Voevodsky, A. Suslin and E. Friedlander), AM **143**, Princeton University Press, 2000.

Part IV

Hodge theoretic invariants of cycles

6

Three lectures on the Hodge conjecture

James D. Lewis

Department of Mathematics, University of Alberta, Edmonton, Alberta,
Canada T6G 2G1

Abstract

The statement of the Hodge conjecture for projective algebraic manifolds is presented in its classical form, as well as the general (Grothendieck amended) version. The intent of these lectures is to focus on some specific examples, rather than present a general survey overview, as can be found in [Lew2] and [Shi]. A number of exercises for the reader are sprinkled throughout the lectures. For background material, the reader is assumed to have some familiarity with the geometry of complex manifolds, such as can be found in chapter 0 of [G-H1].

Keywords: Hodge conjecture, normal function, Abel–Jacobi map, algebraic cycle.
1991 Mathematics subject classification: 14C30, 14C25

6.1 Lecture 1: the statement and some standard examples

6.1.1 Some preliminary material

Let $\mathbb{P}^N = \{\mathbb{C}^{N+1}\backslash\{0\}\}/\mathbb{C}^\times$ be "a" complex projective N-space. A projective algebraic manifold X is a closed embedded submanifold of $\mathbb{P}^N$. By a theorem of Chow, X is cut out by the zeros of a finite number of homogeneous polynomials, satisfying a certain jacobian criterion (so that $X \subset \mathbb{P}^N$ is indeed smooth). The fact that X is projective algebraic implies that X contains 'plenty' of subvarieties. Let $z^k(X)$ be the free abelian group generated by (irreducible) subvarieties of codimension k in X. If $\dim X = n$, then $z^k(X) = z_{n-k}(X)$, the group generated by dimension $n - k$ subvarieties of X. For example, $z_0(X) = \{\sum_{i=1}^r n_i p_i \mid p_1, \ldots, p_r \in X, \ r \geq 1, \ n_i \in \mathbb{Z}\}$ = group of 0-cycles. Now let $E_X^k = \mathbb{C}$-valued C^∞ k-forms on X. We have the decomposition:

$$E_X^k = \bigoplus_{p+q=k} E_X^{p,q}, \qquad \overline{E_X^{p,q}} = E_X^{q,p}, \tag{1.0}$$

Transcendental Aspects of Algebraic Cycles ed. S. Müller-Stach and C. Peters.
© Cambridge University Press 2004.

where $E_X^{p,q}$ are the C^∞ (p,q)-forms which in local holomorphic coordinates $z = (z_1, \ldots, z_n) \in X$, are of the form:

$$\sum\nolimits_{|I|=p, |J|=q} f_{IJ} dz_I \wedge d\bar{z}_J, \qquad \begin{aligned} & I = 1 \le i_1 < \cdots < i_p \le n, \\ & J = 1 \le j_1 < \cdots < j_q \le n \\ & dz_I = dz_{i_1} \wedge \cdots \wedge dz_{i_p}, \\ & d\bar{z}_J = d\bar{z}_{j_1} \wedge \cdots \wedge d\bar{z}_{j_q}. \end{aligned}$$

The differential $d : E_X^k \to E_X^{k+1}$ splits into $d = \partial + \bar{\partial}$, where $\bar{\partial} E_X^{p,q} \subset E_X^{p,q+1}$, $\partial E_X^{p,q} \subset E_X^{p+1,q}$. Since $d^2 = 0$, by Hodge type considerations, $0 = \bar{\partial}^2 = \partial^2 = \partial\bar{\partial} + \bar{\partial}\partial$. The decomposition in (1.0) descends to the cohomological level:

Theorem 1.1 (Hodge decomposition)

$$H_{\text{sing}}^k(X, \mathbb{Q}) \otimes_{\mathbb{Q}} \mathbb{C} \underset{\substack{\text{de Rham} \\ \text{isomorphism}}}{\simeq} H_{\text{DR}}^k(X, \mathbb{C}) = \bigoplus\nolimits_{p+q=k} H^{p,q}(X),$$

$$\|$$

$$\frac{\ker d : E_X^k \to E_X^{k+1}}{dE_X^{k-1}}$$

where $H^{p,q}(X) = d$-*closed* (p,q)-*forms (modulo coboundaries), and* $\overline{H^{p,q}(X)} = H^{q,p}(X)$. *Moreover, all such cohomology groups are finite dimensional. Furthermore,*[1]

$$H^{p,q}(X) \simeq \frac{E_{X,d-\text{closed}}^{p,q}}{\partial\bar{\partial} E_X^{p-1,q-1}}.$$

Now recall dim $X = n$.

Theorem 1.2 (Poincaré and Serre duality) *The following pairings induced by*

$$(w_1, w_2) \mapsto \int_X w_1 \wedge w_2,$$

---

[1] For the reader's convenience, we prove the latter statement here, assuming some knowledge of the Hodge–Kähler identities. Up to a factor of 2, the laplacians Δ_d, Δ_∂, $\Delta_{\bar{\partial}}$ agree, and hence define the same harmonic space. The Green operator G commutes with any operator that commutes with the laplacian. Now suppose $\omega \in E_X^{p,q}$ is an exact form (coboundary), and let $(\ )^*$ stand for adjoint. Then by Hodge type, $\partial\omega = \bar{\partial}\omega = 0$, and $\omega = \bar{\partial}^* \bar{\partial} G_{\bar{\partial}}(\omega) + \bar{\partial}\bar{\partial}^* G_{\bar{\partial}}(\omega)$. But $\bar{\partial} G_{\bar{\partial}}(\omega) = G_{\bar{\partial}}(\bar{\partial}\omega) = 0$. Thus if we set $\eta = \bar{\partial}^* G_{\bar{\partial}}(\omega) \in E_X^{p,q-1}$, we have $\omega = \bar{\partial}\eta$. Next, by the same reasoning, $\omega = \partial\partial^* G_\partial(\omega) = \partial\partial^* G_\partial(\bar{\partial}\eta) = \partial\partial^*\bar{\partial} G_\partial(\eta)$. Now one uses the fact that $[\partial^*, \bar{\partial}] = 0$.

are non-degenerate:

$$H_{DR}^k(X, \mathbb{C}) \times H_{DR}^{2n-k}(X, \mathbb{C}) \to \mathbb{C},$$

$$H^{p,q}(X) \times H^{n-p,n-q}(X) \to \mathbb{C}.$$

Therefore $H^k(X) \simeq H^{2n-k}(X)^\vee$, $H^{p,q}(X) \simeq H^{n-p,n-q}(X)^\vee$.

6.1.2 The cycle class map: the fundamental class

$$\mathrm{cl}_k : z^k(X) \to H_{DR}^{2k}(X, \mathbb{C}) \simeq H_{DR}^{2n-2k}(X, \mathbb{C})^\vee.$$

Let $V \subset X$ be a subvariety of codimension k in X, $\{w\} \in H_{DR}^{2n-2k}(X, \mathbb{C})$. Define $\mathrm{cl}_k(V)(w) = \int_{V^*} w$, and extend to $z^k(X)$ by linearity. (Note that $\dim_{\mathbb{R}} V = 2n - 2k$, where $V^* = V \setminus V_{\mathrm{sing}}$.) We must show that $\int_{V^*} w$ has finite volume, and that the corresponding current is closed, i.e. descends to the cohomology level. All this is essentially a result of $\mathrm{codim}_{V,\mathbb{R}} V_{\mathrm{sing}} \geq 2$; but we can argue differently as follows. One can construct a desingularization $\sigma : \tilde{V} \xrightarrow{\approx} V$, where say $\sigma^{-1}(V_{\mathrm{sing}})$ is a divisor with normal crossings (locally a cut-out of $\tilde{V}$ by $z_1 \cdots z_\ell = 0$). It is obvious then that via $\tilde{V} \to V \hookrightarrow X$, $\sigma^* w$ is a C^∞ form on $\tilde{V}$, and therefore

$$\int_{V^*} w = \int_{\tilde{V}} \sigma^* w$$

has finite volume, since $\tilde{V}$ is compact. Furthermore, if $w = d\eta$ on X, then

$$\int_{V^*} w = \int_{\tilde{V}} \sigma^* d\eta = \int_{\tilde{V}} d(\sigma^* \eta) = \int_{\partial \tilde{V} = 0} \sigma^* \eta = 0,$$

by Stokes' theorem. Suppose $\{w\} \in H^{p,q}(X)$, where $p + q = 2n - 2k$. Then $(p, q) \neq (n - k, n - k) \Rightarrow$ either $p > n - k$ or $q > n - k$. In either case $\int_{V^*} w = 0$, since $\dim V := \dim_{\mathbb{C}} V = n - k$.

Evidently, $\mathrm{cl}_k(z^k(X)) \subset H^{2k}(X, \mathbb{Z})$, thus $\mathrm{cl}_k(z^k(X)) \subset H^{k,k}(X) \cap H^{2k}(X, \mathbb{Z})$ from the above discussion. There are known examples where $\mathrm{cl}_k(z^k(X)) \neq H^{k,k}(X) \cap H^{2k}(X, \mathbb{Z})$, even modulo torsion (see Claim 1.4.2 below); however, the following is still open:

Hodge conjecture 1.3 Hodge$^{k,k}(X, \mathbb{Q})$:

$$\mathrm{cl}_k(z^k(X) \otimes_{\mathbb{Z}} \mathbb{Q}) = H^{k,k}(X, \mathbb{Q}) := H^{k,k}(X) \cap H^{2k}(X, \mathbb{Q}).$$

6.1.3 Some examples[2]

Example I: the Lefschetz $(1, 1)$ **theorem** Hodge$^{1,1}(X, \mathbb{Z})$

The (additive abelian) group $z^1(X)$ is called the group of Weil divisors on X. We introduce the (multiplicative abelian) group of Cartier divisors. Let $\mathbb{C}(X)$ be the rational function field of X, and $\mathbb{C}(X)^\times$ the multiplicative group. Further, let $\mathcal{D}_X = \mathbb{C}(X)^\times / \mathcal{O}_X^\times$, where $\mathcal{O}_X$ is the sheaf of germs of regular functions on X, and $\mathcal{O}_X^\times$ is the corresponding sheaf of nowhere vanishing regular functions. A section $D \in \Gamma(X, \mathcal{D}_X)$ is called a Cartier divisor. In terms of local data (and working in the Zariski topology),

$$D \in \Gamma(X, \mathcal{D}_X) \Leftrightarrow$$
$$D = \left\{ f_\alpha \in \Gamma(U_\alpha, \mathbb{C}(X)^\times) = \mathbb{C}(X)^\times \mid f_\alpha / f_\beta \in \Gamma(U_\alpha \cap U_\beta, \mathcal{O}_X^\times) \right\}.$$

There is a natural divisor map div $: \Gamma(X, \mathcal{D}_X) \to z^1(X)$, which in fact is an isomorphism between these groups. There are two short exact sequences:

$$0 \to \mathcal{O}_X^\times \to \mathbb{C}(X)^\times \to \mathcal{D}_X \to 0,$$

$$0 \to \mathbb{Z} \to \mathcal{O}_{X,\mathrm{an}} \xrightarrow{\exp(2\pi\sqrt{-1}\mathbb{Z})} \mathcal{O}_{X,\mathrm{an}}^\times \to 0,$$

where $\mathcal{O}_{X,\mathrm{an}}$ is the corresponding analytic sheaf.[3] The first sequence yields the short exact sequence:

$$\frac{\Gamma(X, \mathcal{D}_X)}{\Gamma(X, \mathbb{C}(X)^\times)} \simeq H^1(X, \mathcal{O}_X^\times) =: \mathrm{Pic}(X) = \text{ Picard group of } X.$$

(This uses the fact that $H^1(X, \mathbb{C}(X)^\times) = 0$ (use Čech cohomology).) The second sequence yields:

$$0 \to \frac{H^1(X, \mathcal{O}_{X,\mathrm{an}})}{H^1(X, \mathbb{Z})} \to H^1(X, \mathcal{O}_{X,\mathrm{an}}^\times) \xrightarrow{c} H^2(X, \mathbb{Z})$$
$$\to H^2(X, \mathcal{O}_{X,\mathrm{an}}) \to \cdots$$

where c is the first Chern class map, and where $H^1(X, \mathcal{O}_{X,\mathrm{an}}^\times)$ is interpreted as the group of isomorphism classes of holomorphic line bundles over X. The Dolbeault isomorphism theorem gives $H^q(X, \mathcal{O}_{X,\mathrm{an}}) \simeq H^{0,q}(X)$, and by a comparison theorem between analytic and algebraic data (use Chow's theorem or

---

[2] We assume that the reader is familiar with the cohomology of $\mathbb{P}^n$, and more generally the fact that the cohomology of a Grassmannian manifold ($\{\mathbb{C}^k \subset \mathbb{C}^n\}$) is generated by algebraic cocycles.

[3] It is customary to write this as $0 \to \mathbb{Z}(1) \to \mathcal{O}_{X,\mathrm{an}} \xrightarrow{\exp} \mathcal{O}_{X,\mathrm{an}}^\times \to 0$, where $\mathbb{Z}(1) = 2\pi\sqrt{-1}\mathbb{Z}$. This is because there is no canonical choice of $\sqrt{-1}$.

Serre's GAGA), $H^1(X, \mathcal{O}_{X,\mathrm{an}}^\times) = H^1(X, \mathcal{O}_X^\times)$. Putting it all together, there is a diagram:

$$\mathrm{Pic}(X) \xrightarrow{c} H^2(X, \mathbb{Z}) \to H^{0,2}(X)$$

$$\begin{array}{ccccc} \uparrow & \mathrm{cl}_1 \nearrow & & \downarrow & \nearrow \mathrm{Pr}_{0,2} \\ z^1(X) & & H^2(X, \mathbb{C}). & & \end{array}$$

This shows that $\mathrm{cl}_1(z^1(X)) = \{\eta \in H^2(X, \mathbb{Z}) \mid \eta \in H^{1,1}(X)\}$, i.e. the famous Lefschetz $(1, 1)$ theorem holds.

Exercise Recall the following theorems:

Lefschetz theorems 1.4

(i) *(Weak) Let $j : Y \hookrightarrow X$ be a smooth hyperplane section of a smooth X. Then $j_* : H_i(Y, \mathbb{Z}) \to H_i(X, \mathbb{Z})$ is an isomorphism for $i < \dim Y$, and surjective for $i = \dim Y$.*

(ii) *(Hard) Let $L_X : H^i(X, \mathbb{Q}) \to H^{i+2}(X, \mathbb{Q})$ be the operator defined by cupping with the first Chern class of the hyperplane bundle over X, i.e. the cohomology class of a hyperplane section of X. Then for $i \leq n := \dim X$,*

$$L_X^{n-i} : H^i(X, \mathbb{Q}) \xrightarrow{\sim} H^{2n-i}(X, \mathbb{Q})$$

is an isomorphism.

Deduce from this that for $2k \leq n$, $\mathrm{Hodge}^{k,k}(X, \mathbb{Q}) \Rightarrow H^{n-k,n-k}(X, \mathbb{Q})$, where $n = \dim X$. Further deduce that the Hodge conjecture is true for all X of dimension ≤ 3.

Griffiths and Harris introduced a series of conjectures in [G-H2], the weakest of which is the following:

Conjecture 1.4.1 *Let $X \subset \mathbb{P}^4$ be a sufficiently general[4] threefold of degree $d \geq 6$. Then for any curve $C \subset X$, we have $d \mid \deg C$.*

---

[4] General in this case meaning in a transcendental sense, more specifically, where X corresponds to a point in the complement of a *countable* union of proper subvarieties, of the projective space of threefolds in $\mathbb{P}^4$ of a given degree.

Claim 1.4.2 *Assume that Conjecture* 1.4.1 *holds,*[5] *and let* X *be given as in* 1.4.1. *If* β *is a generator for* $H^{2,2}(X, \mathbb{Z}) \simeq \mathbb{Z}$, *then* β *is non-algebraic.*

Proof For our purposes the degree of C can be defined as follows. The image of the fundamental class $\{C\}$ in the map $H_2(C, \mathbb{Z}) \to H_2(\mathbb{P}^4, \mathbb{Z}) = \mathbb{Z} \cdot \mathbb{P}^1$, induced by inclusion, determines a cycle $k \cdot \mathbb{P}^1$ for some integer $k \geq 1$. We define $\deg(C) = k$. The weak Lefschetz theorem in our situation implies that the inclusion map $j : X \hookrightarrow \mathbb{P}^4$ induces an isomorphism $j_* : H_2(X, \mathbb{Z}) \xrightarrow{\sim} H_2(\mathbb{P}^4, \mathbb{Z})$. From Poincaré duality, we deduce that $H_2(X, \mathbb{Z}) \simeq H^4(X, \mathbb{Z}) = H^{2,2}(X, \mathbb{Z})$, i.e. every class in $H_2(X, \mathbb{Z})$ is non-torsion and has Poincaré dual of type $(2, 2)$. Now let us assume the above conjecture is answered in the affirmative. Then $j_*(H_{2,\text{alg}}(X, \mathbb{Z})) = d \cdot H_2(\mathbb{P}^4, \mathbb{Z})$ $(d \geq 6)$, and therefore if β is a generator for $H_2(X, \mathbb{Z})$, then $[\beta] \in H^{2,2}(X, \mathbb{Z})$ and yet $[\beta] \notin H^4_{\text{alg}}(X, \mathbb{Z})$.

Example II: projective bundles and blow-ups

Recall that $\mathbb{P}^N$ describes the family of 1-dimensional subspaces in $\mathbb{C}^{N+1}$, and thus there is an associated tautological (universal) line bundle over $\mathbb{P}^N$, that associates with each point $p \in \mathbb{P}^N$, the corresponding line in $\mathbb{C}^{N+1}$. Let $W \to X$ be a vector bundle of rank r on X, and $\pi : \mathbb{P}[W] \to X$ the corresponding projective bundle. The cohomology of $\mathbb{P}[W]$ is well known, in terms of some Chern class invariants associated to W, and the corresponding tautological line bundle over $\mathbb{P}[W]$. In particular, if μ is the (first) Chern class of the tautological line bundle over $\mathbb{P}[W]$, then identifying $H^\bullet(X)$ with $\pi^* H^\bullet(X)$, we have:

$$H^\bullet(\mathbb{P}[W], \mathbb{Q}) \simeq \frac{H^\bullet(X, \mathbb{Q})[\mu]}{\left(\mu^r - c_1(W)\mu^{r-1} + \cdots + (-1)^r c_r(W)\right)}$$

$$\simeq H^\bullet(X, \mathbb{Q}) \oplus H^{\bullet-2}(X, \mathbb{Q}) \wedge \mu \oplus \cdots \oplus H^{\bullet-2(r-1)}(X, \mathbb{Q}) \wedge \mu^{r-1}.$$

Now let $D \subset X$ be a smooth subvariety of codimension $r \geq 2$ in X, and $N(D/X)$ the corresponding normal bundle (of rank r). Further, let $E = \mathbb{P}[N(D/X)]$, and $\mu \in H^{1,1}(E, \mathbb{Z})$ be the first Chern class of the tautological line bundle over E. Then for example $H^{j-2}(E, \mathbb{Q}) \simeq H^{j-2}(D, \mathbb{Q}) \oplus H^{j-4}(D, \mathbb{Q}) \wedge \mu \oplus \cdots \oplus H^{j-2r}(D, \mathbb{Q}) \wedge \mu^{r-1}$, where $H^*(D, \mathbb{Q})$ is identified with its image $\pi^* : H^*(D, \mathbb{Q}) \to H^*(E, \mathbb{Q})$. Now set $\mathbf{V} = H^{j-2}(D, \mathbb{Q}) \oplus H^{j-4}(D, \mathbb{Q}) \wedge \mu \oplus \cdots \oplus H^{j-2r+2}(D, \mathbb{Q}) \wedge \mu^{r-2}$. Using the fact that the fibers of $\pi : E \to D$ are positive dimensional $(\simeq \mathbb{P}^{r-1}$, recall $r \geq 2)$ and working on the cycle level via Poincaré duality, it follows that cycles represented

---

[5] According to [Ko], the partial results obtained there in the direction of the above conjecture lead to a counterexample to the statement of the Hodge conjecture with $\mathbb{Z}$-coefficients for certain hypersurface threefolds.

by cocyles in $\mathbf{V}$ map to cycles of strictly lower dimension in D. Therefore $\pi_*(\mathbf{V}) = 0$ where $\pi_* : H^{j-2}(E, \mathbb{Q}) \to H^{j-2r}(D, \mathbb{Q})$; moreover since $H^{j-2}(E, \mathbb{Q}) = \mathbf{V} \oplus H^{j-2r}(D, \mathbb{Q}) \wedge \mu^{r-1}$, it follows for dimension reasons that $\pi_* : H^{j-2r}(D, \mathbb{Q}) \wedge \mu^{r-1} \xrightarrow{\simeq} H^{j-2r}(D, \mathbb{Q})$ is an isomorphism; moreover one has the Gysin map $H^{j-2r}(D, \mathbb{Q}) \to H^j(X, \mathbb{Q})$, induced by inclusion. It is easy to verify that

$$H^j(B_D(X), \mathbb{Q}) \simeq \mathbf{V} \oplus H^j(X, \mathbb{Q}).$$

Exercise Let $h : X_1 \to X_2$ be a dominating rational map between fourfolds. Show that $\mathrm{Hodge}^{2,2}(X_1, \mathbb{Q})$ holds $\Rightarrow \mathrm{Hodge}^{2,2}(X_2, \mathbb{Q})$ holds. Hint: First, if h is a morphism, then use the fact that $h_* \circ h^* = \times d$, where d is the degree of h. To reduce to the case of a morphism, use the above results on blow-ups after constructing a 'Hironaka house':

$$
\begin{array}{c}
X_{1,N} \\
\wr \downarrow \tilde{f}_N \\
X_{1,N-1} \\
\wr \downarrow \tilde{f}_{N-1} \\
\vdots \\
\wr \downarrow \qquad\qquad \searrow^{\tilde{f}} \\
\vdots \\
\wr \downarrow \tilde{f}_2 \\
X_{1,1} \\
\wr \downarrow \tilde{f}_1 \\
X_{1,0} \quad = \quad X_1 \xrightarrow{\;h\;} X_2
\end{array}
$$

where (i) $\tilde{f}_j : X_{1,j} \xrightarrow{\approx} X_{1,j-1}$ is a blow-up along a non-singular centre, i.e. $X_{1,j} = B_{D_j}(X_{1,j-1})$ where D_j is a smooth variety of codimension ≥ 2 in $X_{1,j-1}$ and $\tilde{f}_j$ is the corresponding blow-down morphism; (ii) the above diagram is commutative and $\tilde{f} : X_{1,N} \to X_2$ is a morphism (i.e. regular).

Example III: uniruled fourfolds [C-M1]

A variety X is uniruled if it is covered by a family $\{C_b\}_{b \in \Omega}$ of rational curves. One can assume that $\dim \Omega = \dim X - 1$, where Ω is irreducible. If η is the generic point of Ω, then there is a finite field extension L over $\mathbb{C}(\eta)$ such that $C_\eta \times L \simeq \mathbb{P}^1(L)$. This translates to a generically finite to one dominating rational map $h : \Omega' \times \mathbb{P}^1 \to X$ where Ω' is smooth.[6]

---

[6] A *geometric* proof of this fact, that gives a fairly explicit description of such an h, is given in [Lew2, pp. 216–218].

Exercise   Show that if $\dim \Omega' = 3$, then $\text{Hodge}^{2,2}(\Omega' \times \mathbb{P}^1, \mathbb{Q})$ holds.

Note that the Hodge conjecture is true through dimension ≤ 3; moreover by a sequence of blow-ups, we can assume that h is a morphism. Thus $\text{Hodge}^{2,2}(\Omega' \times \mathbb{P}^1, \mathbb{Q}) \Rightarrow \text{Hodge}^{2,2}(X, \mathbb{Q})$. This result includes unirational fourfolds (also see [Mu1]). For example, Conte and Murre show that a smooth complete intersection fourfold with $H^{4,0}(X) = 0$ is uniruled.

Exercise   Show that a smooth projective uniruled variety X of dimension n satisfies $H^{n,0}(X) = 0$.

6.1.4 The general Hodge conjecture

Grothendieck [Gro] was the first to introduce the notion of the coniveau filtration.

Proposition-Definition 1.5   *The (descending) filtration by coniveau on singular cohomology is given by any of the three equivalent definitions below:*

$$N^p H^i(X, \mathbb{Q}) := \ker\left(H^i(X, \mathbb{Q}) \to \varinjlim_{cd_X Y \geq p} H^i(X \backslash Y, \mathbb{Q}) \right)$$

$$:= \text{Image}\left(\sum_{cd_X Y \geq p} H^i_Y(X, \mathbb{Q}) \to H^i(X, \mathbb{Q}) \right)$$

$$:= \text{Gysin Images}\left(\sum_{cd_X Y = r \geq p} H^{i-2r}(\tilde{Y}, \mathbb{Q}) \to H^i(X, \mathbb{Q}) \right),$$

where $\tilde{Y} \to Y$ is a desingularization. (The third characterization follows from the work of Deligne [De, Proposition 8.2.7, Corollary 8.2.8].)

Recall that a Hodge structure of weight $m \in \mathbb{Z}$ is a finite-dimensional vector space $H = H_{\mathbb{Q}}$ with a decomposition $H \otimes \mathbb{C} = \oplus_{p+q=m} H^{p,q}$, where $\overline{H^{p,q}} = H^{q,p}$. It is easy to verify that $N^p H^i(X, \mathbb{Q}) \subset F^p H^i(X, \mathbb{C})$ (exercise!); however, as Grothendieck pointed out [Gro], the inclusion $N^p H^i(X, \mathbb{Q}) \subset F^p H^i(X, \mathbb{C}) \cap H^i(X, \mathbb{Q})$ is not in general an equality, since the left-hand side is a Hodge substructure of $H^i(X, \mathbb{Q})$ by the third characterization of $N^p H^i(X, \mathbb{Q})$ above, whereas the right-hand side need not be (see Section 6.1.8). Let $F^p_H H^i(X, \mathbb{Q})$ be the maximum Hodge substructure of $F^p H^i(X, \mathbb{C}) \cap H^i(X, \mathbb{Q})$. The (Grothendieck amended) general Hodge conjecture (GHC) asserts that

GHC(p, i, X): The inclusion

$$N^p H^i(X, \mathbb{Q}) \subset F_H^p H^i(X, \mathbb{Q})$$

is an equality.

Exercise Let X be a fourfold. Show that GHC(1, 4, X) $\Rightarrow$ Hodge$^{2,2}(X, \mathbb{Q})$ holds, and that GHC(2, 4, X) $\Leftrightarrow$ Hodge$^{2,2}(X, \mathbb{Q})$ holds.

We now recall the Chow group $\mathrm{CH}^k_{\mathrm{alg}}(X)$ of algebraic cycles, algebraically equivalent to zero (see Section 6.2.2). We put $J^k_a(X) = \Phi_k(\mathrm{CH}^k_{\mathrm{alg}}(X))$, where $\Phi_k : \mathrm{CH}^k_{\mathrm{hom}}(X) \to J^k(X)$ is the Abel–Jacobi map (discussed in Section 6.1.5 below).

Proposition 1.6 $J^k_a(X)$ *is an abelian variety* (*even though $J^k(X)$ is in general only a complex torus*).

Proof (Outline) First of all, $\xi \in \mathrm{CH}^k_{\mathrm{alg}}(X) \Rightarrow \xi \in w_*(\mathrm{CH}^1_{\mathrm{alg}}(\Gamma))$ for some smooth curve Γ. But $\mathrm{CH}^1_{\mathrm{alg}}(\Gamma) \simeq J^1(\Gamma)$. One can likewise argue that $\exists \tilde{w} \in z^k(J^1(\Gamma) \times X)$ such that $\xi \in \tilde{w}_*(J^1(\Gamma))$. Further, $\Phi_{k,*} \circ \tilde{w}_* : J^1(\Gamma) \to J^k(X)$ is a homomorphism. Next, assume given an abelian variety A and a homomorphism $w_* : A \to J^k(X)$ induced by a cycle $w \in z^k(A \times X)$, such that $\dim \Phi_k \circ w_*(A)$ is maximal. I claim that $\Phi_k \circ w_*(A) = J^k_a(X)$; otherwise $\exists [\xi] \in J^k_a(X) \backslash \{\Phi_k \circ w_*(A)\}$. But $[\xi] \in \tilde{w}_*(J^1(\Gamma))$ for some smooth curve Γ and cycle $\tilde{w} \in z^k(J^1(\Gamma) \times X)$, hence we can replace A by $A \times J^1(\Gamma)$, w by $\mathrm{Pr}_1^* w + \mathrm{Pr}_2^* \tilde{w}$ in $z^k(A \times J^1(\Gamma) \times X)$. By maximality of dimension, it therefore follows that $\Phi_{k,*} \circ w_*(A) = J^k_a(X)$. Next, by Poincaré's complete reducibility theorem, $\exists$ B, C abelian varieties in A, where $B = $ connected component of $\ker \left(A \to J^k_a(X) \right)$, such that $B + C = A$ and $B \cap C$ is finite. Thus the corresponding map $C \to J^k_a(X)$ is finite, hence the lattices defining C and $J^k_a(X)$ coincide when tensored with $\mathbb{Q}$. Therefore $J^k_a(X)$ is an abelian variety. $\square$

Exercise Show that the Lie algebra, Lie($J^k_a(X)) = N^{k-1} H^{2k-1}(X, \mathbb{Q}) \otimes \mathbb{R}$, where $N^{k-1} H^{2k-1}(X, \mathbb{Q}) \otimes \mathbb{R}$ has a suitable complex structure. Deduce that:

$$\text{GHC}(k - 1, 2k - 1, X) \text{ holds} \Leftrightarrow \text{Lie}(J^k_a(X)) = F_H^{k-1} H^{2k-1}(X, \mathbb{Q}) \otimes \mathbb{R}.$$

Exercise Let X be a threefold. Show that:

$$\text{GHC}(1, 3, X) \text{ holds} \Leftrightarrow \text{Hodge}^{2,2}(\Gamma \times X, \mathbb{Q}) \text{ holds} \ \forall \text{ smooth curves } \Gamma.$$

Exercise   Let $n = \dim X$. Show that the statement:

'GHC$(1, n, X)$ holds for X'

is a birational invariant statement about X.

Exercise   Assume $\dim X = n$, and that the GHC holds. Further, assume that Level$(H^*(X)) \leq \ell$, where Level$(H^*(X)) = \max\{p - q \mid H^{p,q}(X) \neq 0\}$. Then if $[\Delta(p,q)]$, $p + q = 2n$, is the corresponding Künneth component of the diagonal class:

$$H^{2n}(X \times X, \mathbb{Q}) \ni [\Delta]$$

$$= \bigoplus_{p+q=2n} [\Delta(p,q)] \in \bigoplus_{p+q=2n} H^p(X, \mathbb{Q}) \otimes H^q(X, \mathbb{Q}),$$

show that we can arrange for $|\Delta(p,q)| \subset Y_p \times W_q$, where $\mathrm{codim}_X Y_p \geq \frac{p-\ell}{2}$ and $\mathrm{codim}_X W_q \geq \frac{q-\ell}{2}$.

Exercise   Assume that the Hodge conjecture 1.3 holds for projective algebraic manifolds. Show that $\xi \in N^j H^p(X, \mathbb{Q}) \Leftrightarrow \exists$ a smooth projective S of dimension $\ell = p - 2j$, and a cycle $w \in CH^{p-j}(S \times X)$ such that $\xi \in [w]_* H^\ell(S, \mathbb{Q})$. Deduce that there exist S and w as above, such that $N^j H^p(X, \mathbb{Q}) = [w]_* H^\ell(S, \mathbb{Q})$.

Exercise   Assume a given X of dimension n. Further, let

$$\mu : H^1(X, \mathbb{Q}) \to H^{2n-1}(X, \mathbb{Q}),$$

$$\nu : H^{2n-1}(X, \mathbb{Q}) \to H^1(X, \mathbb{Q}),$$

be morphisms of Hodge structures. Show that μ and ν are algebraic cycle induced. Under the assumption of the Hodge conjecture 1.3, give a generalization of this result.

6.1.5 Appendix: the Abel–Jacobi map

Set $z^k_{\mathrm{hom}}(X) = \ker(\mathrm{cl}_k : z^k(X) \to H^{2k}(X, \mathbb{Z}))$. We construct a second cycle class map:

$$\Phi_k : z^k_{\mathrm{hom}}(X) \to J^k(X),$$

where $J^k(X)$ is a certain compact complex torus. Let

$$F^r H^i(X, \mathbb{C}) = \bigoplus_{p+q=i, \, p \geq r} H^{p,q}(X),$$

and note that

$$H^{2k-1}(X, \mathbb{C}) = F^k H^{2k-1}(X, \mathbb{C}) \oplus \overline{F^k H^{2k-1}(X, \mathbb{C})}.$$

Thus

$$J^k(X) := \frac{H^{2k-1}(X, \mathbb{C})}{F^k H^{2k-1}(X, \mathbb{C}) + H^{2k-1}(X, \mathbb{Z})}$$

is a compact complex torus, called the Griffiths jacobian. Serre duality induces the perfect pairing

$$\frac{H^{2k-1}(X, \mathbb{C})}{F^k H^{2k-1}(X, \mathbb{C})} \times F^{n-k+1} H^{2n-2k+1}(X, \mathbb{C}) \to \mathbb{C}.$$

Thus by compatibility of Poincaré and Serre duality, we arrive at:

Corollary 1.7

$$J^k(X) \simeq \frac{F^{n-k+1} H^{2n-2k+1}(X, \mathbb{C})^\vee}{H_{2n-2k+1}(X, \mathbb{Z})},$$

where the denominator (group of periods) is identified with its image

$$H_{2n-2k+1}(X, \mathbb{Z}) \to F^{n-k+1} H^{2n-2k+1}(X, \mathbb{C})^\vee$$

by the formula

$$\{\gamma\} \mapsto \left(\{w\} \in F^{n-k+1} H^{2n-2k+1}(X, \mathbb{C}) \mapsto \int_\gamma w \right),$$

which is a well defined map by Stokes' theorem.

Prescription for Φ_k: let $\xi \in z^k_{\text{hom}}(X)$. Then $\xi = \partial \zeta$ bounds a $2n - 2k + 1$ real dimensional chain ζ in X. Let $\{w\} \in F^{n-k+1} H^{2n-2k+1}(X, \mathbb{C})$. Define:

$$\Phi_k(\xi)(\{w\}) = \int_\zeta w \quad \text{(modulo periods)}.$$

Proposition 1.8 $\Phi_k(\xi)$ *is well defined.*

Outline of proof First, if $\partial \zeta' = \xi$, then set $\varphi = \zeta - \zeta'$, hence $\partial \varphi = \partial \zeta - \partial \zeta' = \xi - \xi = 0$, i.e. $\{\varphi\} \in H_{2n-2k+1}(X, \mathbb{Z})$. Thus $\int_\zeta w - \int_{\zeta'} w = \int_\varphi w \in$ group of periods. Next, we must show that Φ_k depends only on the complex structure of X, i.e. independent of cohomological representative of $\{w\}$, in terms of the Hodge structure. For this, we use a result of Dolbeault:

Lemma 1.9

$$F^k H^\ell(X, \mathbb{C}) = \frac{\ker d : F^k E_X^\ell \to E_X^{\ell+1}}{d(F^k E_X^{\ell-1})}.$$

Therefore, by Lemma 1.9, $\{w\} = \{w'\} \Rightarrow w - w' = d\eta$, $\eta \in F^{n-k+1} E_X^{2n-2k}$. Therefore $\int_\zeta w - \int_\zeta w' = \int_\zeta d\eta \overset{\text{Stokes'}}{=} \int_{\partial\zeta=\xi} \eta$. But ξ involves subvarieties of dimension $n - k$ and $\eta \in F^{n-k+1} E_X^{2n-2k}$. Therefore $\int_\xi \eta = 0$, hence $\int_\zeta w = \int_\zeta w'$.

6.1.6 Appendix: normal functions and the Lefschetz $(1, 1)$ theorem

Lefschetz' original approach to his $(1, 1)$ theorem involved Poincaré normal functions. This appears in [Lef]. He proved his theorem for surfaces. We thus consider the case where X is a smooth surface. Consider a general (Lefschetz) pencil $\{X_t\}_{t \in \mathbb{P}^1}$ of curves in X. Roughly speaking, a holomorphic cross-section:

$$\nu : \mathbb{P}^1 \to \coprod_{t \in \mathbb{P}^1} J(X_t),$$

is called a normal function. Fix a point $p_0 \in \cap_{\mathbb{P}^1} X_t$ (base locus). For smooth X_t, one has a bimeromorphic (= birational) morphism ('Jacobi inversion'):

$$\mathbb{S}^{(g)}(X_t) \to z_{\text{hom}}^1(X_t) \overset{\Phi_1}{\longrightarrow} J^1(X_t), \qquad g = \text{genus of } X_t,$$

given by

$$p_1 + \cdots + p_g \mapsto \Phi_1(p_1 + \cdots + p_g - g \cdot p_0),$$

where $\mathbb{S}^{(g)}(X_t)$ is the gth symmetric product. Thus as $t \in \mathbb{P}^1$ varies, the $\nu(t)$ traces out a codimensional one cycle on X. One way to construct a family of such normal functions is via the restriction:

$$H^{0,1}(X) \to H^{0,1}(X_t) \to J^1(X_t)$$

$$\xi \mapsto \nu_\xi(t).$$

From this construction, Poincaré was able to construct a family of linearly inequivalent divisors on X. This was the original motivation for introducing normal functions. Now consider the holomorphic vector bundle $\mathbb{F} := \coprod_{t \in \mathbb{P}^1} H^{0,1}(X_t)$ with general fiber $H^{0,1}(X_t)$, and the 'sheaf of lattices' $\mathbb{L}_{\mathbb{Z}}$ with fiber $H^1(X_t, \mathbb{Z})$. Further let $\mathbb{J} = \coprod_{t \in \mathbb{P}^1} J(X_t)$. There is a short exact sequence:

$$0 \to \mathbb{L}_{\mathbb{Z}} \to \mathcal{O}_{\mathbb{P}^1}(\mathbb{F}) \to \mathcal{O}_{\mathbb{P}^1}(\mathbb{J}) \to 0.$$

This gives rise to a diagram:

$$\Gamma(\mathbb{P}^1, \mathcal{O}_{\mathbb{P}^1}(\mathbb{J})) \xrightarrow{\delta} H^1(\mathbb{P}^1, \mathbb{L}_\mathbb{Z}) \rightarrow H^1(\mathbb{P}^1, \mathcal{O}_{\mathbb{P}^1}(\mathbb{F})).$$

‖	↑	↑
Group of	Can be identified	This turns out
normal functions	with the key part	to be roughly
	of $H^2(X, \mathbb{Z})$ by	$H^{0,2}(X, \mathbb{C})$
	the Leray spectral	
	sequence associated	
	to $\coprod_{t \in \mathbb{P}^1} X_t \rightarrow \mathbb{P}^1$	

For $v \in \Gamma(\mathbb{P}^1, \mathcal{O}_{\mathbb{P}^1}(\mathbb{J}))$, $\delta(v)$ is the topological invariant associated to v, also called the cohomology class of v. Thus roughly, $H^{1,1}(X, \mathbb{Z}) = \delta(\Gamma(\mathbb{P}^1, \mathcal{O}_{\mathbb{P}^1}(\mathbb{J})))$, i.e. every integral cohomology class of type $(1, 1)$ is the cohomology class of a normal function. Thus the main result in Lefschetz' proof was the identification of the Hodge type of $H^1(\mathbb{P}^1, \mathcal{O}_{\mathbb{P}^1}(\mathbb{F}))$. This was complicated by the fact that although the family of curves $\{X_t\}$ on X are topologically the same, their corresponding analytic structures are not the same. Finally, by using Jacobi inversion, every normal function traces out an algebraic cycle on X which agrees with the original cohomology class in $H^{1,1}(X, \mathbb{Z})$. Thus we again arrive at the Lefschetz $(1, 1)$ theorem for X.

The main supporting point of Lefschetz' approach to his theorem is that the role of normal functions generalizes, in the sense that every (primitive) cohomology class in $H^{k,k}(X, \mathbb{Z})$ is the cohomology class of a normal function. This was the Griffiths' program, which is decribed in [Gr2], as well as in [Z1], together with a modern treatment in [Z2]. The hope was to arrive at an inductive proof of the Hodge conjecture. Unfortunately, Jacobi inversion fails in general. Specifically, Griffiths was the first to exhibit examples (e.g. a general[7] quintic threefold) where $\Phi_2 : z^k_{\text{hom}}(X) \rightarrow J^k(X)$ has at most a countable image.

6.1.7 Appendix: a real regulator on $K_1(X)$ and the Hodge-$\mathcal{D}$ conjecture

Let $X/_\mathbb{C}$ be a projective algebraic manifold of dimension n, let $z^k(X)$ be the free abelian group generated by irreducible subvarieties of codimension k in X,

---

[7] In a transcendental sense; namely, the parameter space of smooth quintic threefolds is not compact, and it is the complement of a certain countable union of proper *analytic* subvarieties that defines the transcendental quintics. See [Gr1, Section 13] for details.

and let

$$z^k(X, 1) = \ker\left[\Gamma\Big(\bigoplus_{\text{codim}_X Z = k-1} \mathbb{C}(Z)^\times\Big) \xrightarrow{\text{div}} z^k(X)\right],$$

where $\mathbb{C}(Z)^\times$ is the multiplicative group of the rational function field $\mathbb{C}(Z)$, and 'div' is the divisor map. Further, let $T^k(X, 1)$ be the subgroup generated by the Tame symbols:

$$T(\{f, g\}) := \sum_{\text{codim}_X Z = k-1} (-1)^{v_Z(f)v_Z(g)} \left(\frac{f^{v_Z(g)}}{g^{v_Z(f)}}\right)_Z$$

where $f, g \in \mathbb{C}(Z)^\times$ as Z runs over all irreducible subvarieties of codimension $k - 2$ on X, and $v_Z(h)$ is the order of vanishing of a rational function h along a codimension one Z. The higher Chow group $\text{CH}^k(X, 1)$, first introduced by Bloch as the homology of a certain simplicial complex, can be defined alternatively as the quotient group:

$$\text{CH}^k(X, 1) = \frac{z^k(X, 1)}{T^k(X, 1)}.$$

Note that the 'usual' kth Chow group is given by

$$\text{CH}^k(X) := \text{coker}\left[\Gamma\Big(\bigoplus_{\text{codim}_X Z = k-1} \mathbb{C}(Z)^\times\Big) \xrightarrow{\text{div}} z^k(X)\right],$$

and that there are induced cycle class maps:

$$\text{cl}_k : \text{CH}^k(X) \to H^{k,k}(X, \mathbb{Z}) \qquad \text{(fundamental class)},$$
$$\Phi_k : \text{CH}^k_{\text{hom}}(X) \to J^k(X) \qquad \text{(Abel–Jacobi map)}.$$

The Riemann–Roch theorem furnishes isomorphisms from K-theory, (via a Chern character map ch):

$$\text{ch} : K_0(X)_\mathbb{Q} \simeq \text{CH}^\bullet(X)_\mathbb{Q} \qquad \text{(Grothendieck)}$$
$$\text{ch} : K_1(X)_\mathbb{Q} \simeq \text{CH}^\bullet(X, 1)_\mathbb{Q} \qquad \text{(Bloch)}.$$

Proposition 1.10 *There is a well defined regulator map*

$$r : \text{CH}^k(X, 1) \to H^{k-1,k-1}(X, \mathbb{R}) \simeq H^{n-k+1,n-k+1}(X, \mathbb{R})^\vee,$$

given by

$$\xi = \sum_i f_i \otimes Z_i \mapsto r(\xi)(\omega) = \sum_i \cdot\int_{Z_i} \omega \log|f_i|,$$

where $\{\omega\} \in H^{n-k+1,n-k+1}(X, \mathbb{R})$.

Idea of proof One first shows that the current $r(\xi)$ in Proposition 1.10 acting on $E_X^{n-k+1,n-k+1}$ is $\partial\bar{\partial}$-closed. This uses the fact that $\sum_i \text{div}(f_i) = 0$, together with a residue argument. Here are some details. Let $\omega \in \partial\bar{\partial}E_X^{n-k,n-k}$. Then we can write $\omega = d\bar{\partial}\eta$ for some $\eta \in E_X^{n-k,n-k}$. Also let

$$\xi = \sum_i f_i \otimes Z_i \in z^k(X, 1)$$

be given as above, and consider the corresponding integral

$$r(\xi)(\omega) = \sum_i \int_{Z_i} \omega \log|f_i|.$$

By Stokes' theorem and a standard calculation (below):

$$\int_{Z_i} (d\bar{\partial}\eta) \log|f_i| = \int_{Z_i} \bar{\partial}\eta \wedge d \log|f_i|$$

$$= \int_{Z_i} \bar{\partial}\eta \wedge \partial \log|f_i| = \int_{Z_i} d\eta \wedge \partial \log|f_i|,$$

where the latter two equalities follow by Hodge type, and the former uses

$$d\big((\bar{\partial}\eta) \log|f|\big) = (d\bar{\partial}\eta) \log|f| - \bar{\partial}\eta \wedge d \log|f|.$$

More specifically, by taking ϵ-tubes about the components $D \subset \text{div}(f)$, and using that $\dim D = n - k$ and $\bar{\partial}\eta \in E_X^{n-k,n-k+1}$, and that $\log|f|$ is locally L^1, we find

$$\lim_{\epsilon \to 0} \int_{\text{Tube}_\epsilon((f))} (\bar{\partial}\eta) \log|f| = 0.$$

Note that $\bar{\partial}\eta \wedge \bar{\partial}\log|f| \in E_{X,L^1}^{n-k,n-k+2}$ (locally L^1 forms) and that $\dim Z_i = n - k + 1$. Thus we are left with an integral of the form $\int_Z d\eta \wedge \partial \log|f|$ as indicated above. Next,

$$d(\eta \wedge \partial \log|f|) = d\eta \wedge \partial \log|f|,$$

since $\bar{\partial}\partial \log|f| = 0$. Thus

$$\int_{Z_i} (d\bar{\partial}\eta) \log|f_i| = \int_{Z_i} d(\eta \wedge \partial \log|f_i|) = \lim_{\epsilon \to 0} \int_{\text{Tube}_\epsilon((f_i))} \eta \wedge \partial \log|f_i|.$$

If we put $Z = Z_i$, for a given i, with z a local coordinate on Z_{reg}, then using the dictionary $|f_\alpha| \leftrightarrow |z|$ we have the residue integral:

$$\lim_{\epsilon \to 0} \int_{|z|=\epsilon} \eta \wedge \partial \log|z|^2 = \lim_{\epsilon \to 0} \int_{|z|=\epsilon} \eta \wedge \frac{dz}{z} = 2\pi\sqrt{-1} \int_{\{z=0\}\cap Z} \eta|_{\{z=0\}\cap Z},$$

where

$$\eta|_{\{z=0\}\cap Z} = \text{Residue}_{\{z=0\}\cap Z}\left(\eta \wedge \frac{dz}{z}\right)$$

(i.e. taking 'tubes' is dual to taking 'residues'). Then by computing the residue, and linearity, we find that

$$r(\xi)(\omega) = 2\pi\sqrt{-1}\sum_D\left[\left(\sum_i v_D(f_i)\right)\int_D \eta\right].$$

(We note that there remains the possibility that $Z = Z_i$ is singular along D. To remedy this, one may pass to a normalization of Z with the same calculations above.) Since $\sum_i \text{div}(f_i) = 0$, we have $r(\xi)(\omega) = 0$. Thus we obtain a well defined map $r : z^k(X, 1) \to H^{k-1,k-1}(X, \mathbb{R})$ by Theorem 1.1. To show that $r\big(T(\{f, g\})\big) = 0$, for $f, g \in \mathbb{C}(Z)^\times$, $\text{codim}_X Z = k - 2$, we consider the following. Set $F := (f, g) : Z \to \mathbb{P}^1 \times \mathbb{P}^1$, and $t, s \in \mathbb{C}(\mathbb{P}^1)^\times$ corresponding to the affine coordinates on $\mathbb{P}^1$. By a proper modification, we can assume that F is a morphism. Then $F^*(\{t, s\}) = \{f, g\}$, and by functoriality and a standard calculation, $r\big(T(\{f, g\})\big) = 0$. $\qquad\square$

Beilinson's Hodge-$\mathcal{D}$ conjecture [Be] asserts that

(1.11) $r : \text{CH}^k(X, 1)_{\mathbb{R}} \to H^{k-1,k-1}(X, \mathbb{R}),$

is surjective. This was recently proven to be *false* (see [MS]), by establishing a Noether–Lefschetz theorem analogue in this setting. It uses a connectedness result of Nori [No]. For example, the Hodge-$\mathcal{D}$ conjecture fails when $X \subset \mathbb{P}^3$ is a general surface of degree ≥ 5, with $k = 2$ [MS]. There remains the possibility, however, that the Hodge-$\mathcal{D}$ conjecture is true for smooth and proper X defined over a number field. It is reasonable to expect the following:

Conjecture 1.12   Let X be a $K3$ surface with general moduli. Then the Hodge-$\mathcal{D}$ conjecture holds for X. Specifically,

$$r : \text{CH}^2(X, 1)_{\mathbb{R}} \to H^{1,1}(X, \mathbb{R}),$$

is onto.

Question 1.13   (Variant of the Hodge conjecture.) Let $X/_{\mathbb{C}}$ be a projective algebraic manifold. Is it the case that the Hodge classes $H^{k-1,k-1}(X, \mathbb{Q})$ are in the image of r in (1.11)?[8]

---

[8] Gordon and Lewis [G-L] introduce a twisted version $\underline{\text{CH}}^k(X, 1)$ of $\text{CH}^k(X, 1)$ involving 'flat line bundles', and construct a corresponding regulator $\underline{r} : \underline{\text{CH}}^k(X, 1) \to H^{k-1,k-1}(X, \mathbb{R})$. In this setting, the corresponding analogue of Question 1.13 seems much more tractable.

6.1.8 Appendix: Grothendieck's counterexample

Let $X = T_1 \times T_2 \times T_3$ where $T_j = \mathbb{C}/\Lambda_j$ and where $\Lambda_j = \mathbb{Z} \oplus \mathbb{Z}\tau_j$ and say $\text{Im}(\tau_j) > 0$ for $j = 1$, 2, 3. Now by the Künneth formula, we have $\dim_{\mathbb{Q}}[\{F^1 H^3(X, \mathbb{C})\} \cap H^3(X, \mathbb{Q})] \equiv \dim_{\mathbb{Q}}[\{F^1 H^3(X, \mathbb{C})\} \cap \{H^1(T_1, \mathbb{Q}) \otimes H^1(T_2, \mathbb{Q}) \otimes H^1(T_3, \mathbb{Q})\}] \bmod 2$. Now set $\omega = dz_1 \wedge dz_2 \wedge dz_3$ and $V = \{\xi \in H_1(T_1, \mathbb{Q}) \otimes H_1(T_2, \mathbb{Q}) \otimes H_1(T_3, \mathbb{Q}) \mid \int_\xi \omega = 0\}$. By Poincaré and Serre duality $\dim_{\mathbb{Q}} V \equiv \dim_{\mathbb{Q}}[\{F^1 H^3(X, \mathbb{C})\} \cap H^3(X, \mathbb{Q})] \bmod 2$. For $j = 1$, 2, 3, let $\{\alpha_j, \beta_j\}$ be the generators of $H_1(T_j, \mathbb{Z}) \simeq \mathbb{Z}^2$ corresponding to $\{1, \tau_j\}$ respectively. Consider the basis of $H_1(T_j, \mathbb{Z}) \otimes H_1(T_2, \mathbb{Z}) \otimes H_1(T_3, \mathbb{Z}) \simeq \mathbb{Z}^8$ given by:

$$\begin{aligned}
\xi_1 &= \alpha_1 \times \alpha_2 \times \alpha_3 & \xi_5 &= \beta_1 \times \beta_2 \times \alpha_3 \\
\xi_2 &= \beta_1 \times \alpha_2 \times \alpha_3 & \xi_6 &= \beta_1 \times \alpha_2 \times \beta_3 \\
\xi_3 &= \alpha_1 \times \beta_2 \times \alpha_3 & \xi_7 &= \alpha_1 \times \beta_2 \times \beta_3 \\
\xi_4 &= \alpha_1 \times \alpha_2 \times \beta_3 & \xi_8 &= \beta_1 \times \beta_2 \times \beta_3.
\end{aligned}$$

We now compute

$$\int_{\xi_1} \omega = 1 \qquad \int_{\xi_2} \omega = \tau_1 \qquad \int_{\xi_3} \omega = \tau_2 \qquad \int_{\xi_4} \omega = \tau_3$$

$$\int_{\xi_5} \omega = \tau_1\tau_2 \qquad \int_{\xi_6} \omega = \tau_1\tau_3 \qquad \int_{\xi_7} \omega = \tau_2\tau_3 \qquad \int_{\xi_8} \omega = \tau_1\tau_2\tau_3.$$

Let $\xi = \sum_j r_j \xi_j$ where $r_j \in \mathbb{Q}$ (for $j = 1, \ldots, 8$). Then:

$$(**) \qquad 0 = \int_\xi \omega = r_1 + r_2\tau_1 + r_3\tau_2 + r_4\tau_3 + r_5\tau_1\tau_2$$
$$+ r_6\tau_1\tau_3 + r_7\tau_2\tau_3 + r_8\tau_1\tau_2\tau_3.$$

Now for example set $\tau_1 = \tau_2 = \tau_3 = (\sqrt[3]{2})e^{2\pi\sqrt{-1}/3} = \tau$ say. Then $(**)$ becomes:

$$(***) \qquad 0 = r_1 + (r_2 + r_3 + r_4)\tau + (r_5 + r_6 + r_7)\tau^2 + r_8\tau^3,$$

and hence

$$\begin{aligned}
r_1 + 2r_8 &= 0 \\
r_2 + r_3 + r_4 &= 0 \\
r_5 + r_6 + r_7 &= 0,
\end{aligned}$$

a rank 3 linear system of equations, and therefore in this case $\dim_{\mathbb{Q}} V = 8 - 3 = 5 \equiv 1 \bmod 2$, a fortiori $\dim_{\mathbb{Q}}[\{F^1 H^3(X, \mathbb{C})\} \cap H^3(X, \mathbb{Q})] \equiv 1$ (2). But any Hodge substructure of $H^3(X, \mathbb{Q})$ must be even dimensional (why?). Hence $N^1 H^3(X, \mathbb{Q}) \neq F^1 H^3(X, \mathbb{C}) \cap H^3(X, \mathbb{Q})$.

6.2 Lecture 2: a geometric approach

6.2.1 The cylinder map for the quintic fourfold

Verifications of the GHC in various geometrically interesting cases, using a cylinder map construction, abound in the literature. We work out a prototypical example situation. We construct the cylinder map associated to the family of lines on a general[9] quintic fourfold X, and deduce Hodge$^{2,2}(X, \mathbb{Q})$ from (2.0) below. This construction is taken from [Lew1], which goes back to some ideas of Clemens (see [T, p. 42], and also [B-M]). For a number of applications of cylinder maps, the reader can also consult for example [C-G], [Co], [Co-M], [Mu2], [Te], [Lew2, Chapter 13], [Lew3], [Lew4], [Sh1]–[Sh4], to name a few. Let $G := \mathrm{Grass}(2, 6) = \{\mathbb{C}^2 \subset \mathbb{C}^6\}$ be the Grassmannian of 2-dimensional subspaces of $\mathbb{C}^6$. Then G can be identified with $\{\mathbb{P}^1 \subset \mathbb{P}^5\}$; moreover $\dim G = 8$. Any hypersurface of degree d in $\mathbb{P}^n$ determines a corresponding point in $\mathbb{P}^N$, where $N = \binom{n+d}{d} - 1$, i.e.

$$\sum_{\substack{\alpha=(\alpha_0,\dots,\alpha_n)\in\mathbb{Z}_+^{n+1} \\ \alpha_0+\cdots+\alpha_n=d}} a_\alpha z_0^{\alpha_0} \cdots z_n^{\alpha_n} = 0 \quad \mapsto \quad [\dots, a_\alpha, \dots] \in \mathbb{P}^N.$$

Let $\mathbb{P}^N$ ($N = 251$) be the projective space of hypersurfaces of degree 5 in $\mathbb{P}^5$, and put $\mathcal{H} = \{(c, t) \in G \times \mathbb{P}^N \mid \mathbb{P}_c^1 \subset X_t\}$, with projection diagram:

$$\mathcal{H}$$
$$\pi_1 \swarrow \qquad \searrow \pi_2$$
$$G \qquad\qquad\qquad \mathbb{P}^N.$$

An elementary topological argument shows that any such quintic fourfold contains a line $\mathbb{P}^1$, and hence as we will see below, at the very least, a 2-dimensional family of lines. (Otherwise, the techniques below will imply that the cylinder map Φ_* introduced below would be zero. Thus $H^4(X, \mathbb{Q}) \simeq \mathbb{Q}$, which is far from being the case! (see [Lew3, p. 194]). Thus π_2 is surjective. On the other hand, π_1 defines a $\mathbb{P}^{N-6}$-bundle over G. (Hint: Up to a PGL action, we can assume $\mathbb{P}^1$ is cut out of $\mathbb{P}^5$ by $\{z_2 = z_3 = z_4 = z_5 = 0\} \subset \mathbb{P}^5$. Then any homogeneous polynomial of degree 5 in the coordinates $[z_0, z_1, z_2, z_3, z_4, z_5]$ cannot involve $\sum_{j=0}^{5} z_0^j z_1^{5-j}$.) Thus $\dim \mathcal{H} = N + 2$, and hence if $X \subset \mathbb{P}^5$ is a general hypersurface of degree 5, and if we set $\Omega_X := \{\mathbb{P}^1$

---

[9] 'General' is interpreted here to mean the following. Given a family of varieties $\{Y_t\}_{t\in S}$, over some base variety S, a general Y_t means that t belongs to some given non-empty Zariski open subset of S, satisfying a number of desirable features (e.g. Y_t smooth, etc.).

$\subset X\} \subset G$, then Ω_X is a smooth surface. It turns out to be irreducible as well. If $Z \subset \mathbb{P}^6$ is a given general quintic hypersurface, then likewise $\Omega_Z := \{\mathbb{P}^1 \subset Z\}$ is smooth (and irreducible) and of dimension 4; moreover in this case through a generic point $p \in Z$ passes 5! lines. Hint: We can assume by a PGL action that $p = [1, 0, 0, 0, 0, 0, 0] \in X \subset \mathbb{P}^6$, and therefore X is the hypersurface defined by $F(z_0, \ldots, z_6) = \sum_{j=1}^5 z_0^{5-j} F_j(z_1, \ldots, z_6)$, where $F_j(z_1, \ldots, z_6)$ is homogeneous of degree j. Thus in the affine coordinates $(x_1, \ldots, x_6) = (z_1/z_0, \ldots, z_6/z_0)$, $p = (0, 0, 0, 0, 0, 0)$, and $Z \cap \mathbb{C}^6$ is cut out by $z_0^{-5} F(z_0, \ldots, z_6) =: f(x_1, \ldots, x_6) = \sum_{j=1}^5 f_j(x_1, \ldots, x_6) = 0$, where

$$f_j(x_1, \ldots, x_6) = z_0^{-5} F_j(z_1, \ldots, z_6),$$

is homogeneous of degree j. Then in affine cordinates, any line ℓ through p in Z, must be described in the form $\ell = \{t \cdot v \mid t \in \mathbb{C}\}$, for some non-zero $v \in \mathbb{C}^6$. Thus $\ell \subset Z \Leftrightarrow [v] \in \Sigma_p := \{f_1 = f_2 = f_3 = f_4 = f_5 = 0\} \subset \mathbb{P}^5$. Note that generically, Σ_p consists of 5! points by Bezout's theorem. Let

$$P(X) = \{(c, x) \in \Omega_X \times X \mid x \in \mathbb{P}_c^1\}$$
$$P(Z) = \{(c, z) \in \Omega_Z \times Z \mid z \in \mathbb{P}_c^1\}.$$

Note that $P(X)$ is the projectivization of the pull-back to Ω_X of the universal rank 2 vector bundle over G, namely the bundle over G that associates to each point $c \in G$, the corresponding 2-dimensional subspace $\mathbb{C}_c^2 \subset \mathbb{C}^6$; similarly for $P(Z)$. Hence $P(X)$ and $P(Z)$ are smooth and irreducible. Note that the projection $\pi_Z : P(Z) \to Z$ is onto, since through every point of Z, there passes a line. In particular, $\deg \pi_Z = 5!$ By Bertini's theorem, $\tilde{X} := \pi_Z^{-1}(X)$ is smooth, since X is general. One has an obvious diagram of projections (or restrictions of projections), and inclusions:

$$
\begin{array}{ccccc}
P(X) & \xrightarrow{\pi_X} & & X & \\
& \searrow^{i_1} & \pi \nearrow & & \searrow^{j} \\
& & \tilde{X} & & Z \\
& & \searrow^{i} & \pi_Z \nearrow & \\
\rho_X \downarrow & \rho \downarrow & & P(Z) & \\
& & \swarrow \rho_Z & & \\
\Omega_X & \hookrightarrow & \Omega_Z. & &
\end{array}
$$

Note that any line $\mathbb{P}^1 \subset Z \subset \mathbb{P}^6$ either meets $X = \mathbb{P}^5 \cap Z$ at a single point, or lies entirely in X. Thus it is reasonably obvious that $\tilde{X} = B_{\Omega_X}(\Omega_Z) =$ blow-up of Ω_Z along Ω_X. We introduce the *cylinder map* $\Phi_* := \pi_{X,*} \circ \rho_X^*$. For a cycle

$\gamma \subset \Omega_X$, $\Phi_*(\gamma) = \{\cup_{c \in \gamma} \mathbb{P}^1_c\}$ determines a corresponding (co)cycle on X. Now for any cycle $\xi \subset X$, $\pi_* \circ \pi^*(\xi) = (\deg \pi)\xi$. (Since $\deg \pi = 5!$ is *finite*.) Now recall that $H^4(\tilde{X}, \mathbb{Q}) \simeq H^4(\Omega_Z, \mathbb{Q}) \oplus H^2(\Omega_X, \mathbb{Q})$, the isomorphism is in fact given by $\rho_* \oplus \rho_{X,*} \circ i_1^*$, with inverse $\rho^* + i_{1,*} \circ \rho_X^*$. Thus under this decomposition of $H^4(\tilde{X}, \mathbb{Q})$, the surjective morphism π_* is a sum of two morphisms:

$$\pi_* : \rho^*\big(H^4(\Omega_Z, \mathbb{Q})\big) \to H^4(X, \mathbb{Q}),$$

$$\pi_* : i_{1,*} \circ \rho_X^*\big(H^2(\Omega_X, \mathbb{Q})\big) \to H^4(X, \mathbb{Q}).$$

But $\pi_* \circ \rho^* = \pi_* \circ \{\rho_Z \circ i\}^* = \{\pi_* \circ i^*\} \circ \rho_Z^* = j^* \circ \{\pi_{Z,*} \circ \rho_Z^*\}$, hence $\pi_* \circ \rho^*\big(H^4(\Omega_Z, \mathbb{Q})\big) \subset j^* H^4(Z, \mathbb{Q})$. In fact, by the weak Lefschetz theorem for cohomology, $H^4(Z, \mathbb{Z}) \simeq H^4(\mathbb{P}^6, \mathbb{Z}) \simeq \mathbb{Z}$. In particular, $\pi_* \circ \rho^*\big(H^2(\Omega_X, \mathbb{Q})\big) = \mathbb{Q}\omega \wedge \omega$, where ω is a Kähler class (first Chern class of the hyperplane bundle) of X. Next $\pi_* \circ i_{1,*} \circ \rho_X^* = \pi_{X,*} \circ \rho_X^* = \Phi_*$. Thus $\Phi_* : H^2(\Omega_X, \mathbb{Q}) \to H^4(X, \mathbb{Q})/\mathbb{Q}\omega \wedge \omega$ is onto. This, together with the Noether–Lefschetz theorem implies that:

(2.0) $\qquad\qquad \Phi_* : H^2(\Omega_X, \mathbb{Q}) \to H^4(X, \mathbb{Q})$ is onto.

In particular, Hodge$^{2,2}(X, \mathbb{Q})$ holds for X.

6.2.2 Appendix

In this part, we need to introduce the following terminology. Let ξ_1, $\xi_2 \in z^k(X)$. We say that ξ_1 and ξ_2 are rationally equivalent to each other, and write $\xi_1 \sim_{\text{rat}} \xi_2$, if there is a cycle $w \in z^k(\mathbb{P}^1 \times X)$ in general position, such that $\xi_1 - \xi_2 = w(0) - w(\infty)$. We say that ξ_1 and ξ_2 are algebraically equivalent to each other, and write $\xi_1 \sim_{\text{alg}} \xi_2$, if there are a smooth connected curve Γ, points P, $Q \in \Gamma$, and a cycle $w \in z^k(\Gamma \times X)$ in general position, such that $\xi_1 - \xi_2 = w(P) - w(Q)$. The subgroups of cycles rationally, algebraically and homologically equivalent to zero are denoted by $z^k_{\text{rat}}(X)$, $z^k_{\text{alg}}(X)$, $z^k_{\text{hom}}(X)$. We put $\text{CH}^k(X) := z^k(X)/z^k_{\text{rat}}(X)$, $\text{CH}^k_{\text{alg}}(X) := z^k_{\text{alg}}(X)/z^k_{\text{rat}}(X)$, and $\text{Griff}^k(X) := z^k_{\text{hom}}(X)/z^k_{\text{alg}}(X)$ (the Griffiths group). (Note: by a norm argument, this definition of $\text{CH}^k(X)$ agrees with the earlier definition given in Section 6.1.7.) Now let $X \subset \mathbb{P}^{n+1}$ be any smooth hypersurface of degree d cut out by an irreducible homogeneous polynomial $F(z_0, \ldots, z_{n+1})$ ($\deg F = d$). Set $G(z_0, \ldots, z_{n+2}) = F + z_{n+2}^d$, $Z = \{G = 0\} \subset \mathbb{P}^{n+2}$ and note that $X = \mathbb{P}^{n+1} \cap Z$, where $\mathbb{P}^{n+1} = \{z_{n+2} = 0\} \subset \mathbb{P}^{n+2}$. One can easily check that Z is smooth. Let $j : X \hookrightarrow Z$ be the inclusion and $\nu : Z \to \mathbb{P}^{n+1}$ the projection from $[0, \ldots, 0, 1] \in \mathbb{P}^{n+2}$. (Explicitly, $\nu\big([z_0, \ldots, z_{n+2}]\big) = [z_0, \ldots, z_{n+1}]$.) We will also consider the inclusion

$i : X \hookrightarrow \mathbb{P}^{n+1}$. The following result is a Chow analogue of the weak Lefschetz theorem:

Proposition 2.1 [Lew4] *The following diagram is commutative:*

$$CH^{\bullet}(Z) \xrightarrow{dj^*} CH^{\bullet}(X)$$

$$v_* \searrow \qquad \uparrow i^*$$

$$CH^{\bullet}(\mathbb{P}^{n+1}).$$

Using Proposition 2.1, the results of this lecture generalize as follows.

Theorem 2.2 [Lew3] *Let X be a general hypersurface of degree $d \geq 3$ in $\mathbb{P}^{n+1}$, and let $k = [\frac{n+1}{d}]$. Let us further assume that $k(n+2-k)+1 - \binom{d+k}{k} \geq 0$. Then there is a smooth subvariety $\Omega_X(k) \subset \text{Grass}\{\mathbb{P}^k \subset X\}$ of dimension $n - 2k$ such that the following hold for the cylinder map Φ_*.*

(i) *If n is odd, then $\Phi_* : H^{n-2k}(\Omega_X(k), \mathbb{Z}) \to H^n(X, \mathbb{Z})$ is onto.*
(ii) *If n is even, then $\Phi_* : H^{n-2k}(\Omega_X(k), \mathbb{Q}) \to H^n(X, \mathbb{Q})$ is onto.*
(iii) *$\Phi_* : CH^{\bullet-k}(\Omega_X) \otimes \mathbb{Q} \to \{CH^{\bullet}(X)/\mathbb{Z}H_X^{\bullet}\} \otimes \mathbb{Q}$ is onto, where $H_X := \mathbb{P}^n \cap X$ is a hyperplane section of X.*
(iv) *$\Phi_* : CH_{\text{alg}}^{\bullet-k}(\Omega_X) \to CH_{\text{alg}}^{\bullet}(X)$ is onto.*
(v) *$\Phi_* : CH_{\text{hom}}^{\bullet-k}(\Omega_X) \otimes \mathbb{Q} \to CH_{\text{hom}}^{\bullet}(X) \otimes \mathbb{Q}$ is onto.*
(vi) *$\Phi_* : \text{Griff}^{\bullet-k}(\Omega_X) \otimes \mathbb{Q} \to \text{Griff}^{\bullet}(X) \otimes \mathbb{Q}$ is onto.*

Corollary 2.3 [Lew3] *Let X be given as in Theorem 2.2. Then the following hold.*

(i) *$CH_{\text{alg}}^r(X) = 0$ for $r \leq k$ and for $r \geq n - k + 1$.*
(ii) *$CH_{\text{hom}}^r(X)_{\mathbb{Q}} = 0$ for $r \leq k$ and for $r \geq n - k + 1$.*
(iii) *$CH_{\text{alg}}^{k+1}(X)$ is finite dimensional.[10]*
(iv) *$\text{Griff}^{k+1}(X) \otimes \mathbb{Q} = \text{Griff}^{n-k}(X) \otimes \mathbb{Q} = 0$.*

Exercise Let $Q_n \subset \mathbb{P}^{n+1}$ be a smooth quadric hypersurface (i.e. $\deg Q_n = 2$). Show that $H^{\bullet}(Q_n, \mathbb{Q})$ is generated by algebraic cocyles. (Hint: Let $v_p : Q_n \to \mathbb{P}^n$ be the projection from a general point $p \in Q_n$. Describe the cohomology of the blow-up $B_p(Q_n)$ in terms of the cohomology of $\mathbb{P}^n$ and the variety of lines $\{\mathbb{P}^1 \subset Q_n \mid p \in \mathbb{P}^1\}$ through p.)

---

[10] Finite dimensional means that there is a cycle induced surjective homomorphism $A \to CH_{\text{alg}}^{k+1}$ (X), for some abelian variety A.

Problem Note that Theorem 2.2 establishes GHC(k, n, X) for X, k given there, satisfying the numerical condition in Theorem 2.2. In fact, a deformation argument allows one to deduce the same result, even if X is not general (provided, however, that X is smooth). Investigate GHC(k, n, X) for those smooth X which do *not* satisfy the numerical condition. For example, the *first* five cases of X where this condition fails are the following:

n	d	k
9	5	2
11	4	3
11	6	2
12	4	3
12	6	2

6.2.3 Appendix

Another generalization of Lecture 2 is the following.

Theorem 2.4 [Lew1] *Let $X \subset Z$ be an inclusion of algebraic manifolds with respective dimensions* 4, 5. *Let* $\{C_b\}_{b \in \Omega}$ *be a family of curves covering Z where* $\dim \Omega = 4$, *and set* $\Omega_X = \{b \in \Omega \mid \dim C_b \cap X = 1\}$. *Assume:*

 (i) *X meets the generic curve C_b transversally in a single point;*
 (ii) *codim$_\Omega \Omega_X \geq 2$.*

   Then Hodge$^{2,2}(Z, \mathbb{Q}) \Rightarrow$ Hodge$^{2,2}(X, \mathbb{Q})$.

Remark In the case $\{C_b\}_{b \in \Omega}$ is a family of rational curves, condition (ii) can be dropped.

6.3 Lecture 3: the method via the diagonal class

Motivation Let X be a projective algebraic manifold of dimension n, and consider the diagonal class $\Delta \in \mathrm{CH}^n(X \times X)$. By collecting the coefficients of the polynomials cutting out X in some $\mathbb{P}^N$ (a finite number by Hilbert's basis theorem), we can assume that $X = X/_k$ is defined over a field k, of finite transcendence degree over $\mathbb{Q}$. Let $\eta \in \mathrm{Spec}(X/_k)$ be the generic point. Then $\mathcal{O}_{X,\eta} = k(\eta) = k(X)$, the function field of X. Set $L = k(\eta)$. Then clearly we can view Δ as defining a family of 0-cycles in X, parameterized by X, with generic fiber the generic point $\eta = \Delta_\eta = \Delta_*(\eta) \in CH^n(X_L)$. The idea of studying the Chow group via the generic 0-cycle $\{\eta\} \in CH^n(X_L)$ was suggested by Colliot-Thélène, and exploited by Bloch and Srinivas (see [Blo] and [B-S]). We explain the ideas below.

Lemma 3.0

$$CH^\bullet(X_L) = \varinjlim_{\substack{U \subset X/k \\ \text{Zariski open}}} CH^\bullet(X/_k \times_k U).$$

Proof First of all,

$$(X_L)^\bullet = \varinjlim_{\substack{U \subset X/k \\ \text{Zariski open}}} (X/_k \times_k U)^\bullet,$$

where $(\cdots)^\bullet = $ points of codimension $\bullet$, and

$$CH^\bullet(X_L) = \operatorname{coker}\left(\coprod_{x \in (X_L)^{\bullet-1}} L(x)^\times \xrightarrow{\text{div}} \coprod_{x \in (X_L)^\bullet} \mathbb{Z} \right),$$

$$CH^\bullet(X \times U) = \operatorname{coker}\left(\coprod_{y \in (X \times U)^{\bullet-1}} k(y)^\times \xrightarrow{\text{div}} \coprod_{y \in (X \times U)^\bullet} \mathbb{Z} \right).$$

Now take the direct limit of the latter expression to get the former.

Now consider an embedding $L \hookrightarrow \mathbb{C}$.

Lemma 3.1 *The kernel of the pull-back* $CH^\bullet(X_L) \to CH^\bullet(X_\mathbb{C})$ *is torsion.*

Proof Step I Suppose that K/L is a finite extension. Then one has a finite proper map $X_K \to X_L$, hence a norm $CH^\bullet(X_K) \to CH^\bullet(X_L)$. Thus the kernel of the pull-back $CH^\bullet(X_L) \to CH^\bullet(X_K)$ is torsion.

Step II Let $\overline{L}$ be the algebraic closure of L in $\mathbb{C}$. Then the kernel of the pull-back $CH^\bullet(X_L) \to CH^\bullet(X_{\overline{L}})$ is torsion.

Proof Suppose $\xi \in CH^\bullet(X_L)$ has the property that $\xi = 0$ in $CH^\bullet(X_{\overline{L}})$. Then by collecting the coefficients in $\overline{L}$ of polynomials defining $\xi \sim_{\text{rat}} 0$, we can assume that $\xi = 0$ in $CH^\bullet(X_K)$, for some finite extension K/L. By Step I, we are done.

Step III The kernel of the pull-back $\mathrm{CH}^\bullet(X_{\overline{L}}) \to \mathrm{CH}^\bullet(X_{\mathbb{C}})$ is zero.

Proof We can write:

$$\mathrm{CH}^\bullet(X_{\mathbb{C}}) = \varinjlim_{U/\overline{L} \text{ finite type}} \mathrm{CH}^\bullet(X_{\overline{L}} \times_{\overline{L}} U).$$

But $\overline{L}$ is algebraically closed, hence $U(\overline{L}) \neq \emptyset$. But an $\overline{L}$-point of U gives a section of $\mathrm{CH}^\bullet(X_{\overline{L}}) \to \mathrm{CH}^\bullet(X_{\overline{L}} \times_{\overline{L}} U)$. Thus $\xi \mapsto 0 \in \mathrm{CH}^\bullet(X_{\mathbb{C}}) \Rightarrow \xi \mapsto 0 \in \mathrm{CH}^\bullet(X_{\overline{L}} \times_{\overline{L}} U)$ for some $U/_{\overline{L}}$. Hence $\xi = 0$ in $\mathrm{CH}^\bullet(X_{\overline{L}})$.

Now let $D \subset X$ be a subvariety, and assume that $\mathrm{CH}^n(X \backslash D)_{\mathbb{Q}} = 0$. We can view X, D as defined over a given field k of finite transcendence degree over $\mathbb{Q}$, as mentioned earlier. Thus:

$$\eta = \Delta(\eta) \in \mathrm{CH}^n(X_L) \to \mathrm{CH}^n((X \backslash D)_L)$$
$$\downarrow$$
$$\mathrm{CH}^n((X \backslash D)_{\mathbb{C}})_{\mathbb{Q}} = 0.$$

Hence $N\{\eta\} \mapsto 0 \in \mathrm{CH}^n((X \backslash D)_L)$ for some $N \in \mathbb{N}$ by Lemma 3.1. By Lemma 3.0,

$$\mathrm{CH}^n((X \backslash D)_L) = \varinjlim_{U \subset X/k} \mathrm{CH}^n((X \backslash D)/_k \times_k U),$$

hence $N\Delta \mapsto 0 \in \mathrm{CH}^n((X \backslash D)/_k \times_k U)$ for some $U \subset X/_k$. Set $E := X \backslash U$. By using the exact sequence:

$$\left\{ \begin{array}{l} \text{Cycles supported on} \\ (D \times X) \cup (X \times E) \end{array} \right\} \to \mathrm{CH}^n(X/_k \times_k X/_k)$$
$$\to \mathrm{CH}^n(U \times_k (X \backslash D)/_k) \to 0,$$

it follows that $N\Delta \sim_{\mathrm{rat}} \Gamma_1 + \Gamma_2$, where Γ_1 is supported on $D \times X$ and Γ_2 is supported on $X \times E$.

Corollary 3.2 [B-S] *Given the notation above:*

(i) *If* $\dim D \leq 3$, *then* $\mathrm{Hodge}^{2,2}(X, \mathbb{Q})$ *holds.*

(ii) *If* $\dim D \leq 2$, *then the Abel–Jacobi map (see Section 6.1.5)* $\Phi_2 : z^2_{\mathrm{hom}}(X) \to J^2(X)$ *is surjective; and more specifically,* $J^2_a(X) = J^2(X)$, *i.e.* $\Phi_2(z^2_{\mathrm{alg}}(X)) = J^2(X)$. *In particular,* $\mathrm{GHC}(1, 3, X)$ *holds.*

Proof We have

$$H^\bullet(X, \mathbb{C}) = N[\Delta]_* H^\bullet(X, \mathbb{C}) = [\Gamma_1]_* H^\bullet(X, \mathbb{C}) + [\Gamma_2]_* H^\bullet(X, \mathbb{C}).$$

We work with the diagrams below, where

$$(\tilde{E} \xrightarrow{\sigma} E \hookrightarrow X, \ \tilde{D} \xrightarrow{\sigma} D \hookrightarrow X),$$

are desingularizations, together with the projection formula

$$\tilde{D} \times X$$
$$\downarrow {\scriptstyle \sigma \times 1}$$

$$\pi_1 \swarrow \qquad X \times X \qquad \searrow \pi_2$$
$$\text{Pr}_1 \swarrow \qquad\qquad \searrow \text{Pr}_2$$
$$\tilde{D} \xrightarrow{j} X \qquad\qquad\qquad X$$

$$X \times \tilde{E}$$
$$\downarrow {\scriptstyle 1 \times \sigma}$$

$$\pi_1 \swarrow \qquad X \times X \qquad \searrow \pi_2$$
$$\text{Pr}_1 \swarrow \qquad\qquad \text{Pr}_2 \searrow$$
$$X \qquad\qquad\qquad X \xleftarrow{j} \tilde{E}$$

$$\left[\tilde{\Gamma}_1 \xmapsto{(\sigma \times 1)_*} \Gamma_1 \quad ; \quad \tilde{\Gamma}_2 \xmapsto{(1 \times \sigma)_*} \Gamma_2 \right]$$

Step I

$$[\Gamma_1]_* H^\bullet(X) = \text{Pr}_{2,*}\big([\Gamma_1] \bullet \text{Pr}_1^* H^\bullet(X)\big)$$

$$= \text{Pr}_{2,*}\big((\sigma \times 1)_* [\tilde{\Gamma}_1] \bullet \text{Pr}_1^* H^\bullet(X)\big)$$

$$= \text{Pr}_{2,*} \circ (\sigma \times 1)_* \big([\tilde{\Gamma}_1] \bullet (\sigma \times 1)^* \circ \text{Pr}_1^* H^\bullet(X)\big)$$

$$= \pi_{2,*}\big([\tilde{\Gamma}_1] \bullet \pi_1^* \circ j^* H^\bullet(X)\big), \ \text{use} \begin{cases} j \circ \pi_1 = \text{Pr}_1 \circ (\sigma \times 1), \\ \pi_2 = \text{Pr}_2 \circ (\sigma \times 1) \end{cases}$$

$$=: [\tilde{\Gamma}_1]_* (j^* H^\bullet(X)).$$

But $j^* H^\bullet(X) \subset H^\bullet(\tilde{D})$, hence $[\Gamma_1]_* H^\bullet(X) \subset [\tilde{\Gamma}_1]_* H^\bullet(\tilde{D})$.

(For Step II below, we can assume that $E \subset X$ is a divisor.)

Step II

$$[\Gamma_2]_* H^\bullet(X) = \mathrm{Pr}_{2,*}\left([\Gamma_2]\bullet \mathrm{Pr}_1^* H^\bullet(X)\right)$$

$$= \mathrm{Pr}_{2,*}\left((1 \times \sigma)_*[\tilde{\Gamma}_2]\bullet \mathrm{Pr}_1^* H^\bullet(X)\right)$$

$$= \mathrm{Pr}_{2,*} \circ (1 \times \sigma)_*\left([\tilde{\Gamma}_2]\bullet(1 \times \sigma)^* \circ \mathrm{Pr}_1^* H^\bullet(X)\right)$$

$$= j_* \circ \pi_{2,*}\left([\tilde{\Gamma}_2]\bullet\pi_1^* H^\bullet(X)\right), \text{ use} \begin{cases} j \circ \pi_2 = \mathrm{Pr}_2 \circ(1 \times \sigma), \\ \pi_1 = \mathrm{Pr}_1 \circ(1 \times \sigma) \end{cases}$$

$$= j_*[\tilde{\Gamma}_2]_* H^\bullet(X) \subset j_* H^{\bullet-2}(\tilde{E}),$$

hence $[\Gamma_2]_* H^\bullet(X) \subset j_* H^{\bullet-2}(\tilde{E})$.

Now set $\bullet = 4$ in Steps I and II above. Since the Hodge conjecture holds for projective algebraic manifolds of dimension ≤ 3, together with the Lefschetz $(1, 1)$ theorem, it follows that $\mathrm{Hodge}^{2,2}(X, \mathbb{Q})$ holds, i.e. Corollary 3.2(i) holds. Corollary 3.2(ii) is a consequence of the above and some functorial properties of the Abel–Jacobi mapping, together with some classical results on divisors and 0-cycles. This will be left to the reader.

Exercise Let X be a projective algebraic manifold of dimension n. Suppose that $\mathrm{CH}^n(X\backslash D)_{\mathbb{Q}} = 0$ for some divisor $D \subset X$. Show that $\mathrm{GHC}(1, n, X)$ holds. Hence using remark (1) following Proposition 3.3 in Section 6.3.1 below, deduce $\mathrm{GHC}(1, n, X)$ for any smooth complete intersection $X \subset \mathbb{P}^{n+r}$ of dimension n, which satisfies $H^{n,0}(X) = 0$.

6.3.1 Appendix

Proposition 3.3 (See [Ro, Section 4]) *Let $X \subset \mathbb{P}^{n+1}$ be a smooth hypersurface of degree $d \leq n + 1$. (Note: $d \leq n + 1 \Leftrightarrow H^{n,0}(X) = 0$.) Then $\mathrm{CH}^n(X\backslash\{p\})_{\mathbb{Q}} = 0$, for any $p \in X$.*

Remarks

(1) A similar story holds for smooth complete intersections X of dimension n in $\mathbb{P}^{n+r}$ with $H^{n,0}(X) = 0$ [Ro, Section 4] and in fact Roitman proves in this case that $\mathrm{CH}^n_{\mathrm{alg}}(X) = 0$.

(2) Proposition 3.3 can also be deduced from Section 6.2.2, so long as X is general.

Proof of Proposition 3.3 It suffices to show that $N \cdot (\{p - q\}) = 0$ for any points $p, q \in X$ and some integer $N \neq 0$. First fix a point $p \in X$.

By a PGL action, we may assume that $p = [1, 0, \ldots, 0]$ in the homogeneous coordinates $[z_0, \ldots, z_{n+1}] \in \mathbb{P}^{n+1}$. Write $X = \{F = 0\} \subset \mathbb{P}^{n+1}$, $F = F(z_0, \ldots, z_{n+1})$ homogeneous of degree d. Since $p \in X$, it follows that $F(z_0, \ldots, z_{n+1}) = \sum_{j=1}^{d} z_0^{d-j} F_j(z_1, \ldots, z_{n+1})$, where F_j is homogeneous of degree j in $(z_1, \ldots, z_{n+1})$. Now consider the local affine coordinates $(x_1, \ldots, x_{n+1}) = (\frac{z_1}{z_0}, \ldots, \frac{z_{n+1}}{z_0})$, where p now corresponds to $(0, \ldots, 0)$. Write $f = f(x_1, \ldots, x_{n+1}) = \frac{F}{z_0^d} = \sum_{j=1}^{d} f_j(x_1, \ldots, x_{n+1})$, $f_j = F_j / z_0^j$ homogeneous of degree j in $(x_1, \ldots, x_{n+1})$. Any line $\ell := \mathbb{P}^1 \subset \mathbb{P}^{n+1}$ passing through $p \in X$, corresponds in affine coordinates to a line $\mathbb{C} \subset \mathbb{C}^{n+1}$ passing through $(0, \ldots, 0) \in \{f = 0\} \subset \mathbb{C}^{n+1}$. Thus $\ell_{[v]}$ through $(0, \ldots, 0)$ looks like $\ell(t) = \{tv | t \in \mathbb{C}\}$, where $v \in \mathbb{C}^{n+1} \backslash \{0\}$, i.e. $[v] \in \mathbb{P}^n$. Note that $f(tv) = t f_1(v) + \cdots + t^d f_d(v)$. Thus $\ell \subset X \Leftrightarrow [v] \in \{f_1 = \cdots = f_d = 0\} \subset \mathbb{P}^n$, and either $\{\ell \subset X$ or $\ell \cap X = dp\} \Longleftrightarrow [v] \in \{f_1 = \cdots = f_{d-1} = 0\} \subset \mathbb{P}^n$. Set $\Sigma_p = \{f_1 = \cdots = f_{d-1} = 0\} \subset \mathbb{P}^n$, and note that $\dim \Sigma_p \geq n - (d - 1) = n + 1 - d \geq 0$, since $d \leq n + 1$; moreover, $\deg \Sigma_p \leq M$, where $M = (d - 1)!$ (Bezout's theorem). By taking hyperplane sections, we can assume that $\dim \Sigma_p = 0$. Set $Z_p = \cup_{[v] \in \Sigma_p} \ell_{[v]}$. Then Z_p is a cone with vertex p, and any point $p' \in Z_p$ has the property that $p' \sim_{\mathrm{rat}} p$. Let $M(p) = \deg Z_p$, $M(q) = \deg Z_q$, for any given $p, q \in X$. Then $M(p)$, $M(q) \leq (d - 1)!$ and $M(p)Z_q - M(q)Z_p \sim 0 = CH_1(\mathbb{P}^{n+1})_{\deg 0}$. Thus $N(\{p - q\}) = 0$, for some $N \leq d((d - 1)!)^2$. (Note: Since N is bounded, it likewise follows that $\mathrm{CH}_{\mathrm{alg}}^n(X) = 0$, using the well known divisibility of $\mathrm{CH}_{\mathrm{alg}}^n(X)$.)

6.3.2 Appendix

The method of [B-S] generalizes as follows (see [Pa] for details).

Proposition 3.4 *Let X be a projective algebraic manifold, $\dim X = n$, and assume given subvarieties $Y_j \subset X$ for each $j \in \{0, \ldots, m\}$ such that $\mathrm{CH}_j(X \backslash Y_j)_{\mathbb{Q}} = 0$ for each j. Then for each j, we have cycles $\Gamma_j \in \mathrm{CH}^n(X \times X)_{\mathbb{Q}}$ such that $|\Gamma_j| \subset Y_j \times X$, and a cycle $\Gamma^{m+1} \in \mathrm{CH}^n(X \times X)_{\mathbb{Q}}$ such that $|\Gamma^{m+1}| \subset X \times W$, where $W \subset X$ has pure codimension $m + 1$, so that the diagonal class $\Delta_X \in \mathrm{CH}^n(X \times X)_{\mathbb{Q}}$ decomposes into:*

$$\Delta_X = \Gamma_0 + \cdots + \Gamma_m + \Gamma^{m+1}.$$

Proof By induction on $m \geq -1$, the statement is clearly true for $m = -1$, where $\Delta_X = \Gamma^0$. By induction, we can assume given

$$\Delta_X = \Gamma_0 + \cdots + \Gamma_{m-1} + \Gamma^m,$$

where accordingly, $|\Gamma^m| \subset X \times W'$, W' having pure codimension m on X. Further, we assume that $\mathrm{CH}_m(X \backslash Y_m)_\mathbb{Q} = 0$ for some subvariety $Y_m \subset X$. There is no loss of generality in assuming that W' is irreducible, and that X, W', Γ^m are defined over a subfield $k \subset \mathbb{C}$ with $\mathrm{trdeg}_\mathbb{Q} k < \infty$. Note that Γ^m, being supported on $X \times W'$, determines a corresponding cycle class $\tilde{\Gamma}^m \in \mathrm{CH}^{n-m}(X \times W')$, and thus a class in $\mathrm{CH}^{n-m}(X_L)$, by restriction to the generic point of W'/k, where $L = k(W')$. Next, $\mathrm{CH}^{n-m}(X_L) = \mathrm{CH}_m(X_L)$ and $\mathrm{CH}_m((X \backslash Y_m)_\mathbb{C})_\mathbb{Q} = 0$. Now fix an embedding $L \hookrightarrow \mathbb{C}$. Using the injection $\mathrm{CH}_m((X \backslash Y_m)_L)_\mathbb{Q} \hookrightarrow \mathrm{CH}_m((X \backslash Y_m)_\mathbb{C})_\mathbb{Q}$ (Lemma 3.1), it follows by the same arguments in Lecture 3, that in $\mathrm{CH}^n(X \times X)_\mathbb{Q}$, $\Gamma^m = \Gamma_m + \Gamma^{m+1}$ where $|\Gamma_m| \subset Y_m \times X$, and $|\Gamma^{m+1}| \subset X \times W$, where W has pure codimension 1 in W', i.e. has pure codimension $m + 1$ in X. $\square$

Exercise Using Proposition 3.4, deduce the following:

Corollary 3.5 *Assume a given projective algebraic manifold X. Suppose that $\mathrm{CH}_\ell(X)_\mathbb{Q} \simeq \mathbb{Q}$ for all $\ell \in \{0, \ldots, m\}$. Then $H^p(X, \mathbb{Q}) = N^{m+1} H^p(X, \mathbb{Q})$ for all $p \geq 2m + 1$. In particular,* $\mathrm{GHC}(m + 1, p, X)$ *holds for all $p \geq 2m + 1$.*

6.4 Appendix: the Tate conjecture, absolute Hodge cycles, and some recent developments

We recall the statement of the Hodge conjecture (classical version), namely: for a smooth projective variety of dimension n over $\mathbb{C}$, the cycle class map

$$\mathrm{cl}_k : \mathrm{CH}^k(X) \otimes \mathbb{Q} \to H^{2k}(X, \mathbb{Q}) \cap H^{k,k}(X),$$

is surjective. By Poincaré duality, the space $H^{2k}(X, \mathbb{Q}) \cap H^{k,k}(X)$ is identified with

$$\left\{ \gamma \in H_{2n-2k}(X, \mathbb{Q}) \ \Big| \ \int_\gamma \omega = 0, \forall \, \omega \in F^{n-k+1} H^{2n-2k}(X, \mathbb{C}) \right\}.$$

In other words, the space of analytic $=$ algebraic (co)cycles can be 'computed' using Hodge theory. Actually, as indicated earlier (in a footnote, regarding the exponential short exact sequence), it is more natural to throw in twists by introducing the pure Hodge structure $\mathbb{Q}(m) := \mathbb{Q}(2\pi \sqrt{-1})^m$ of type $(-m, -m)$, for any given integer m, and in which case the cycle map becomes

$(*)$ $\mathrm{cl}_k : \mathrm{CH}^k(X) \otimes \mathbb{Q} \to H^{2k}(X, \mathbb{Q}(k)) \cap H^{k,k}(X),$

which is $(2\pi\sqrt{-1})^k$ times the fundamental class of an algebraic cycle. Thus the image of the cycle class map in $(*)$ lies in the $\mathbb{Q}(k)$-classes of Hodge type $(0,0)$. (Note $\mathbb{Q}(1)$ is called the Tate twist, but more generally see below.) The Hodge conjecture has an arithmetical analogue, namely the Tate conjecture, which we will state briefly. First, we will assume that the reader has some familiarity with ℓ-adic cohomology, although only the formal properties of this cohomology are needed here. A good reference for this is the book by Milne [Mi1]. As only the formal properties of ℓ-adic cohomology are needed, and for a description of these properties, the reader can also consult Hartshorne's book [Har, Appendix C]. Let X be a smooth projective variety over a field L, with algebraic closure $\overline{L}$, $G = \mathrm{Gal}(\overline{L}/L)$ the Galois group, and $\overline{X} = X \times_L \overline{L}$. For a prime $\ell \neq \mathrm{char}(L)$, we consider the ℓ-adic field $\mathbb{Q}_\ell$, namely the quotient field of the ℓ-adic integers

$$\mathbb{Z}_\ell := \lim_{\leftarrow} \mathbb{Z}/\ell^r \mathbb{Z},$$

and the ℓ-adic cohomology

$$H^i_{\mathrm{et}}(\overline{X}, \mathbb{Q}_\ell) = \left(\lim_{\leftarrow} H^i_{\mathrm{et}}(\overline{X}, \mathbb{Z}/\ell^r \mathbf{Z}) \right) \otimes \mathbb{Q}_\ell.$$

As in the transcendental case, there is a cycle class map:

$$\mathrm{cl}_k : \mathrm{CH}^k(X) \otimes \mathbb{Q}_\ell \to H^{2k}_{\mathrm{et}}(\overline{X}, \mathbb{Q}_\ell),$$

and likewise, it is more natural to modify the weights by introducing the Tate twist[11] (see [Ta1]), i.e.

$$\mathrm{cl}_k : \mathrm{CH}^k(X) \otimes \mathbb{Q}_\ell \to H^{2k}_{\mathrm{et}}(\overline{X}, \mathbb{Q}_\ell(k)),$$

whose image lies in the subspace $H^{2k}_{\mathrm{et}}(\overline{X}, \mathbb{Q}_\ell(k))^G$ of classes invariant under G. The celebrated Tate conjecture [Ta1] asserts that for a finitely generated field L (i.e. finitely generated over the prime field), the cycle class map

$$\mathrm{cl}_k : \mathrm{CH}^k(X) \otimes \mathbb{Q}_\ell \to H^{2k}_{\mathrm{et}}(\overline{X}, \mathbb{Q}_\ell(k))^G,$$

is surjective.

---

[11] Set $\mathbb{Q}_\ell(1) := \mathbb{Z}_\ell(1) \otimes_{\mathbb{Z}_\ell} \mathbb{Q}_\ell$ where

$$\mathbb{Z}_\ell(1) := \lim_{\leftarrow} \mu_{\ell^n},$$

and where μ_{ℓ^n} are the ℓ^nth roots of unity in $\overline{L}$, and which has a natural G action. Then the definition of twisted ℓ-adic cohomology can be taken to be: $H^{2k}_{\mathrm{et}}(\overline{X}, \mathbb{Q}_\ell(k)) = H^{2k}_{\mathrm{et}}(\overline{X}, \mathbb{Q}_\ell) \otimes \mathbb{Q}_\ell(1)^{\otimes k}$. Note: $\mathbb{Q}_\ell(m)$ for $m < 0$ is also defined [Ta1], although we do not need it here.

Exercise Let $X \subset \mathbb{P}^5$ be a general quintic fourfold over $\mathbb{C}$. Note that X can be defined over a field of finite transcendence degree over $\mathbb{Q}$. Formulate a version of the Tate conjecture for X in terms of the Fano variety of lines on X. (Warning: Unlike the Hodge conjecture, the Tate conjecture is not known in general for surfaces.)

A survey of the status of the Tate conjecture as of 1994 can be found in [Ta2]. Some special cases mentioned there are the following. In the case of *divisors* (i.e. codimension 1 cycles), the conjecture has been verified in a number of instances, such as for abelian varieties (Tate, Zarhin in characteristic $p > 0$; Faltings in characteristic 0, as part of his proof of the Mordell conjecture), for $K3$ surfaces in characteristic 0 (and with some restrictions in characteristic $p > 0$), and for various types of modular surfaces and threefolds. For higher codimensional cycles, the conjecture has been verified for a number of Fermat hypersurfaces (due to Tate, Shioda) as well as for many classes of abelian varieties (due to Tate, Tankeev, Murty, Shioda *et al.*). The reader is encouraged to consult the references cited there [Ta2].

It is general 'yoga' that the Hodge and Tate conjectures are like opposite sides of the same coin. That yoga abounds, for indeed there are a number of examples in the literature where the Hodge and Tate conjectures can be verified simultaneously. For example, from the work of [P], it is known that for an abelian variety X defined over a finitely generated subfield $L \subset \mathbb{C}$, the Tate conjecture implies the Hodge conjecture. Also, as another example, it is shown in [Mi3] that if the Hodge conjecture holds for all abelian varieties of CM type over $\mathbb{C}$, then the Tate conjecture holds for all abelian varieties defined over the algebraic closure of a finite field. Another case in point is the comparison of both conjectures in [Hu, Section 8.9], making use of the absolute Hodge cycles, which we will now describe.

One way of incorporating the analytic and arithmetic points of view is via Deligne's notion of absolute Hodge cycles. Let us now assume that $\mathrm{char}(L) = 0$, with X defined over L, and from now on we will assume that $L = \overline{L}$ is algebraically closed and of finite transcendence degree over $\mathbb{Q}$. Thus there is an embedding $\sigma : L \hookrightarrow \mathbb{C}$. The comparison isomorphism theorems give

(A.0) $H^i(X \times_{L,\sigma} \mathbb{C}, \mathbb{Q}) \otimes_{\mathbb{Q}} \mathbb{Q}_\ell \simeq H^i_{\mathrm{et}}(X \times_{L,\sigma} \mathbb{C}, \mathbb{Q}_\ell) \simeq H^i_{\mathrm{et}}(X, \mathbb{Q}_\ell),$

and

(A.1) $$H^i(X \times_{L,\sigma} \mathbb{C}, \mathbb{Q}) \otimes_{\mathbb{Q}} \mathbb{C} \simeq H^i_{\mathrm{DR}}(X \times_{L,\sigma} \mathbb{C}, \mathbb{C})$$
$$\simeq H^i_{\mathrm{DR}}(X/L) \otimes_{L,\sigma} \mathbb{C},$$

where $H^i(X \times_{L,\sigma} \mathbb{C}, \mathbb{Q})$ is singular cohomology, and $H^i_{DR}(X/L)$ is algebraic de Rham cohomology (see [DMOS] for a definition), and where $H^i_{DR}(X/\mathbb{C}) :=$ $H^i_{DR}(X/\mathbb{C}, \mathbb{C})$ agrees with ordinary de Rham cohomology. Now consider the products

$$(A.2) \qquad H^i_{DR}(X \times_{L,\sigma} \mathbb{C}) \times \left(\prod_{\ell} H^i_{et}(X \times_{L,\sigma} \mathbb{C}, \mathbb{Q}_{\ell}) \right),$$

$$(A.3) \qquad H^i_{DR}(X/L) \otimes_{L,\sigma} \mathbb{C} \times \left(\prod_{\ell} H^i_{et}(X, \mathbb{Q}_{\ell}) \right).$$

From (A.0) and (A.1), one has the following data.

(1) An isomorphism which we will denote by $\sigma^* = \sigma^*_{DR} \times \sigma^*_{et}$:

$$\sigma^* : H^i_{DR}(X/L) \otimes_{L,\sigma} \mathbb{C} \times \left(\prod_{\ell} H^i_{et}(X, \mathbb{Q}_{\ell}) \right)$$

$$\xrightarrow{\sim} H^i_{DR}(X \times_{L,\sigma} \mathbb{C}) \times \left(\prod_{\ell} H^i_{et}(X \times_{L,\sigma} \mathbb{C}, \mathbb{Q}_{\ell}) \right),$$

(2) a diagonal embedding

$$H^i(X \times_{L,\sigma} \mathbb{C}, \mathbb{Q}) \hookrightarrow H^i_{DR}(X \times_{L,\sigma} \mathbb{C}) \times \left(\prod_{\ell} H^i_{et}(X \times_{L,\sigma} \mathbb{C}, \mathbb{Q}_{\ell}) \right).$$

Now roughly speaking, an absolute Hodge cycle relative to σ is a class $\xi \in H^{2k}_{DR}(X/L) \times \left(\prod_{\ell} H^{2k}_{et}(X, \mathbb{Q}_{\ell}) \right)$, for which $\sigma^*(\xi)$ belongs to the rational subspace $H^{2k}(X \times_{k,\sigma} \mathbb{C}, \mathbb{Q})$ and for which the de Rham component of $\sigma^*(\xi)$ lies in $F^k H^{2k}_{DR}(X \times_{L,\sigma} \mathbb{C})$. A precise statement of an absolute Hodge cycle relative to σ involves twists. Thus to state this precisely, we briefly introduce these twists. For an integer $m \geq 1$, let $\mu_m(L) = \{ \zeta \in L \mid \zeta^m = 1 \}$, and put

$$\mathbf{A}^f(1) = \left(\varprojlim \mu_m(L) \right) \otimes_{\mathbf{Z}} \mathbb{Q}.$$

One sets $\mathbf{A}^f(m) = \mathbf{A}^f(1)^{\otimes m}$. There is a natural extension of $\mathbf{A}^f(m)$ for negative m, but we will omit this here [DMOS, p. 19]. (Here $\mathbf{A}^f(m)$ is a module over the so-called ring $\mathbf{A}^f$ of finite adèles for $\mathbb{Q}$.) Now put

$$H^i_{et,f}(X) = \prod_{\ell} H^i_{et}(X, \mathbb{Q}_{\ell}) \otimes_{\mathbf{Z}} \mathbb{Q},$$

and

$$H^i_{et,f}(X)(m) = H^i_{et,f}(X) \otimes_{\mathbf{A}^f} \mathbf{A}^f(m).$$

Now with twists thrown into the picture, we have the following data:

(1) An isomorphism

$$\sigma^* : \left(H^i_{\mathrm{DR}}(X/_L)(m) \otimes_{L,\sigma} \mathbb{C}\right) \times H^i_{\mathrm{et,f}}(X)(m)$$

$$\xrightarrow{\sim} \left(H^i_{\mathrm{DR}}(X \times_{L,\sigma} \mathbb{C})(m) \otimes_{L,\sigma} \mathbb{C}\right) \times H^i_{\mathrm{et,f}}(X \times_{L,\sigma} \mathbb{C})(m),$$

(2) an embedding

$$H^i(X \times_{k,\sigma} \mathbb{C}, \mathbb{Q}(m)) \hookrightarrow \left(H^i_{\mathrm{DR}}(X \times_{L,\sigma} \mathbb{C})(m) \otimes_{L,\sigma} \mathbb{C}\right)$$

$$\times H^i_{\mathrm{et,f}}(X \times_{L,\sigma} \mathbb{C})(m).$$

Definition A.4 A class

$$\xi \in \left(H^{2k}_{\mathrm{DR}}(X/_L)(m)\right) \times H^i_{\mathrm{et,f}}(X)(m)$$

is called a Hodge cycle relative to σ, if $\sigma^*(\xi)$ lies in the rational subspace $H^{2k}(X \times_{L,\sigma} \mathbb{C}, \mathbb{Q}(k))$, and the de Rham component of ξ is contained in $F^0 H^{2k}_{\mathrm{DR}}(X \times_{L,\sigma} \mathbb{C})(k) = F^k H^{2k}_{\mathrm{DR}}(X \times_{L,\sigma} \mathbb{C})$. If ξ is a Hodge cycle relative to all embeddings $\sigma : L \hookrightarrow \mathbb{C}$, then it is called an absolute Hodge cycle.

Note that any algebraic cycle on $X/_L$ is an absolute Hodge cycle and conversely the Hodge conjecture implies that all absolute Hodge cycles are algebraic cycles.

It is anticipated that every Hodge cycle is an absolute Hodge cycle, i.e. if ξ is a Hodge cycle relative to one embedding $\sigma : L \hookrightarrow \mathbb{C}$, then it is a Hodge cycle relative to all embeddings σ. So far, this has been proven by Deligne in the case where X is an abelian variety [DMOS].

Next, we want to discuss some recent developments on the Hodge conjecture since the appearance of the survey book [Lew2]. For recent works on abelian varieties, the reader can consult for example [Mur], [Ge], [Ge-V], [Mi2], [Mi3], [M-Z1], [M-Z2], [Ha] and [Ab1]–[Ab6]. One of the more interesting developments that we wish to explain is the recent work of Abdulali [Ab6]. To explain his result, we introduce the following terminology. Let $V = V_\mathbb{Q}$ be a finite dimensional $\mathbb{Q}$-vector space with Hodge structure

$$V_\mathbb{C} := V_\mathbb{Q} \otimes_\mathbb{Q} \mathbb{C} = \bigoplus_{p+q=N} V^{p,q},$$

of weight N. The Hodge structure V is said to be *geometric* if it is isomorphic to a Hodge substructure of the cohomology of a smooth projective variety over $\mathbb{C}$, and *effective* if $V^{p,q} = 0$ unless $p, q \geq 0$. Let m be an integer. The twisted Hodge structure $V(m) = V_\mathbb{Q} \otimes \mathbb{Q}(2\pi\sqrt{-1})^m$ is the Hodge structure of weight $N - 2m$ with $V(m)^{p,q} = V^{p-m,q-m}$. The Grothendieck amended Hodge conjecture implies that the twist $V(m)$ of any geometric Hodge structure V is still geometric provided that $V(m)$ is effective. We say that a smooth complex projective variety X is *dominated* by a class of varieties $\mathcal{Y}$ if, for every irreducible Hodge structure W occurring in the cohomology of X, the twist $W(r)$ occurs in the cohomology of some member of $\mathcal{Y}$, where if N is the weight of the Hodge structure W, then

$$r := \min\{p \mid W^{p,N-p} \neq 0\},$$

is the *height* of the Hodge structure. As observed by Grothendieck [Gro], the (always assumed amended) general Hodge conjecture for X is implied by the existence of a class of varieties $\mathcal{Y}$ which dominate X, together with the classical Hodge conjecture for $Y \times X$ for all $Y \in \mathcal{Y}$.

Exercise Prove this!

Abdulali shows in a series of papers [Ab2], [Ab4], [Ab5] that certain abelian varieties are dominated by subclasses of the class of all abelian varieties, and he used this to deduce the general Hodge conjecture in some cases. This raises the question as to whether every abelian variety is dominated by the class of all abelian varieties. Abdulali [Ab6] shows that the answer to this question is *no*. As indicated in his paper, this suggests that those extraordinary Hodge structures that he constructs can be used as a testing ground for the general Hodge conjecture.

Acknowledgement

I would like to thank Chris Peters and Stefan Müller-Stach for their superb organizational skills in putting together this wonderful conference, and for making our stay in Grenoble very agreeable. I would also like to thank the referee for his or her careful reading and constructive comments on this paper. Finally, I am very grateful to Noriko Yui for her constructive comments regarding Section 6.4 on the Tate conjecture. This work was partially supported by a grant from the Natural Sciences and Engineering Research Council of Canada.

References

[Ab1]   Abdulali, S. Algebraic cycles in families of Abelian varieties, *Can. J. Math.* **46** (1994) 1121–1134.

[Ab2]   Abelian varieties and the general Hodge conjecture, *Compositio Math.* **109** (1997) 341–355.

[Ab3]   Abelian varieties of type III and the Hodge conjecture, *Int. J. Math.* **10** (1999) 667–675.

[Ab4]   Filtrations on the cohomology of Abelian varieties. In: *The Arithmetic and Geometry of Algebraic Cycles, Banff, 1998* (B. B. Gordon, J. D. Lewis, S. Müller-Stach, S. Saito and N. Yui, eds.), *CRM Proc. Lecture Notes* **24**, pp. 3–12, AMS, Providence, RI, 2000.

[Ab5]   Hodge structures on Abelian varieties of CM-type, *J. reine angew. Math.* **534** (2001) 33–39.

[Ab6]   Hodge structures on Abelian varieties of type III, *Ann. Math.* **155** (2002) 925–928.

[Be]   Beilinson, A. Higher regulators and values of *L*-functions, *J. Sov. Math.* **30** (1985) 2036–2070.

[Blo]   Bloch, S. On an argument of Mumford in the theory of algebraic cycles. In: *Algebraic Geometry, Angers, 1979* (A. Beauville, ed.), pp. 217–221, Nijthoff & Noordhoff, Rockville MD, 1980.

[B-M]   Bloch, S. and Murre, J. On the Chow groups of certain types of Fano threefolds, *Compositio Math.* **39** (1979) 47–105.

[B-S]   Bloch, S. and Srinivas, V. Remarks on correspondences and algebraic cycles, *Am. J. Math.* **105** (1983) 1235–1253.

[C-G]   Clemens, C. H. and Griffiths, P. A. The intermediate jacobian of the cubic threefold, *Ann. Math.* **95** (2) (1972) 281–356.

[Co]   Collino, A. The Abel–Jacobi isomorphism for the cubic fivefold, *Pac. J. Math.* **122** (1) (1986) 43–55.

[Co-M]   Collino, A. and Murre, J. P. The intermediate jacobian of a cubic threefold with one ordinary double point (I) and (II), *Proc. Koninklijke Ned. Akad. Wetenschap., Amst.* **81** (1978) 43–71.

[C-M1]   Conte A. and Murre, J. P. The Hodge conjecture for hypersurfaces admitting a covering by rational curves, *Math. Ann.* **238** (1) (1978) 79–88.

[C-M2]   The Hodge conjecture for Fano complete intesections of dimension four. In: *Algebraic Geometry, Angers, 1979* (A. Beauville, ed.), pp. 129–141, Nijthoff & Noordhoff, Germantown, MD, 1980.

[De]   Deligne, P. Théorie de Hodge. III, *Inst. Hautes Études Sci. Publ. Math.* **44** (1974) 5–77.

[DMOS]   Deligne, P., Milne, J., Ogus, A. and Shih, K. *Hodge Cycles, Motives, and Shimura Varieties, Lecture Notes in Mathematics* **900**, Springer-Verlag, Berlin, 1989.

[Ge]   van Geemen, B. Kunga-Satake varieties and the Hodge conjecture. In: *The Arithmetic and Geometry of Algebraic Cycles, Banff, 1998* (B. B. Gordon, J. D. Lewis, S. Müller-Stach, S. Saito and N. Yui, eds.), *NATO Sci. Ser. C Math. Phys. Sci.* **548**, pp. 51–82, Kluwer, Dordrecht, 2000.

[Ge-V]   van Geemen, B. and Verra, A. Quaternionic pryms and Hodge classes, *Topology*, in press.

[G-L]   Gordon, B. and Lewis, J. D. Regulators of higher Chow cycles, in preparation.

[Gr1]   Griffiths, P. A. On the periods of certain rational integrals: I and II, *Ann. Math.* **90** (3) (1969) 460–541.

[Gr2]   A theorem about normal functions associated to Lefschetz pencils on algebraic varieties, *Am. J. Math.* **101** (1) (1979) 94–131.

[G-H1]   Griffiths, P. A. and Harris, J. *Principles of Algebraic Geometry*, Wiley, 1978.

[G-H2]   On the Noether–Lefschetz theorem and some remarks on codimension two cycles, *Math. Ann.* **271** (1985) 31–51.

[Gro]   Grothendieck, A. Hodge's general conjecture is false for trivial reasons, *Topology* **8** (1969) 299–303.

[Har]   Hartshorne, R. *Algebraic Geometry, Graduate Texts in Mathematics*, Springer-Verlag, New York, 1987.

[Ha]   Hazama, F. General Hodge conjecture for Abelian varieties of CM-type, *Proc. Jpn. Acad. Ser. A Math. Sci.* **78** (2002) 72–75.

[Hu]   Hulsbergen, W. Conjectures in arithmetic algebraic geometry. In: *Aspects of Mathematics* **18**, Vieweg & Sohn, Braunschweig, 1992.

[J1]   Jannsen, U. *Mixed Motives and Algebraic K-Theory, Lecture Notes in Mathematics* **1400**, Springer-Verlag, Berlin, 1989.

[J2]   Deligne homology, Hodge $\mathcal{D}$-conjecture and motives. In: *Beilinson's Conjectures on Special Values of L-Functions*, Academic Press, New York, 1988.

[Ko]   Kollár, J. *et al.*, *Trento Examples, Springer Lecture Notes* **1515**, pp. 134–135, Springer, 1992.

[Lef]   Lefschetz, S. *L'Analysis Situs et la Géométrie Algébrique*, Gauthier-Villars, Paris, 1924.

[Lew1]   Lewis, J. D. The Hodge conjecture for a certain class of fourfolds, *Math. Ann.* **268** (1) (1984) 85–90.

[Lew2]   A Survey of the Hodge Conjecture, *CRM Monograph Series (Am. Math. Soc.)* **10**, 1999.

[Lew3]   Cylinder homomorphisms and Chow groups, *Math. Nachr.* **160** (1993) 205–221.

[Lew4]   The cylinder correspondence for hypersurfaces of degree n in $\mathbb{P}^n$, *Am. J. Math.* **110** (1988) 77–114.

[Mi1]   Milne, J. *Étale Cohomology, Princeton Mathematical Series* **33**, Princeton University Press, 1980.

[Mi2]   Lefschetz classes on Abelian varieties, *Duke Math. J.* **96** (1999) 639–675.

[Mi3]   Lefschetz motives and the Tate conjecture. Preprint 1999.

[M-Z1]   Moonen, B. J. and Zarhin, Y. Hodge classes on abelian varieties of low dimension, *Math. Ann.* **315** (1999) (4) 711–733.

[M-Z2]   Weil classes and Rosati involutions on complex abelian varieties. In: *Recent Progress in Algebra, Taejon/Seoul, 1997, Contemp. Math.* **224**, pp. 229–236, AMS, Providence, RI, 1999.

[MS]   Müller-Stach, S. Constructing indecomposable motivic cohomology classes on algebraic surfaces, *J. Algebric Geom.* **6** (1997) 513–543.

[Mu1]   Murre, J. P. On the Hodge conjecture for unirational fourfolds, *Indag. Math.* **39** (3) (1977) 230–232.

[Mu2]   Algebraic equivalence modulo rational equivalence on a cubic threefold, *Compositio Math.* **25** (1972) 161–206.

[Mur]   Murty, V. K. Hodge and Weil classes on Abelian varieties. In: *The Arithmetic and Geometry of Algebraic Cycles, Banff, 1998* (B. B. Gordon, J. D.

Lewis, S. Müller-Stach, S. Saito and N. Yui, eds.), *NATO Sci. Ser. C Math. Phys. Sci.* **548**, pp. 83–115, Kluwer, Dordrecht, 2000.

[No]   Nori, M. Algebraic cycles and Hodge theoretic connectivity, *Invent. Math.* **111** (1993) 349–373.

[Pa]   Paranjape, K. Cohomological and cycle theoretic connectivity, Preprint.

[P]   Pjateckii-Sapiro, I. I. Interrelations between the Tate and Hodge conjectures for Abelian varieties, *Math. USSR-Sb.* **14** (1971) 615–625.

[Ro]   Roitman A. The torsion group of 0-cycles modulo rational equivalence, *Ann. Math.* **111** (1980) 553–569.

[Sh1]   Shimada, I. On the cylinder isomorphism associated to the family of lines on a hypersurface, *J. Fac. Sci. Univ. Tokyo Sect. IA Math.* **37** (3) (1990) 703–719.

[Sh2]   On the cylinder homomorphisms of Fano complete intersections, *J. Math. Soc. Jpn.* **42** (4) (1990) 719–738.

[Sh3]   On the cylinder homorphisms II, *UTYO-MATH* **90-23** (1990).

[Sh4]   On the cylinder homomorphism for a family of algebraic cycles, *Duke Math. J.* **64** (1) (1991) 201–205.

[Shi]   Shioda, T. What is known about the Hodge conjecture? *Algebraic Varieties and Analytic Varieties, Adv. Stud. in Pure Math.*, **1**, pp. 55–68, North-Holland, Amsterdam, 1983.

[Ta1]   Tate, J. Algebraic cycles and poles of zeta functions. In: *Arithmetic Algebraic Geometry* (O. F. G. Schilling, ed.), pp. 93–111, Harper & Row, 1965.

[Ta2]   Conjectures on Algebraic Cycles in ℓ-adic cohomology, *Proc. Symp. Pure Math.* **55** (1) (1994) 77–83.

[Te]   Terasoma, T. Hodge conjecture for cubic 8-folds, *Math. Ann.* **288** (1990) 9–19.

[T]   Tyurin, A. Five lectures on three dimensional varieties, *Russ. Math. Surv.* **27** (1972) 1–53.

[Z1]   Zucker, S. Generalized intermediate jacobians and the theorem on normal functions, *Invent. Math.* **33** (3) (1976) 185–222.

[Z2]   Hodge theory with degenerating coefficients: L_2 cohomology in the Poincaré metric, *Ann. Math.* **109** (2) (1979) 415–476.

7
Lectures on Nori's connectivity theorem

J. Nagel

Université Lille 1, Mathématiques, Bâtiment MZ, 596SS
Villeneuve d'Ascq Cedex, France

Introduction

The aim of these lectures is to discuss Nori's connectivity theorem and its applications to the theory of algebraic cycles. I have tried to clarify some of the underlying ideas by emphasizing the relationship of Nori's theorem with the theorems of Griffiths and Green–Voisin on the image of the Abel–Jacobi map for hypersurfaces in projective space.

The notes are divided into six sections that roughly correspond to my six lectures at the summer school in Grenoble. I have added a short Appendix on Deligne cohomology.

7.1 Normal functions

Let X be a smooth projective variety over $\mathbb{C}$. The main invariants used to study the Chow group $CH^p(X)$ of codimension p cycles on X are the *cycle class map*

$$cl_X^p : CH^p(X) \to H^{2p}(X, \mathbb{Z})$$

and the *Abel–Jacobi map*

$$\psi_X^p : CH^p_{\mathrm{hom}}(X) \to J^p(X),$$

which is defined on the kernel of cl_X^p; its target is the intermediate Jacobian

$$J^p(X) = H^{2p-1}(X, \mathbb{C})/F^p H^{2p-1}(X, \mathbb{C}) + H^{2p-1}(X, \mathbb{Z})/\mathrm{tors}$$
$$\cong F^{n-p+1} H^{2n-2p+1}(X, \mathbb{C})^{\vee}/H_{2n-2p+1}(X, \mathbb{Z}).$$

Transcendental Aspects of Algebraic Cycles ed. S. Müller-Stach and C. Peters.
© Cambridge University Press 2004.

The Hodge conjecture gives a conjectural description of the image of cl_X^p. Little is known about the kernel and the image of ψ_X^p, except in special cases such as curves and Fano threefolds. A natural idea is to consider a family of smooth projective varieties $\{X_s\}_{s \in S}$ and to study holomorphic sections of the fiber space of intermediate Jacobians over S to obtain information about cycles on the general fiber. Such sections are called normal functions; they were introduced by Poincaré for families of curves on algebraic surfaces and by Griffiths for algebraic cycles on higher-dimensional varieties.

Let $f : X \to S$ be a smooth projective morphism of quasi-projective varieties. Suppose that the fibers $X_s = f^{-1}(s)$ have dimension $2m - 1$. The intermediate Jacobians $J^m(X_s)$ of the fibers fit together to give a holomorphic fiber space of complex tori

$$J^m(X/S) = \cup_{s \in S} J^m(X_s).$$

The sheaf $H_\mathbb{Z}^{2m-1} = R^{2m-1} f_* \mathbb{Z}/\text{tors}$ is a local system of abelian groups. The associated Hodge bundle $\mathcal{H}^{2m-1} = H_\mathbb{Z}^{2m-1} \otimes_\mathbb{Z} \mathcal{O}_S$ is a vector bundle that carries a flat connection ∇, the Gauss–Manin connection. The Hodge bundle is filtered by holomorphic subbundles $\mathcal{F}^p$. The sheaf

$$\mathcal{J}^m = \mathcal{H}^{2m-1}/\mathcal{F}^m + H_\mathbb{Z}^{2m-1}$$

is the sheaf of sections of the fibration $\pi : J^m(X/S) \to S$.

Definition 7.1.1 *A normal function is a holomorphic section $\nu \in H^0(S, \mathcal{J}^m)$.*

The exact sequence of sheaves

$$0 \to H_\mathbb{Z}^{2m-1} \to \mathcal{H}^{2m-1}/\mathcal{F}^m \to \mathcal{J}^m \to 0$$

induces a map $\partial : H^0(S, \mathcal{J}^m) \to H^1(S, H_\mathbb{Z}^{2m-1})$ that associates to a normal function ν its *cohomological invariant* $\partial(\nu) \in H^1(S, H_\mathbb{Z}^{2m-1})$. The Gauss–Manin connection $\nabla : \mathcal{H}^{2m-1} \to \Omega_S^1 \otimes \mathcal{H}^{2m-1}$ restricts to a map

$$\nabla : \mathcal{F}^m \to \Omega_S^1 \otimes \mathcal{F}^{m-1}$$

by Griffiths transversality. As $\nabla(H_\mathbb{Z}^{2m-1}) = 0$ we obtain an induced map

$$\overline{\nabla} : \mathcal{J}^m \to \Omega_S^1 \otimes \mathcal{H}^{2m-1}/\mathcal{F}^{m-1}$$

whose kernel is denoted by $\mathcal{J}_h^m$. The holomorphic sections of this sheaf are called *quasi-horizontal* normal functions.

Definition 7.1.2 *Let $f : X \to S$ be a smooth projective morphism. The group $Z^p(X/S)$ of relative codimension p cycles over S is the free abelian group* *generated by irreducible subvarieties $Z \subset X$ that are flat of relative dimension* $\dim X - \dim S - p$ *over S.*

We write $Z^m_{\hom}(X/S)$ for the subgroup of relative codimension m cycles whose restriction to every fiber X_s is homologically equivalent to zero. Given $Z \in Z^m_{\hom}(X/S)$, it is possible to choose a family of $(2m - 1)$-chains $\gamma = (\gamma_s)$ such that $\partial\gamma_s = Z_s$ for all $s \in S$. We then obtain a C^∞ section $\tilde{\nu} \in \Gamma(S, (\mathcal{F}^m)^\vee)$ by setting

$$\langle \tilde{\nu}(s), \omega_s \rangle = \int_{\gamma_s} \omega_s$$

for a C^∞ section $\omega = (\omega_s) \in \Gamma(S, \mathcal{F}^m)$. By projection to $\mathcal{J}^m$ we obtain a section $\nu_Z \in \Gamma(S, \mathcal{J}^m)$.

Let ξ be a vector field on X. There exists a *contraction map*

$$i_\xi : A^k(X) \to A^{k-1}(X)$$

that is defined as the unique derivation that equals the evaluation map on 1-forms and is linear for C^∞ functions. Let ξ be a local lifting of the vector field $\frac{\partial}{\partial s}$. Locally, the flow associated to ξ defines diffeomorphisms $\varphi_s : X_0 \to X_s$ (Ehresmann's fibration theorem). The *Lie derivative* of a family $\omega = (\omega_s)$ of differential k-forms with respect to ξ is defined by

$$L_\xi\omega = \frac{\partial}{\partial s}\Big|_{s=0} \varphi_s^* \omega_s.$$

A similar formula defines $L_{\bar{\xi}}\omega$ if $\bar{\xi}$ is a local lifting of $\frac{\partial}{\partial \bar{s}}$. The Lie derivative can be computed using the *Cartan formula*

$$L_\xi\omega = di_\xi\Omega + i_\xi d\Omega$$

where Ω is a k-form on X such that $\Omega|_{X_s} = \omega_s$ for all $s \in S$. By definition, the Gauss–Manin derivative along ξ of a family of closed forms ω is given by

$$\nabla_\xi\omega = [L_\xi\omega].$$

This is well defined and maps families of closed forms to families of closed forms since $[L_\xi, d] = 0$ by the Cartan formula.

Proposition 7.1.3 *The section $\nu = \nu_Z \in \Gamma(S, \mathcal{J}^m)$ associated to a relative cycle $Z \in Z^m_{\hom}(X/S)$ is a quasi-horizontal normal function.*

Proof It suffices to cover S by contractible open subsets and to verify the assertion locally. Hence we may assume that S is a disc. We shall sketch the proof under the assumption that the relative cycle $Z = \sum_{i=1}^{k} n_i Z_i$ satisfies the following two properties:

(i) Z_i is smooth and $f|_{Z_i} : Z_i \to S$ is smooth for $i = 1, \ldots, k$;
(ii) $Z_i \cap Z_j = \emptyset$ if $i \neq j$.

Under these assumptions there exists a C^∞ trivialization $X \cong X_0 \times S$ that induces trivializations $Z_i \cong Z_{i,0} \times S$ for $i = 1, \ldots, k$ and a C^∞ decomposition

$$T_X|_{X_s} \cong T_{X_s} \oplus f^* T_{S,s}$$

for all $s \in S$. See [30] for a more detailed discussion. Let $\mathcal{A}^1(X/S) = \mathcal{A}^1(X)/f^*\mathcal{A}^1(S)$ be the sheaf of relative differentiable 1-forms. The sheaf $\mathcal{A}^{2m-1}(X/S) = \bigwedge^{2m-1}\mathcal{A}^1(X/S)$ admits a decomposition

$$\mathcal{A}^{2m-1}(X/S) = \bigoplus_{p+q=2m-1} \mathcal{A}^{p,q}(X/S).$$

Set $\mathcal{F}^m \mathcal{A}^{2m-1}(X/S) = \bigoplus_{p \geq m} \mathcal{A}^{p,2m-1-p}(X/S)$, and let $\omega = (\omega_s)_{s \in S}$ be a C^∞ section of $\mathcal{F}^m \mathcal{A}^{2m-1}(X/S)$ such that ω_s is closed for all $s \in S$. There exists $\Omega \in F^m \mathcal{A}^{2m-1}(X)$ such that $\Omega|_{X_s} = \omega_s$ for all $s \in S$. The form Ω is uniquely determined if we impose the condition that $i_\chi \Omega = 0$ for every horizontal vector field χ. Choose $\gamma_0 \in C_{2m-1}(X_0)$ such that $\partial \gamma_0 = Z_0$ and define $\gamma_s = (\varphi_s)_* \gamma_0$. By construction we have $\partial \gamma_s = Z_s$ for all $s \in S$. Set $h(s) = \langle \tilde{\nu}(s), \omega_s \rangle$. Let ξ be a lifting of the vector field $\frac{\partial}{\partial s}$, and let $\bar{\xi}$ be a lifting of $\frac{\partial}{\partial \bar{s}}$. To show that h is holomorphic we compute

$$\frac{\partial}{\partial \bar{s}}\bigg|_{s=0} h(s) = \frac{\partial}{\partial \bar{s}}\bigg|_{s=0} \int_{\varphi_{s*}\gamma_0} \omega_s = \frac{\partial}{\partial \bar{s}}\bigg|_{s=0} \int_{\gamma_0} \varphi_s^* \omega_s = \int_{\gamma_0} \frac{\partial}{\partial \bar{s}}\bigg|_{s=0} \varphi_s^* \omega_s$$
$$= \int_{\gamma_0} L_{\bar{\xi}} \omega|_{X_0}.$$

Using the Cartan formula we find that $L_{\bar{\xi}} \omega = i_{\bar{\xi}} d\Omega$ (remember that $i_{\bar{\xi}}\Omega = 0$). Since contraction with $\bar{\xi}$ cannot annihilate any of the dz we find that $L_{\bar{\xi}} \omega|_{X_0} \in F^m A^{2m-1}(X_0)$. If ω is a holomorphic section of $\mathcal{F}^m$ then $\nabla_{\bar{\xi}} \omega = 0$, hence $L_{\bar{\xi}} \omega$ is exact. An important consequence of Hodge theory is that the differential $d : A^k(X) \to A^{k+1}(X)$ is *strictly compatible* with $F^\bullet$. This means that we can find $\eta \in F^m A^{2m-2}(X_0)$ such that $L_{\bar{\xi}} \omega = d\eta$. Using the Stokes formula we find that

$$\frac{\partial}{\partial \bar{s}}\bigg|_{s=0} h(s) = \int_{Z_0} \eta = 0$$

since the complex dimension of Z_0 is $m - 1$. This proves that ν is a holomorphic section of $\mathcal{J}^m$. To check the quasi-horizontality property, we use the Leibniz rule

$$\left.\frac{\partial}{\partial s}\right|_{s=0} h(s) = \langle \nabla_\xi \tilde{v}(0), \omega_0 \rangle + \langle \tilde{v}(0), \nabla_\xi \omega(0) \rangle.$$

If we can show that

$$(*) \qquad\qquad \langle \nabla_\xi \tilde{v}(0), \omega_0 \rangle = 0$$

for every section $\omega \in \Gamma(S, \mathcal{F}^{m+1})$ it follows that

$$\nabla_\xi \tilde{v}(0) \in F^{m+1} H^{2m-1}(X_0)^\perp = F^{m-1} H^{2m-1}(X_0)$$

which means that $\overline{\nabla} v(0) = 0$. To verify $(*)$ we compute

$$\left.\frac{\partial}{\partial s}\right|_{s=0} h(s) = \int_{\gamma_0} L_\xi \omega|_{X_0}.$$

As the closed form $L_\xi \omega|_{X_0} = i_\xi d\Omega|_{X_0} \in F^m A^{2m-1}(X_0)$ represents $\nabla_\xi \omega$ we have

$$\nabla_\xi \omega = i_\xi d\Omega|_{X_0} + d\eta$$

where η can be chosen in $F^m A^{2m-1}(X_0)$, again by strict compatibility of d with the Hodge filtration. The Stokes formula then shows that

$$\int_{\gamma_0} d\eta = 0,$$

hence

$$\left.\frac{\partial}{\partial s} h(s)\right|_{s=0} = \int_{\gamma_0} i_\xi d\Omega|_{X_0} = \langle \tilde{v}(0), \nabla_\xi \omega(0) \rangle$$

and this equality is equivalent to $(*)$. $\qquad\qquad\qquad\qquad\qquad\qquad\square$

Let Y be a smooth projective variety of even dimension $2m$ and let $\{X_t\}_{t \in T}$ be a family of smooth hypersurface sections of Y. Let X_T be the total space of the family $\{X_t\}$ with inclusion $r : X_T \to Y \times T$. Suppose there exists a codimension m cycle Z on Y such that $Z \cap X_t$ is homologically equivalent to zero for all $t \in T$. The normal function associated to the relative cycle $r^*(Z \times T) \in Z^m_{\text{hom}}(X_T/T)$ is denoted by v_Z. By construction $v_Z(t)$ is the image of Z_t under the Abel–Jacobi map on X_t. Using Deligne cohomology (see Section 7.7) it is possible to associate a normal function to Hodge classes on Y. Define

$$\text{Hdg}^m(Y)_0 = \ker(i^* : \text{Hdg}^m(Y) \to \text{Hdg}^m(X))$$

and consider the commutative diagram

$$0 \to J^m(Y) \to H_{\mathcal{D}}^{2m}(Y, \mathbb{Z}(m)) \to \mathrm{Hdg}^m(Y) \to 0$$

$$\downarrow{i^*} \qquad\qquad \downarrow{i^*} \qquad\qquad \downarrow{i^*}$$

$$0 \to J^m(X) \to H_{\mathcal{D}}^{2m}(X, \mathbb{Z}(m)) \to \mathrm{Hdg}^m(X) \to 0.$$

Given $\xi \in \mathrm{Hdg}^m(Y)_0$, choose a lifting $\tilde{\xi} \in H_{\mathcal{D}}^{2m}(Y, \mathbb{Z}(m))$. As $i^*\tilde{\xi}$ maps to zero in $\mathrm{Hdg}^m(X)$, it belongs to $J^m(X)$. To get a well-defined map we have to pass to the quotient

$$J_{\mathrm{var}}^m(X) = J^m(X)/i^* J^m(Y).$$

Define $\nu(s) = i_s^*\tilde{\xi}$. One can show that $\nu \in H^0(S, \mathcal{J}_{\mathrm{var}}^m)$ is a quasi-horizontal normal function, cf. [12, Lecture 6].

Examples 7.1.4 Some other examples of normal functions:

(1) Let C be a non-hyperelliptic curve of genus 4. It is known that the canonical image $\varphi_K(C) \subset \mathbb{P}^3$ is the complete intersection of a quadric Q and a cubic F. Let ℓ_1, ℓ_2 be two lines from the different rulings of the quadric and set $D = \ell_1 \cap F - \ell_2 \cap F \in Z_{\mathrm{hom}}^1(C)$. If $C_S = \{C_s\}_{s \in S}$ is a family of non-hyperelliptic genus 4 curves, we obtain a relative cycle $D_S = \cup_{s \in S} D_s \in Z_{\mathrm{hom}}^1(C_S/S)$ and a normal function ν_D.

(2) Let C be a smooth curve of genus $g \geq 3$. The choice of a base point $x \in C$ defines an embedding $i_x : C \to J(C)$ of C into the g-dimensional abelian variety $J(C)$. There is an involution i on $J(C)$ given by multiplication by -1. Define $C_x^+ = i_x(C), C_x^- = i_*(C_x^+)$ and $Z_{C,x} = C_x^+ - C_x^-$. As i_* acts as the identity on $H^{2g-2}(J(C), \mathbb{Z}) = \bigwedge^{2g-2} H^1(J(C), \mathbb{Z})$, we have $[Z_{C,x}] \in \mathrm{CH}_{\mathrm{hom}}^{g-1}(J(C))$. We obtain a normal function $\tilde{\nu}$ over an open subset of the moduli space $M_{g,1}$ of pointed genus g curves associated to the relative cycle $Z = \cup_{(C,x)} Z_{C,x}$. Let $P H_3(J(C), \mathbb{Z})$ be the cokernel of the map

$$H_1(J(C), \mathbb{Z}) \to H_3(J(C), \mathbb{Z})$$

given by Pontryagin product with $[C_x] \in H_2(J(C), \mathbb{Z})$. One can verify that the projection of the Abel–Jacobi image of $C_x^+ - C_x^-$ to the primitive intermediate Jacobian

$$J_{\mathrm{pr}}^{g-1}(J(C)) = F^2 H_{\mathrm{pr}}^3(J(C), \mathbb{C})^\vee / P H_3(J(C), \mathbb{Z})$$

does not depend on the choice of the base point x. Hence $\tilde{\nu}$ descends to a section $\nu \in H^0(U, \mathcal{J}_{\mathrm{pr}}^{g-1})$ defined over a Zariski open subset $U \subset M_g$.

A *local lifting* of a normal function ν is a local section of the Hodge bundle that projects to ν. For normal functions that satisfy the quasi-horizontality property we can define a new invariant that measures the obstruction for the existence of *flat* local liftings. To define this invariant, consider the de Rham complex for the Hodge bundle $\mathcal{H} = \mathcal{H}^{2m-1}$

$$\Omega^\bullet(\mathcal{H}) = (\mathcal{H} \xrightarrow{\nabla} \Omega^1_S \otimes \mathcal{H} \xrightarrow{\nabla} \Omega^2_S \otimes \mathcal{H} \to \dots)$$

and its subcomplex

$$\Omega^\bullet(\mathcal{F}^m) = (\mathcal{F}^m \xrightarrow{\nabla} \Omega^1_S \otimes \mathcal{F}^{m-1} \xrightarrow{\nabla} \Omega^2_S \otimes \mathcal{F}^{m-2} \to \cdots).$$

Let $\tilde{\Omega}^\bullet(\mathcal{F}^m) = \Omega^\bullet(\mathcal{F}^m) \oplus H_\mathbb{Z}$ be the complex obtained by adding the local system $H_\mathbb{Z}$ in degree zero. The quotient of the de Rham complex by this modified subcomplex is the complex

$$\Omega^\bullet(\mathcal{H}/\mathcal{F}^m) = (\mathcal{J}^m \xrightarrow{\overline{\nabla}} \Omega^1_S \otimes \mathcal{H}/\mathcal{F}^{m-1} \to \cdots).$$

The connecting homomorphism

$$\mathcal{J}^m_h = \mathcal{H}^0(\Omega^\bullet(\mathcal{H}/\mathcal{F}^m)) \to \mathcal{H}^1(\tilde{\Omega}^\bullet(\mathcal{F}^m)) = \mathcal{H}^1(\Omega^\bullet(\mathcal{F}^m))$$

induces a map

$$\delta : H^0(S, \mathcal{J}^m_h) \to H^0(S, \mathcal{H}^1(\Omega^\bullet(\mathcal{F}^m))).$$

We call $\delta\nu$ the *infinitesimal invariant* of ν. It is obtained as follows: choose an open covering $\mathcal{U} = \{U_\alpha\}$ of S and apply ∇ to a local lifting $\tilde{\nu}_\alpha$ of ν_α; by quasi-horizontality, $\nabla\tilde{\nu}_\alpha$ comes from a local section of $\Omega^1_S \otimes \mathcal{F}^{m-1}$, which is annihilated by ∇ and is well defined modulo sections in the image of $\nabla :$ $\mathcal{F}^m \to \Omega^1_S \otimes \mathcal{F}^{m-1}$. The local sections $\delta\nu_\alpha = [\nabla\tilde{\nu}_\alpha]$ patch together to give a global section $\delta\nu \in \Gamma(S, \mathcal{H}^1(\Omega^\bullet(\mathcal{F}^m)))$.

Lemma 7.1.5 $\delta\nu = 0 \iff \nu$ *has flat local liftings.*

Proof If $\delta\nu = 0$ then there exists locally a section f of $\mathcal{F}^m$ such that $\nabla\tilde{\nu} = \nabla f$, hence $\tilde{\nu} - f = \nabla(\lambda)$ where λ is locally constant and $\hat{\nu} = \tilde{\nu} - f$ is a flat local lifting of ν. The other direction of the equivalence is clear. $\square$

The complex $\Omega^\bullet(\mathcal{F}^m)$ is filtered by subcomplexes $F^p\Omega^\bullet(\mathcal{F}^m) = \Omega^\bullet(\mathcal{F}^p)$ $(p \geq m)$ with graded pieces

$$\mathrm{Gr}^p_F \Omega^\bullet(\mathcal{F}^m) = (\mathcal{H}^{p,m-p-1} \xrightarrow{\overline{\nabla}} \mathcal{H}^{p-1,m-p} \to \cdots).$$

Note that the differential $\overline{\nabla}$ in these complexes is $\mathcal{O}_S$-linear. There is a natural map

$$H^0(S, \mathcal{H}^1(\Omega^\bullet(\mathcal{F}^m))) \to H^0(S, \mathcal{H}^1(\mathrm{Gr}_F^m \, \Omega^\bullet(\mathcal{F}^m))).$$

The image $\delta_1 \nu$ of $\delta \nu$ under this map is the infinitesimal invariant of normal functions defined by Griffiths.

Remark 7.1.6

(i) If $\{Z(t)\}_{t \in T}$ is a family of codimension m cycles on X, we obtain a relative cycle on the trivial family $X \times T$. In this case, the sheaf $\mathcal{J}^m$ is the constant sheaf $J^m(X)$ and the associated normal function is a map $\nu : T \to J^m(X)$, defined after the choice of a base point t_0 by $\nu(t) = \psi(Z(t) - Z(t_0))$. The infinitesimal invariant $\delta_1 \nu$ is an element of

$$\Omega_T^1 \otimes H^{m-1,m}(X) \cong \mathrm{Hom}(T, H^{m-1,m}(X)).$$

It coincides with the differential ν_* of ν, which is called the infinitesimal Abel–Jacobi map.

(ii) The infinitesimal invariant $\delta_1 \nu$ often carries geometric information. For genus 4 curves, Griffiths [16] showed that it determines the cubic containing the canonical image of the curve. For genus 3 curves, Collino and Pirola [7] showed that it determines the canonical equation of the curve.

Remark 7.1.7 Griffiths defined the *fixed part* of $\mathcal{J}^m$ as the sheaf

$$\mathcal{J}_{\mathrm{fix}}^m = H_{\mathbb{C}}/\mathcal{F}^m \cap H_{\mathbb{C}} + H_{\mathbb{Z}}.$$

The reason for this terminology is that if $\{X_s\}$ is a family of hypersurface sections of Y, the fixed part can be identified with the constant sheaf $J^m(Y)$ using a monodromy argument. (We shall see this in Section 7.3 for $Y = \mathbb{P}^{2m}$.) We have an exact sequence

$$0 \to \mathcal{J}_{\mathrm{fix}}^m \to \mathcal{J}_h^m \to \mathcal{H}^1(\Omega^\bullet(\mathcal{F}^m)) \to 0,$$

hence

$$\delta \nu = 0 \iff \nu \in H^0(S, \mathcal{J}_{\mathrm{fix}}^m).$$

To conclude this lecture, we mention without proof two important theorems on normal functions. Suppose that S is a smooth curve and $f : X \to S$ admits a compactification $\bar{f} : \bar{X} \to \bar{S}$ such that the fibers $\bar{f}^{-1}(s)$ over the points $s \in \bar{S} \setminus S$ are divisors with simple normal crossings. By work of Schmid and Steenbrink it is possible to extend the Hodge bundle and its subbundles to vector bundles $\bar{\mathcal{H}}$ and $\bar{\mathcal{F}}^m$ on $\bar{S}$. Let $j : S \to \bar{S}$ be the inclusion map. The *Zucker extension* $\bar{\mathcal{J}}^m$

of $\mathcal{J}^m$ is the sheaf

$$\bar{\mathcal{J}}^m = \mathcal{H}/\bar{\mathcal{F}}^m + j_* H_{\mathbb{Z}}.$$

Strictly speaking, normal functions should be defined as global sections of $\bar{\mathcal{J}}^m$ (i.e. sections of $\mathcal{J}^m$ that extend over the singular fibers). There exists a map

$$\partial : H^0(\bar{S}, \bar{\mathcal{J}}^m) \to H^1(\bar{S}, j_* H_{\mathbb{Q}}).$$

Theorem 7.1.8 (Zucker) *The group $H^1(\bar{S}, j_* H_{\mathbb{Q}})$ carries a Hodge structure of weight $2m$ and the image of ∂ coincides with the set of Hodge classes $\mathrm{Hdg}^m H^1(\bar{S}, j_* H_{\mathbb{Q}})$.*

There exists a criterion for extendability of normal functions. For every $s \in \bar{S} \setminus S$, let $\Delta^*(s)$ be a punctured disc centred at s and let $\partial_s(v) \in H^1(\Delta^*(s), H_{\mathbb{Z}})$ be the cohomological invariant of $v|_{\Delta^*(s)}$.

Theorem 7.1.9 (El Zein–Zucker) *Let v be a normal function. If $\partial_s(v) = 0$ for all $s \in \bar{S} \setminus S$, then v extends to a section $\bar{v} \in H^0(\bar{S}, \bar{\mathcal{J}}^m)$.*

Bibliographical hints A good introduction to normal functions is Zucker's paper [33]; see also [15]. The proof of Proposition 7.1.3 is taken from [26]. Zucker's theorem on normal functions can be found in [31] and [32]. For a discussion of extendability of normal functions, see [8]. The invariant $\delta_1 v$ was discovered by Griffiths [16]. The definition of δv is due to Green [11].

7.2 Griffiths's theorem

Let X be a smooth projective variety. The group $\mathrm{CH}^p_{\mathrm{hom}}(X)$ of codimension p cycles homologically equivalent to zero contains as a subgroup the group $\mathrm{CH}^p_{\mathrm{alg}}(X)$ of cycles algebraically equivalent to zero. For divisors ($p = 1$) and zero-cycles ($p = \dim X$) both groups coincide but in general they may be different. The quotient group

$$\mathrm{Griff}^p(X) = \mathrm{CH}^p_{\mathrm{hom}}(X) / \mathrm{CH}^p_{\mathrm{alg}}(X)$$

is called the *Griffiths group* of codimension p cycles.

In 1969 Griffiths showed that there exist quintic threefolds $X \subset \mathbb{P}^4$ such that the difference of two lines on X is not algebraically equivalent to zero. This follows from the theorem below. Recall that a property (P) is said to hold for a *very general* point of a topological space T if the subset of elements that do not satisfy (P) is a countable union of proper closed subsets of T.

Theorem 7.2.1 (Griffiths) *Let Y be a smooth projective variety of even dimension $2m$ and let $\{X_t\}_{t \in \mathbb{P}^1}$ be a Lefschetz pencil of hyperplane sections of Y. Suppose that $H^{2m-1}(Y) = 0$ and that*

$$H^{2m-1}(X_t, \mathbb{C}) \neq H^{m,m-1}(X_t) \oplus H^{m-1,m}(X_t). \tag{1}$$

If $Z \in Z^m(Y)$ and $Z \cap X_t$ is algebraically equivalent to zero for very general $t \in \mathbb{P}^1$, then Z is homologically equivalent to zero.

Remark 7.2.2 The assumption $H^{2m-1}(Y) = 0$ is only included to simplify the proof; it can be omitted. In a later lecture we shall explain Nori's generalization of Theorem 7.2.1.

Before we start with the proof of Theorem 7.2.1 we introduce some notation. Let $B = X_0 \cap X_\infty$ be the base locus of the pencil, and let $\tilde{Y}$ be the blow-up of Y along B. Let U be the complement of the discriminant locus $\Delta \subset \mathbb{P}^1$. We have a diagram

$$
\begin{array}{ccccc}
X_U & \xrightarrow{\ j\ } & \tilde{Y} & \xrightarrow{\ \pi\ } & Y \\
\downarrow{\scriptstyle f} & & \downarrow{\scriptstyle \tilde{f}} & & \\
U & \xrightarrow{\ j\ } & \mathbb{P}^1. & &
\end{array}
$$

To prove Theorem 7.2.1 we need several lemmas. Let

$$J_{\mathrm{alg}}^m(X) = \mathrm{im}(\psi_X^m : \mathrm{CH}_{\mathrm{alg}}^m(X) \to J^m(X))$$

be the algebraic part of the intermediate Jacobian of $X = X_t$.

Lemma 7.2.3 *With the hypotheses of Theorem 7.2.1 we have $J_{\mathrm{alg}}^m(X_t) = 0$ if $t \in U$ is very general.*

Proof Let $X = X_t$ be a general hyperplane section of Y. Recall that a cycle $z \in Z^m(X)$ is algebraically equivalent to zero if there exist a variety S, a relative cycle $\mathcal{Z} \in Z^m(X \times S/S)$ and two points s_0, s_1 in S such that $z = \mathcal{Z}(s_0) - \mathcal{Z}(s_1)$. We may assume that S is a smooth irreducible curve. Define a map

$$g : S \to J^m(X)$$

by $g(s) = \psi_X(\mathcal{Z}(s_0) - \mathcal{Z}(s))$. As $\psi_X(z) = g(s_1)$ it suffices to study the image of g. As g is holomorphic, it factorizes over a map

$$h : J(S) \to J^m(X)$$

by the universal property of the Jacobian. By Poincaré duality, the induced map on homology groups

$$h_* : H_1(J(S), \mathbb{Z}) = H_1(S, \mathbb{Z}) \to H_1(J^m(X), \mathbb{Z}) = H_{2m-1}(X, \mathbb{Z})$$

gives a map

$$h_* : H^1(S) \to H^{2m-1}(X).$$

This map is a morphism of Hodge structures of type $(m-1, m-1)$ induced by the correspondence $[\mathcal{Z}] \in \mathrm{CH}^m(X \times S)$. As it induces the map h by passage to the quotient, we find that $J^m_{\mathrm{alg}}(X)$ is the intermediate Jacobian associated to a sub-Hodge structure

$$H_{\mathrm{alg}} \subset H^{m-1,m}(X) \oplus H^{m,m-1}(X) \cap H^{2m-1}(X, \mathbb{Q}).$$

Let

$$\rho : \pi_1(U) \to \mathrm{Aut}\, H^{2m-1}(X, \mathbb{Q})$$

be the monodromy representation, and let $\Gamma = \mathrm{im}\,\rho$ be the monodromy group. By Picard–Lefschetz theory we know that $H^{2m-1}(X, \mathbb{Q})$ is an irreducible Γ-module. If $t \in U$ is very general, it is possible to 'spread out' every cycle on X_t to a relative cycle over U (this will be explained in more detail in the next lecture), hence $H_{\mathrm{alg}} \subset H^{2m-1}(X, \mathbb{Q})$ is a Γ-submodule. If $H_{\mathrm{alg}} \neq 0$ we would get $H_{\mathrm{alg}} = H^{2m-1}(X, \mathbb{Q})$, which is impossible by condition (1) of Theorem 7.2.1. $\qquad\square$

The Leray spectral sequence for the map $f : X_U \to U$ defines a filtration $L^\bullet$ on $H^{2m}(X_U)$. We have

$$L^1 H^{2m}(X_U, \mathbb{Q}) = \ker(H^{2m}(X_U, \mathbb{Q}) \to H^0(U, R^{2m} f_* \mathbb{Q})).$$

Note that the primitive cohomology

$$H^{2m}_{\mathrm{pr}}(Y, \mathbb{Q}) = \ker(\cup c_1(\mathcal{O}_Y(1)) : H^{2m}(Y, \mathbb{Q}) \to H^{2m+2}(Y, \mathbb{Q})).$$

coincides with the kernel of the restriction map $i^* : H^{2m}(Y, \mathbb{Q}) \to H^{2m}(X_t, \mathbb{Q})$ if X_t is smooth. Hence, if $\alpha \in H^{2m}_{\mathrm{pr}}(Y, \mathbb{Q})$ then $\pi^*\alpha \in L^1 H^{2m}(X_U, \mathbb{Q})$. As $f : X_U \to U$ is a smooth morphism, the sheaf $R^{2m-1} f_* \mathbb{Q}$ is locally constant; we denote it by $H^{2m-1}_{\mathbb{Q}}$. Since $U = \mathbb{P}^1 \setminus \Delta$ is an affine curve, we have $H^2(U, R^{2m-2} f_* \mathbb{Z}) = 0$. Hence

$$L^1 H^{2m}(X_U) = \mathrm{Gr}^1_L H^{2m}(X_U) \cong H^1(U, H^{2m-1}_{\mathbb{Q}}).$$

Recall that we have maps $\pi : \tilde{Y} \to Y$ and $r : X_U \to \tilde{Y}$. Define the *Griffiths homomorphism*

$$\mathrm{Griff} : H^{2m}_{\mathrm{pr}}(Y, \mathbb{Q}) \to H^1(U, H^{2m-1}_{\mathbb{Q}})$$

by $\mathrm{Griff}(\alpha) = (\pi \circ r)^* \alpha$.

Lemma 7.2.4 *If* $Z \in Z^m(Y)$ *and* $[Z] \in H^{2m}_{pr}(Y, \mathbb{Q})$ *then* $\mathrm{Griff}[Z]$ *coincides with the cohomological invariant* $\partial(\nu_Z)$ *of the normal function* ν_Z *associated to* Z.

Proof Set $Z_U = r^*\pi^*Z \in Z^m_{hom}(X_U/U)$ and choose a covering of U by contractible open subsets U_i, $i \in I$. We can choose a family of cochains $\gamma^i(t)$ such that $\delta\gamma^i(t) = Z_U(t)$ for all $t \in U_i$. The 1-cocycle $\{\gamma^{ij}(t)\}$ defined by $\gamma^{ij}(t) = \gamma^i(t) - \gamma^j(t)$ represents the cycle class $\mathrm{cl}(Z_U) \in L^1 H^{2m}(X_U, \mathbb{Q}) \cong H^1(U, H^{2m-1}_{\mathbb{Q}})$. We can perform a similar construction in homology: choose chains $\gamma_i(t)$ such that $\partial\gamma_i(t) = Z_i(t)$ and define $\gamma_{ij}(t) = \gamma_i(t) - \gamma_j(t)$. Let ω be a section of $(\mathcal{F}^m)^\vee$. We obtain a local lifting $\tilde{\nu}_i$ of ν_Z by setting

$$\langle \tilde{\nu}_i(t), \omega(t) \rangle = \int_{\gamma_i(t)} \omega(t).$$

The corresponding 1-cocycle $\{\tilde{\nu}_{ij}\} \in \check{C}^1(\mathcal{U}, (\mathcal{F}^m)^\vee)$ is given by integration along $\gamma_{ij}(t)$, hence it comes from $\{\gamma_{ij}\} \in \check{C}^1(U, H^{\mathbb{Z}}_{2m-1})$. Using Poincaré duality to identify the local systems $H^{\mathbb{Q}}_{2m-1}$ and $H^{2m-1}_{\mathbb{Q}}$ we find that $\partial(\nu_Z)$ is represented by the 1-cocycle $\{\gamma^{ij}\} \in \check{C}^1(U, H^{2m-1}_{\mathbb{Q}})$. $\square$

Lemma 7.2.5 *The Griffiths homomorphism is injective if* $H^{2m-1}_{\mathbb{Q}} \neq 0$.

Proof As U is an affine curve, the Leray spectral sequence induces an isomorphism

$$H^{2m}(X_U, \mathbb{Q}) \cong H^0(U, R^{2m} f_*\mathbb{Q}) \oplus H^1(U, H^{2m-1}_{\mathbb{Q}}).$$

As $[Z]$ maps to zero in the first summand, we have

$$\mathrm{Griff}[Z] = 0 \iff \pi^*[Z] \in \ker(r^* : H^{2m}(\tilde{Y}) \to H^{2m}(X_U)).$$

Set $\Sigma = \bar{f}^{-1}(\Delta) \subset \tilde{Y}$. The exact sequence

$$H^{2m}_\Sigma(\tilde{Y}) \xrightarrow{\tau_*} H^{2m}(\tilde{Y}) \xrightarrow{r^*} H^{2m}(X_U)$$

shows that $\mathrm{Griff}[Z] = 0$ if and only if $\pi^*[Z] \in \mathrm{im}\,\tau_*$. By Poincaré–Lefschetz duality we have

$$H^{2m}_\Sigma(\tilde{Y}, \mathbb{Q}) \cong H_{2m}(\Sigma, \mathbb{Q}) \cong \oplus_{s \in \Delta} H_{2m}(X_s, \mathbb{Q}).$$

Let X_0 be a smooth fiber. By Picard–Lefschetz theory we have an exact sequence

$$0 \to H_{2m}(X_0) \to H_{2m}(X_s) \to \mathbb{Z} \xrightarrow{\partial} H_{2m-1}(X_0) \to H_{2m-1}(X_s) \to 0$$

and $\partial(1) = \delta_s$ is the vanishing cycle associated to the singular fiber X_s. As $H_{2m-1}(X_0)$ is generated by vanishing cycles and the vanishing cycles are conjugate under the action of the monodromy group, the hypothesis of the lemma shows that $\delta_s \neq 0$ for all $s \in \Delta$, hence $H_{2m}(X_s) \cong H_{2m}(X_0)$ for all $s \in \Delta$. Since $H_{2m}(X_0) \cong H^{2m-2}(X_0) \cong H^{2m-2}(Y)$ by Poincaré duality and the Lefschetz hyperplane theorem, $H_{2m}(X_s)$ can be identified with $H^{2m-2}(Y)$. Under this identification $\pi_* \circ \tau_* : H^{2m-2}(Y) \to H^{2m}(Y)$ is identified with the Lefschetz operator L_Y, which is given by cup product with $c_1(\mathcal{O}_Y(1))$. As $\pi_* \pi^*[Z] = [Z]$ we find

$$\pi^*[Z] \in \operatorname{im} \tau_* \Longleftrightarrow [Z] \in \operatorname{im} L_Y$$
$$\Longleftrightarrow 0 = [Z] \in H^{2m}_{\mathrm{pr}}(Y).$$

$\square$

We can now finish the proof of Griffiths's theorem.

Proof (Theorem 7.2.1) Suppose that $Z \in Z^m(Y)$ is a cycle such that $Z_t = Z \cap X_t$ is algebraically equivalent to zero for very general t. Then $\psi(Z_t) = 0$ for very general t by Lemma 7.2.3, hence the normal function ν_Z is zero. This implies that $\partial(\nu_Z) = 0$, hence $\operatorname{Griff}[Z] = 0$ by Lemma 7.2.4 and $[Z] = 0$ by Lemma 7.2.5. $\square$

We mention two applications of Theorem 7.2.1. Note that the theorem also applies to hypersurface sections (use a Veronese embedding to reduce to the case of hyperplane sections).

Corollary 7.2.6 *Let $Y \subset \mathbb{P}^{2m+1}$ ($m \geq 2$) be a smooth quadric and let X be the intersection of Y with a hypersurface of degree $d \geq 4$. If X is very general then $\operatorname{Griff}^m(X) \otimes \mathbb{Q} \neq 0$.*

Proof The quadric Y contains two families of m-planes. Let L_1, L_2 be two m-planes from the two different families. If $d \geq 4$ the Hodge structure on $H^{2m-1}(X)$ is not of type $\{(m-1, m), (m, m-1)\}$ and we can apply Theorem 7.2.1 to the cycle $Z = L_1 - L_2$. (If $m \geq 3$ it suffices to take $d \geq 3$.) $\square$

Corollary 7.2.7 *Let Y be a smooth quintic fourfold such that $\operatorname{im}(\operatorname{cl}^2_Y) \cap H^4_{\mathrm{pr}}(Y, \mathbb{Q}) \neq 0$ and let $\{X_t\}_{t \in \mathbb{P}^1}$ be a Lefschetz pencil of hyperplane sections of Y. If t is very general then $\operatorname{Griff}^2(X_t) \otimes \mathbb{Q} \neq 0$.*

The Fermat quintic $Y \subset \mathbb{P}^5$ is an example of a quintic fourfold that satisfies the condition of Corollary 7.2.7. It contains two planes P_1, P_2 such that

$0 \neq [P_1 - P_2] \in H^4_{pr}(Y, \mathbb{Q})$. By Corollary 7.2.7, the difference of the two lines $L_{1,t} = P_1 \cap H_t$ and $L_{2,t} = P_2 \cap H_t$ on the quintic threefold $X_t = Y \cap H_t$ is a non-torsion element of $\mathrm{Griff}^2(X_t)$ if t is very general. Note that the set of quintic fourfolds Y that satisfy the condition of Corollary 7.2.7 is a countable union of proper closed subsets of $\mathbb{P}H^0(\mathbb{P}^5, \mathcal{O}_\mathbb{P}(5))$ by the Noether–Lefschetz theorem.

Bibliographical hints Griffiths's theorem appears in [14, Theorem 14.1]. A detailed study of the Griffiths homomorphism can be found in [18]. For the proof of Griffiths's theorem we have followed the arguments of Voisin [26].

7.3 The theorem of Green–Voisin

Let $S \subset \mathbb{P}^3$ be a surface of degree $d \geq 4$. If S is very general then $\mathrm{Pic}(S) \cong \mathbb{Z}$ by the Noether–Lefschetz theorem. Using the Lefschetz hyperplane theorem and the exponential sequence one easily proves that $\mathrm{Pic}(X) \cong \mathbb{Z}$ for every smooth hypersurface $X \subset \mathbb{P}^{n+1}$ of dimension $n \geq 3$. Griffiths and Harris asked whether $\mathrm{CH}^2(X) \cong \mathbb{Z}$ if X is a very general threefold of degree $d \geq 6$ in $\mathbb{P}^4$. If this question has an affirmative answer, it would follow that the image of the Abel–Jacobi map ψ_X is zero if $d \geq 6$. Modulo torsion, this has been proved by Green and Voisin.

Theorem 7.3.1 (Green–Voisin) *Let $X \subset \mathbb{P}^{2m}$ be a smooth hypersurface of degree d and dimension $2m - 1$ ($m \geq 2$). If X is very general and if $d \geq 2 + \frac{4}{m-1}$ then the image of*

$$\psi^m_X : \mathrm{CH}^m_{\mathrm{hom}}(X) \to J^m(X)$$

is contained in the torsion points of $J^m(X)$.

The proof of this result is obtained by a careful study of the infinitesimal invariant of a normal function. Again we need several lemmas, the first of which is a technique known as 'spreading out' an algebraic cycle.

Lemma 7.3.2 *Let $f : X \to S$ be a smooth projective morphism. If $Z_0 \in Z^m_{\mathrm{hom}}(X_{s_0})$ and $s_0 \in S$ is very general, then there exist a finite covering $g : T \to S$, a point $t_0 \in g^{-1}(s_0)$ and a relative cycle $Z_T \in Z^m_{\mathrm{hom}}(X_T/T)$ such that $Z_T(t_0) = Z_0$.*

Proof There exist a relative Chow variety $\mathrm{Chow}^m(X/S)$ that parametrizes algebraic cycles of relative codimension m over S and a dominant map

$$p : \mathrm{Chow}^m(X/S) \to S.$$

Let $\Sigma \subset S$ be the image of the irreducible components of $\mathrm{Chow}^m(X/S)$ that do not dominate S. The subset $\Sigma \subset S$ is a countable union of Zariski closed subsets of S. If $s_0 \in S \setminus \Sigma$ then there exists a finite morphism $g : T \to S$ such that $\mathrm{Chow}^m(X \times_S T/T)$ admits a rational section that passes through Z_0; let Z_T be its image. By construction the cycles $Z(t)$ are algebraically equivalent, hence homologically equivalent. $\qquad\square$

Remark 7.3.3 By shrinking S and T, we may assume that the relative Chow variety $\mathrm{Chow}^m(X \times_S T/T)$ admits a section and that $g : T \to S$ is a finite étale morphism.

Set $V = H^0(\mathbb{P}^{2m}, \mathcal{O}_{\mathbb{P}}(d))$ and let $U \subset \mathbb{P}(V)$ be the complement of the discriminant locus. Consider the universal family of hypersurfaces

$$X_U = \{(x, F) \in \mathbb{P}^{2m} \times U \,|\, F(x) = 0\}.$$

Let $0 \in U$ be a base point and let $Z \in Z^m_{\mathrm{hom}}(X_0)$. If the base point is very general, we can apply Lemma 7.3.2 to the morphism $f : X_U \to U$ to 'spread out' Z_0 to a relative cycle $Z_T \in Z^m_{\mathrm{hom}}(X_T/T)$ over a new base T. After deleting the branch locus of the finite morphism $g : T \to U$ we may assume that g is étale. Let $\nu \in H^0(T, \mathcal{J}^m)$ be the normal function associated to Z_T. Write $S = \mathbb{C}[X_0, \dots, X_{2m}]$. Let $f \in S_d$ be a homogeneous polynomial of degree d, and let $J_f = (\frac{\partial f}{\partial X_0}, \dots, \frac{\partial f}{\partial X_{2m}})$ be the Jacobian ideal of f. The quotient ring $R_f = S/J_f$ is a graded ring, called the Jacobi ring of f. Griffiths proved that the cohomology of $X = V(f)$ in the middle dimension can be described by the Jacobi ring:

$$H^{p,q}(X) \cong R_{(q+1)d-2m-1} \qquad (p + q = 2m - 1).$$

The following result is known as the *symmetrizer lemma*. We shall not prove it here, as we shall prove a more general result in the lectures on Nori's theorem.

Lemma 7.3.4 (Donagi–Green) *The Koszul complex*

$$\textstyle\bigwedge^2 S_d \otimes S_{a-d} \to S_d \otimes S_a \to S_{a+d} \to 0$$

is exact if $a - d > 0$.

We shall use the symmetrizer lemma to study the cohomology sheaves of the complex

$$\Omega^\bullet(\mathcal{F}^m) = (\mathcal{F}^m \to \Omega^1_T \otimes \mathcal{F}^{m-1} \to \cdots).$$

Lemma 7.3.5 *If $d \geq 2 + \frac{4}{m-1}$ then $\mathcal{H}^0(\Omega^\bullet(\mathcal{F}^m)) = \mathcal{H}^1(\Omega^\bullet(\mathcal{F}^m)) = 0$.*

Proof By a spectral sequence argument, it suffices to verify the assertion for the graded pieces

$$\operatorname{Gr}_F^p \Omega^\bullet(\mathcal{F}^m) = (\mathcal{H}^{p,2m-1-p} \to \Omega_T^1 \otimes \mathcal{H}^{p-1,2m-p} \to \cdots).$$

To this end, it suffices to show that the complex

$$0 \to H^{p,2m-p-1}(X_t) \to \Omega_{T,t}^1 \otimes H^{p-1,2m-p}(X_t)$$
$$\to \Omega_{T,t}^2 \otimes H^{p-2,2m-p+1}(X_t)$$

is exact as far as written for all $p \geq m$ and for all $t \in T$. The dual complex is

$$\bigwedge^2 T_t \otimes H^{2m-p+1,p-2}(X_t) \to T_t \otimes H^{2m-p,p-1}(X_t)$$
$$\to H^{2m-p-1,p}(X_t) \to 0.$$

As $g : T \to U$ is étale, we can identify the tangent space T_t with the tangent space to U at $g(t)$, which is isomorphic to S_d. Using Griffiths's description of the cohomology groups of X_t we can identify the dual complex with the complex

$$\bigwedge^2 S_d \otimes R_{(p-1)d-2m-1} \to S_d \otimes R_{pd-2m-1} \to R_{(p+1)d-2m-1}.$$

A diagram chase shows that this complex is exact at the middle term if

(i)
$$\bigwedge^2 S_d \otimes S_{(p-1)d-2m-1} \to S_d \otimes S_{pd-2m-1} \to S_{(p+1)d-2m-1}$$

   is exact at the middle term;

(ii) the map $S_d \otimes J_{pd-2m-1} \to J_{(p+1)d-2m-1}$ is surjective.

By the symmetrizer lemma, (i) holds for all $p \geq m$ if $(m-1)d \geq 2m+2$, which translates into the condition of the lemma. As the Jacobian ideal J_f is generated in degree $d-1$, (ii) holds if $(m-1)d - 2m - 1 \geq d - 1$; this condition is weaker than the first condition. The multiplication map

$$S_d \otimes R_{pd-2m-1} \to R_{(p+1)d-2m-1}$$

is surjective for all $p \geq m$ if $md \geq 2m+1$; again, this condition is weaker than the first condition. $\qquad\square$

The proof of Theorem 7.3.1 is finished by the following monodromy argument.

Lemma 7.3.6 *If* $\mathcal{H}^0(\Omega^\bullet(\mathcal{F}^m)) = \mathcal{H}^1(\Omega^\bullet(\mathcal{F}^m)) = 0$ *then* ν *is a torsion section of* $\mathcal{J}^m$.

Proof As $\delta\nu$ is a global section of $\mathcal{H}^1(\Omega^\bullet(\mathcal{F}^m))$, we have $\delta\nu = 0$. Hence ν has flat local liftings. The vanishing of $\mathcal{H}^0(\Omega^\bullet(\mathcal{F}^m))$ implies that these flat local liftings are unique up to sections of the local system $H_\mathbb{Z} = R^{2m-1} f_* \mathbb{Z}/\mathrm{tors}$. Let

$$\rho : \pi_1(T, t_0) \to \mathrm{Aut}\, H^{2m-1}(X_0, \mathbb{C})$$

be the monodromy representation. By the uniqueness property of flat local liftings obtained above, it follows that

$$(*) \qquad\qquad \rho(\gamma)(\tilde{\nu}(0)) - \tilde{\nu}(0) \in H^{2m-1}(X_0, \mathbb{Z})$$

for all $\gamma \in \pi_1(T, t_0)$. To show that $\nu(0) \in J^m(X_0)$ is torsion, we have to prove that $\tilde{\nu}(0) \in H^{2m-1}(X_0, \mathbb{Q})$. We shall verify that this follows from $(*)$. Let $\{\gamma_i\}$ be a set of generators of $\pi_1(U, 0)$ coming from a Lefschetz pencil $L \subset U$, and let $\{\delta_i\}$ be the corresponding set of vanishing cocycles in $H^{2m-1}(X_0, \mathbb{Z})$. Let N be the index of the subgroup $g_*\pi_1(T, t_0) \subset \pi_1(U, 0)$. We have $\gamma_i^N = g_*\tilde{\gamma}_i$ some $\tilde{\gamma}_i \in \pi_1(T, t_0)$. By the Picard–Lefschetz formula we have

$$\rho(\tilde{\gamma}_i)(\tilde{\nu}(0)) - \tilde{\nu}(0) = \varepsilon N \langle \tilde{\nu}(0), \delta_i \rangle \delta_i$$

where $\varepsilon \in \{-1, 1\}$. Using $(*)$ we obtain $\langle \tilde{\nu}(0), \delta_i \rangle \in \mathbb{Q}$ for all i. As the vanishing cocycles generate $H^{2n-1}(X, \mathbb{Q})$, it follows that $\tilde{\nu}(0) \in H^{2m-1}(X_0, \mathbb{Q})$. $\qquad\square$

Remark 7.3.7

(i) Apart from the known exceptions $X = V(d) \subset \mathbb{P}^4$, $d \le 5$, the only exceptions to the Green–Voisin theorem are cubic fivefolds and cubic sevenfolds. In both cases the image of the Abel–Jacobi map is non-zero modulo torsion; see [4] and [1].

(ii) Griffiths and Harris have shown that there are no non-zero normal functions over the complement $U \subset \mathbb{P}H^0(\mathbb{P}^4, \mathcal{O}_\mathbb{P}(d))$ of the discriminant locus if $d \ge 3$ (this is also true for $d = 2$, since the intermediate Jacobian of a quadric is zero). We have seen that there can be non-zero normal functions if $d \le 5$, but they are multivalued sections of $\mathcal{J}^m$ that only become well defined after passing to a finite covering $T \to U$.

Bibliographical hints Theorem 7.3.1 was proved independently by Green and Voisin. Green's proof has appeared in [11]; see also [26]. The conjectures of Griffiths and Harris have appeared in [17]. The monodromy argument in Lemma 7.3.6 is taken from [27, Lecture 4].

7.4 Nori's connectivity theorem

Nori's connectivity theorem is a far-reaching generalization of the theorem of Green–Voisin, based on the observation that their result can be interpreted in terms of the cohomology of the universal family of hypersurfaces in projective space.

Notation Let $(Y, \mathcal{O}_Y(1))$ be a smooth polarized variety over $\mathbb{C}$ and let $X \subset Y$ be a smooth complete intersection of multi-degree $(d_0, \ldots, d_r)$ and of dimension n. Set

$$E = \mathcal{O}_Y(d_0) \oplus \cdots \oplus \mathcal{O}_Y(d_r)$$

and define $S = \mathbb{P}H^0(Y, E)$. Let $U \subset S$ be the complement of the discriminant locus. Over S we have the universal family

$$X_S = \{(y, s) \in Y \times S \mid s(y) = 0\} \subset Y_S = Y \times S.$$

Given a morphism $T \to S$ we obtain induced families

$$X_T = X \times_S T, \qquad Y_T = Y \times T$$

over T by base change. Let $r : X_T \to Y_T$ be the inclusion map, and let $p : Y_T \to T$, $f = p \circ r : X_T \to T$ be the projection maps. Nori's main result is the following connectivity theorem for the pair (Y_T, X_T).

Theorem 7.4.1 (Nori) *If* $\min(d_0, \ldots, d_r) \gg 0$ *then for every smooth morphism* $T \to S$ *we have* $H^{n+k}(Y_T, X_T, \mathbb{Q}) = 0$ *for all* $k \leq n$.

Remark 7.4.2

(i) For every base change $T \to S$ we have $H^k(Y_T, X_T, \mathbb{Z}) = 0$ for all $k \leq n$ ($T \to S$ is not necessarily smooth). To see this, note that the Lefschetz hyperplane theorem shows that the restriction map $i^* R^q p_* \mathbb{Z} \to R^q f_* \mathbb{Z}$ is an isomorphism if $q < n$ and is injective if $q = n$. As an exercise, the reader may verify that this implies that the restriction map $H^k(Y_T, \mathbb{Z}) \to H^k(X_T, \mathbb{Z})$ is an isomorphism if $k \leq n - 1$ and is injective if $k = n$ by comparing the Leray spectral sequence

$$E_2^{p,q} = H^p(T, R^q f_* \mathbb{Z}) \Rightarrow H^{p+q}(X_T, \mathbb{Z})$$

to the Künneth spectral sequence

$$\tilde{E}_2^{p,q} = H^p(T, \mathbb{Z}) \otimes H^q(Y, \mathbb{Z}) \Rightarrow H^{p+q}(Y_T, \mathbb{Z}).$$

If $k \geq 1$ one can construct examples where $H^{n+k}(Y_T, X_T, \mathbb{Z})$ is non-zero even if $\min(d_0, \ldots, d_r) \gg 0$.

(ii) Set $\mathcal{E} = p_Y^* E \otimes p_S^* \mathcal{O}_S(1)$. The variety $X_S \subset Y_S$ is the zero locus of the tautological section $\tau \in H^0(Y_S, \mathcal{E})$. As $\mathcal{E}$ is an ample vector bundle on the projective variety Y_S, it follows that $H^k(Y_S, X_S, \mathbb{Z}) = 0$ for all

$p \leq \dim X_S$ [19]. This connectivity result is stronger than the statement in Nori's theorem for $T = S$, but it is not invariant under base change.

(iii) The local system $R^n f_* \mathbb{Q}$ splits as a direct sum of a fixed part $i^* R^n p_* \mathbb{Q}$ and a variable part $\mathbb{V}$. Nori's connectivity theorem is equivalent to a statement about cohomology with values in the local system $\mathbb{V}$. This follows from Deligne's theorem on the degeneration at E_2 of the Leray spectral sequence for f with $\mathbb{Q}$-coefficients. Define

$$H^k(Y)_0 = \ker(i^* : H^k(Y) \to H^k(X)).$$

Note that this group is zero if $k \leq n$. By Deligne's theorem, we have

$$H^{n+k}(X_T, \mathbb{Q}) \cong \bigoplus_{p+q=n+k} H^p(T, R^q f_* \mathbb{Q}).$$

Comparing the decomposition of $H^{n+k}(X_T)$ with the Künneth decomposition of $H^{n+k}(Y_T)$ we find that $H^{n+k}(Y_T, X_T) = 0$ for all $k \leq c$ if and only if the map

$$\bigoplus_{i=0}^{k-1} H^i(T, \mathbb{Q}) \otimes H^{n+k-i}(Y, \mathbb{Q})_0 \to H^k(T, \mathbb{V})$$

is an isomorphism for all $k < c$ and is injective for $k = c$.

(iv) Nori's theorem does not necessarily hold if we omit the smoothness assumption on the base change. For instance, let $L \subset S$ be a Lefschetz pencil of hypersurface sections on a smooth projective variety Y such that $\dim Y = 2m$ and $H^{2m}_{\mathrm{pr}}(Y, \mathbb{Q}) \neq 0$. Let $T = U \cap L$ be the smooth part of the Lefschetz pencil. Using (iii), the vanishing of $H^{2m}(Y_T, X_T)$ is equivalent to the vanishing of $H^0(T, \mathbb{V})$ and the injectivity of the map

$$\psi_1 : H^{2m}_{\mathrm{pr}}(Y, \mathbb{Q}) \to H^1(T, \mathbb{V}).$$

By Picard–Lefschetz theory we have $H^0(T, \mathbb{V}) = 0$. As the map ψ_1 coincides with the Griffiths homomorphism discussed in Section 7.2, it follows from Griffiths's theorem that $H^{2m}(Y_T, X_T) = 0$ if $d \gg 0$. Again using (iii), the vanishing of $H^{2m+1}(Y_T, X_T, \mathbb{Q})$ would imply that there is an injective map

$$\psi_2 : H^{2m+1}(Y, \mathbb{Q})_0 \oplus H^1(T, \mathbb{Q}) \otimes H^{2m}_{\mathrm{pr}}(Y, \mathbb{Q}) \to H^2(T, \mathbb{V}).$$

As T is affine, we have $H^2(T, \mathbb{V}) = 0$. As $H^1(T, \mathbb{Q})$ is non-zero if $d \gg 0$, the left-hand side is non-zero for $d \gg 0$; hence ψ_2 cannot be injective and $H^{2m+1}(Y_T, X_T, \mathbb{Q}) \neq 0$ for all $d \gg 0$. A similar argument shows that $H^{n+c}(Y_T, X_T) \neq 0$ if $L \subset S$ is a general linear subspace of dimension $e < c$. By analogy with Griffiths's theorem, Nori [24, Conjecture 7.4.1] conjectures that $H^{n+k}(Y_T, X_T) = 0$ for all $k \leq c$ if L is a general linear subspace of dimension c in S.

The following result shows that Nori's theorem implies the theorem of Green–Voisin.

Theorem 7.4.3 *Let X be a very general smooth complete intersection in Y of multi-degree $(d_0, \ldots, d_r)$ and dimension n, with inclusion map $i : X \to Y$. If $\min(d_0, \ldots, d_r) \gg 0$ then*

$$\mathrm{im}(\mathrm{cl}^p_{\mathcal{D}, X}) \subset i^* H^{2p}_{\mathcal{D}}(Y, \mathbb{Q}(p))$$

for all $p < n$.

Proof Let $\Delta \subset S$ be the discriminant locus. If $s_0 \in S$ is very general, there exist for every $z_0 \in \mathrm{CH}^p_{\mathrm{hom}}(X_{s_0})$ a subset $U' \subset S \setminus \Delta$ containing s_0, a finite étale covering $g : T \to U'$, a relative cycle $Z \in \mathrm{CH}^p_{\mathrm{hom}}(X_T/T)$ and $t_0 \in g^{-1}(s_0)$ such that $Z_T(t_0) = z_0$. Consider the commutative diagram

$$
\begin{array}{ccc}
H^{2p}_{\mathcal{D}}(Y_T, \mathbb{Q}(p)) & \xrightarrow{r^*} & H^{2p}_{\mathcal{D}}(X_T, \mathbb{Q}(p)) \\
\downarrow{\scriptstyle k^*} & & \downarrow{\scriptstyle j^*} \\
H^{2p}_{\mathcal{D}}(Y, \mathbb{Q}(p)) & \xrightarrow{i^*} & H^{2p}_{\mathcal{D}}(X_{s_0}, \mathbb{Q}(p)).
\end{array}
$$

It follows from Theorem 7.4.1 and the long exact sequence of Deligne cohomology (Section 7.7) that

$$H^{n+k}_{\mathcal{D}}(Y_T, X_T, \mathbb{Q}(p)) = 0$$

for all $k \leq n$. Hence the map r^* is surjective for all $p \leq n - 1$. Choose $\xi \in H^{2p}_{\mathcal{D}}(Y_T, \mathbb{Q}(p))$ such that $r^*\xi = \mathrm{cl}_{\mathcal{D}}(Z_T)$ and put $\eta = k^*\xi \in H^{2p}_{\mathcal{D}}(Y, \mathbb{Q}(p))$. By construction we have $i^*\eta = j^* \mathrm{cl}_{\mathcal{D}}(Z_T) = \mathrm{cl}_{\mathcal{D}}(z_0)$. $\square$

Remark 7.4.4

(i) By the Lefschetz hyperplane theorem, the statement of the theorem is only non-trivial if $n = 2p$ or $n = 2p - 1$. Let us take $Y = \mathbb{P}^N$. If we apply the theorem with $n = 2p$, we find that the image of cl^p_X is isomorphic to $\mathbb{Z}$ (note that $H^{2p}(X, \mathbb{Z})$ is torsion free), which is the general form of the Noether–Lefschetz theorem. If we apply the theorem with $n = 2p - 1$, we find that the image of the Abel–Jacobi map ψ^p_X is contained in the torsion points of $J^p(X)$, which is the Green–Voisin theorem. Note that we only obtain asymptotic versions of these theorems. Paranjape [25] has obtained an effective version of Nori's theorem. It leads to the bound $d \geq 2p + 2$ in both cases. This bound is optimal if $n = 2p$ and $p = 1$

and if $n = 2p - 1$ and $p = 2$, but not in general: we obtained the better bound $d \geq 2 + \frac{4}{\cdot p-1}$ for the Green–Voisin theorem in Section 7.3. We shall see later how to obtain effective degree bounds for Theorem 7.4.1 that do give the optimal bounds for the theorems of Noether–Lefschetz and Green–Voisin.

(ii) Theorem 7.4.3 shows that we cannot expect to obtain a connectivity result for the pair (Y_T, X_T) without conditions on the degrees. For instance we cannot have $H^5(\mathbb{P}_T^4, X_T) = 0$ if $d \leq 5$ since the image of the Abel–Jacobi map on a very general quintic hypersurface $X \subset \mathbb{P}^4$ is not contained in the torsion points of $J^2(X)$ by Griffiths's theorem.

(iii) Theorem 7.4.3 does not hold for zero cycles ($p = n$). Consider for example a family $C_T \to T$ of smooth plane curves of degree d ($n = 1$). If Theorem 7.4.3 could be applied to this case, we would find that the image of the Abel–Jacobi map for a very general plane curve is contained in the torsion points of the Jacobian. This is clearly false by the Jacobi inversion theorem.

Bibliographical hints Nori's theorem has appeared in [24]; see also [12, Lecture 8]. Paranjape's effective version of Nori's theorem can be found in [25]. Green and Müller-Stach have obtained a more precise version of Theorem 7.4.3, see [13].

7.5 Sketch of proof of Nori's theorem

In this section we sketch the proof of an effective version of Nori's theorem along the lines of the proof of Green–Voisin. Our condition on the base change is more restrictive than the one in Nori's original result: we consider smooth morphisms $T \to S$ such that the induced map $X_T \to T$ is smooth, i.e. we demand that the morphism $T \to S$ factors through the complement of the discriminant locus. This suffices for the geometric applications that we shall discuss in the next lecture. We start by recalling a little bit of mixed Hodge theory. The cohomology groups of a quasi-projective variety do not always carry a pure Hodge structure (HS). Consider for example the variety $U = \mathbb{P}^1(\mathbb{C}) \setminus \{0, \infty\}$. As U is homotopically equivalent to a circle its first Betti number is 1, hence $H^1(U)$ cannot carry a pure HS. (A similar result holds for the complement of two points in a smooth compact curve of arbitrary genus.) Deligne has proved that the cohomology groups of smooth quasi-projective variety X always carry a *mixed Hodge structure* (MHS). This means that for every $k \geq 0$ there exist a decreasing filtration $F^\bullet$ on $H^k(X, \mathbb{C})$ (the Hodge filtration) and an increasing filtration $W_\bullet$ on $H^k(X, \mathbb{Q})$ (the weight filtration) such that $F^\bullet$ induces a pure HS of weight m on the graded pieces

$$\mathrm{Gr}_m^W H^k(X) = W_m H^k(X)/W_{m-1} H^k(X)$$

of the weight filtration. By construction, the weight filtration on the cohomology of a smooth quasi-projective variety satisfies

$$\mathrm{Gr}_m^W H^k(X) = 0 \qquad \text{if } m < k.$$

A morphism of MHS $f : H_1 \to H_2$ is a homomorphism of abelian groups that is compatible with the filtrations $F^\bullet$ and $W_\bullet$. Deligne has shown that such morphisms are *strictly compatible* with the Hodge and weight filtrations, i.e.

$$F^p H_2 \cap \mathrm{im}\, f = f(F^p H_1)$$
$$W_m H_2 \cap \mathrm{im}\, f = f(W_m H_1).$$

The proof of Nori's theorem proceeds in several steps.

Step 1: mixed Hodge theory As we usually want to apply Nori's theorem to a quasi-projective base T, we have to deal with the cohomology of the quasi-projective varieties X_T and Y_T. By Deligne's theorem the groups $H^{n+k}(Y_T)$ and $H^{n+k}(X_T)$ carry a MHS. Using a cone construction one can also put a MHS on the relative cohomology $H^{n+k}(Y_T, X_T)$. As the long exact sequence

$$H^{n+k-1}(X_T) \to H^{n+k}(Y_T, X_T) \to H^{n+k}(Y_T) \to H^{n+k}(X_T)$$

is an exact sequence of MHS and morphisms of MHS are strictly compatible with the weight filtration, we have

$$\mathrm{Gr}_m^W H^{n+k}(Y_T, X_T) = 0 \qquad \text{if } m < n + k - 1.$$

The presence of an MHS on $H^{n+k}(Y_T, X_T)$ implies that this vector space vanishes if a large enough subspace of it vanishes.

Lemma 7.5.1 *Suppose there exists a natural number $m \leq [\frac{n+k}{2}]$ such that $F^m H^{n+k}(Y_T, X_T) = 0$. Then $H^{n+k}(Y_T, X_T) = 0$.*

Proof Suppose that $H^{n+k}(Y_T, X_T) \neq 0$. Then there exists $i \geq n + k - 1$ such that $\mathrm{Gr}_i^W H^{n+k}(Y_T, X_T) \neq 0$. We have a Hodge decomposition

$$\mathrm{Gr}_i^W H^{n+k}(Y_T, X_T) = \oplus_{p+q=i} H^{p,q}$$

such that $H^{q,p} = \overline{H}^{p,q}$ (Hodge symmetry). If $H^{p,q} \neq 0$ then $p \leq m - 1$ by the hypothesis of the lemma, hence also $q \leq m - 1$ by Hodge symmetry. But then $i = p + q \leq 2m - 2 \leq n + k - 2$, contradiction. $\square$

It is difficult to apply the previous lemma to prove the vanishing of $H^{n+k}(Y_T, X_T)$ since the Hodge filtration on $H^{n+k}(Y_T, X_T)$ is defined in a complicated way. One starts by choosing compatible *good compactifications* $\bar{Y}_T$ and $\bar{X}_T$ of Y_T and X_T with boundary divisors $\tilde{D}_T = \bar{Y}_T \setminus Y_T$, $D_T = \bar{X}_T \setminus X_T$. Let $j : \bar{X}_T \to \bar{Y}_T$ be the inclusion map (its restriction to X_T is also denoted by j) and let $C^\bullet(\alpha)$ be the cone of the (surjective) map

$$\alpha : \Omega^\bullet_{\bar{Y}_T}(\log \tilde{D}_T) \to j_* \Omega^\bullet_{\bar{X}_T}(\log D_T).$$

One can show that $\mathbb{H}^{n+k}(C^\bullet(\alpha)) \cong H^{n+k}(Y_T, X_T)$. The Hodge filtration on $H^{n+k}(Y_T, X_T)$ is defined by

$$F^p H^{n+k}(Y_T, X_T) = \mathrm{im}(\mathbb{H}^{n+k}(\sigma_{\geq p} C^\bullet(\alpha)) \to \mathbb{H}^{n+k}(C^\bullet(\alpha))).$$

There is another, easier way to put a filtration on $H^{n+k}(Y_T, X_T)$. Define

$$\Omega^\bullet_{Y_T, X_T} = \ker(\beta : \Omega^\bullet_{Y_T} \to j_* \Omega^\bullet_{X_T}).$$

It follows from Grothendieck's algebraic de Rham theorem and the five lemma that $\mathbb{H}^{n+k}(\Omega^\bullet_{Y_T, X_T}) \cong H^{n+k}(Y_T, X_T)$. Hence we can define a second filtration $G^\bullet$ on $H^{n+k}(Y_T, X_T)$ by

$$G^p H^{n+k}(Y_T, X_T) = \mathrm{im}(\mathbb{H}^{n+k}(\sigma_{\geq p} \Omega^\bullet_{Y_T, X_T}) \to \mathbb{H}^{n+k}(\Omega^\bullet_{Y_T, X_T})).$$

As β is surjective, the complex $\Omega^\bullet_{Y_T, X_T}$ is quasi-isomorphic to $C^\bullet(\beta)$. The restriction from $\bar{Y}_T$ to Y_T induces a quasi-isomorphism $C^\bullet(\alpha) \to C^\bullet(\beta)$. Hence

$$F^p H^{n+k}(Y_T, X_T) \subseteq G^p H^{n+k}(Y_T, X_T),$$

so it suffices to show that $G^m H^{n+k}(Y_T, X_T) = 0$ for some $m \leq [\frac{n+k}{2}]$. The advantage of working with $G^\bullet$ is that we can work on Y_T and X_T and do not have to pass to a compactification.

Step 2: semi-continuity There exists a spectral sequence (induced by the filtration bête)

$$E_1^{a,b} = H^a(Y_T, \Omega^b_{Y_T, X_T}) \Rightarrow \mathbb{H}^{a+b}(\Omega^\bullet_{Y_T, X_T}).$$

Using this spectral sequence we find that $G^m H^{n+k}(Y_T, X_T) = 0$ if

$$H^a(Y_T, \Omega^b_{Y_T, X_T}) = 0$$

for all (a, b) such that $a + b \leq n + k$, $b \geq m$.

Let $p : Y_T \to T$ be the projection map. By the Leray spectral sequence it suffices to show that $H^i(T, R^j p_* \Omega^b_{Y_T, X_T}) = 0$ for all (i, j) such that $i + j = a$.

This is certainly true if

$$R^j p_* \Omega^b_{Y_T, X_T} = 0$$

for all $j \leq a$. Let $i_t : Y \to Y_T$ be the inclusion map defined by $i_t(y) = (y, t)$. As the maps $p : Y_T \to T$ and $f : X_T \to T$ are flat and the sheaves $\Omega^b_{Y_T}$ and $\Omega^b_{X_T}$ are locally free, it follows that these sheaves are flat over $\mathcal{O}_T$. Hence $\Omega^b_{Y_T, X_T}$ is flat over $\mathcal{O}_T$. By semi-continuity, $R^j p_* \Omega^b_{Y_T, X_T} = 0$ if

$$H^j(Y, i_t^* \Omega^b_{Y_T, X_T}) = 0$$

for all $t \in T$.

Step 3: Leray filtration Suppose that $f : X_T \to T$ is smooth. In this case we have an exact sequence

$$0 \to f^* \Omega^1_T \to \Omega^1_{X_T} \to \Omega^1_{X_T/T} \to 0.$$

The *Leray filtration* $L^\bullet$ on $\Omega^b_{X_T}$ is defined by

$$L^p \Omega^b_{X_T} = \mathrm{im}(f^* \Omega^p_T \otimes \Omega^{b-p}_{X_T} \to \Omega^b_{X_T}).$$

Its graded pieces are

$$\mathrm{Gr}^p_L \Omega^b_{X_T} \cong f^* \Omega^p_T \otimes \Omega^{b-p}_{X_T/T}.$$

The spectral sequence associated to the induced filtration on $\Omega^b_{X_T} \otimes \mathcal{O}_{X_t}$ is

$$E_1^{p,q} = \Omega^p_{T,t} \otimes H^{p+q}(X_t, \Omega^{b-p}_{X_t}) \Rightarrow H^{p+q}(X_t, \Omega^b_{X_T} \otimes \mathcal{O}_{X_t}).$$

One can show that the d_1 map in this spectral sequence is the differential of the period map; it is given by cup product with the Kodaira–Spencer class. The Leray filtration on $\Omega^\bullet_{Y_T}$ splits, as Y_T is a product. Define a filtration $L^\bullet$ on $\Omega^\bullet_{Y_T, X_T}$ by

$$L^p \Omega^b_{Y_T, X_T} = \ker(L^p \Omega^b_{Y_T} \to L^p j_* \Omega^b_{X_T}).$$

Set $\Omega^b_{(Y_T, X_T)/T} = \ker(\Omega^b_{Y_T/T} \to j_* \Omega^b_{X_T/T})$. We have

$$\mathrm{Gr}^p_L \Omega^b_{Y_T, X_T} \cong f^* \Omega^p_T \otimes \Omega^{b-p}_{(Y_T, X_T)/T}.$$

If we restrict $L^\bullet$ to the fiber $Y \times \{t\}$ we obtain a spectral sequence

$$E_1^{p,q}(b) = \Omega^p_{T,t} \otimes H^{p+q}(Y, \Omega^{b-p}_{Y, X_t}) \Rightarrow H^{p+q}(Y, i_t^* \Omega^b_{Y_T, X_T}).$$

Using the semi-continuity result from Step 2, we find that $H^{n+k}(Y_T, X_T) = 0$

if

$$E_\infty^{p,q}(b) = 0$$

for all (p, q, b) such that $p + q + b \le n + k$, $b \ge m$.

Step 4: reduction to hypersurfaces There exists a trick using projective bundles to reduce questions about complete intersections to hypersurfaces. It was used by Terasoma and Konno to define Jacobi rings for complete intersections in projective space. We can apply a similar trick to the relative cohomology of the pair (Y_T, X_T). Set

$$\mathcal{E} = p_Y^* E \otimes p_T^* \mathcal{O}_T(1)$$

and let $P_T = \mathbb{P}(\mathcal{E}^\vee)$ be the projective bundle associated to $\mathcal{E}^\vee$ with projection map $\pi_T : P_T \to Y_T$. On P_T there exists a tautological line bundle ξ such that $H^0(P_T, \xi) \cong H^0(Y_T, \mathcal{E})$. We know that $X_T \subset Y_T$ is the zero locus of a section $\sigma \in H^0(Y_T, \mathcal{E})$. Let $\tilde{\sigma}$ be the corresponding section of ξ, and let $\tilde{X}_T \subset P_T$ be its zero locus.

Lemma 7.5.2 *For all $k \ge 0$ there is an isomorphism*

$$H^k(Y_T, X_T) \cong H^{k+2r}(P_T, \tilde{X}_T).$$

Proof Consider the diagram

$$\pi_T^{-1}(X_T) = \mathbb{P}(\mathcal{E}^\vee|_{X_T}) \subset \tilde{X}_T \subset P_T \supset P_T \setminus \tilde{X}_T$$
$$\downarrow{\scriptstyle \pi_T} \qquad \downarrow{\scriptstyle \pi_T}$$
$$X_T \subset Y_T \supset Y_T \setminus X_T.$$

As the line bundle ξ restricts to $\mathcal{O}_\mathbb{P}(1)$ on each fiber of π_T, the induced map

$$\pi_T : P_T \setminus \tilde{X}_T \to Y_T \setminus X_T$$

is a fiber bundle with fiber $\mathbb{A}^r$. Hence $(\pi_T)_*$ induces an isomorphism

$$H_c^{k+2r}(P_T \setminus \tilde{X}_T) \cong H_c^k(Y_T \setminus X_T).$$

By Poincaré–Lefschetz duality we find an isomorphism $H^{k+2r}(P_T, \tilde{X}_T) \cong H^k(Y_T, X_T)$. $\qquad \square$

By Lemma 7.5.2 it suffices to prove Nori's theorem for a family of hypersurface sections $X_T \subset Y_T$ defined by sections of a very ample line bundle $L = \mathcal{O}_Y(d)$.

Step 5: base change We say that Nori's condition (N_c) holds if

$$R^a p_* \Omega^b_{Y_T, X_T} = 0$$

for all (a, b) such that $a + b \le n + c$, $b \ge m$. If (N_c) holds, then $H^{n+k}(Y_T, X_T) = 0$ for all $k \le c$.

Lemma 7.5.3 (Nori) *Let U and T be smooth quasi-projective varieties and let $g : T \to U$ be a smooth morphism.*

(i) *If (N_c) holds for U, then (N_c) holds for T;*
(ii) *if g is smooth and surjective and (N_c) holds for T, then (N_c) holds for U.*

For the proof of Lemma 7.5.3, see [24, Lemma 2.2]. Define $V = H^0(Y, L)$, $S = \mathbb{P}(V)$. Let $\Delta \subset S$ and $\Delta' \subset V$ be the discriminant loci and let $U = S \setminus \Delta$, $U' = V \setminus \Delta'$ be their complements. Let $i : U' \to V \setminus \{0\}$ be the inclusion, and let $\pi : V \setminus \{0\} \to \mathbb{P}(V)$ be the projection. The composed map $\pi \circ i : U' \to S$ is a smooth morphism with image U. If (N_c) holds for the base U', then (N_c) holds for every base T such that $T \to U$ is a smooth morphism by Lemma 7.5.3. Hence it suffices to prove that (N_c) holds for one particular choice of the base T, namely $T = U'$. In this case, the tangent space T_t to T at every point t can be identified with V. Hence

$$E^{p,q}_1(b) \cong \bigwedge\nolimits^p V^\vee \otimes H^{p+q}(\Omega^{b-p}_{Y, X_t})$$

for all $t \in T$. There exists a perfect pairing

$$\Omega^p_{Y, X_t} \otimes \Omega^{n+1-p}_Y(\log X_t) \to K_Y$$

given by wedge product. Using this pairing we can identify $K_Y \otimes (\Omega^p_{Y, X_t})^\vee$ with $\Omega^{n+1-p}_Y(\log X_t)$. By Serre duality the dual of $E^{p,q}_1(b)$ is

$$E^{-p, n+1-q}_1(b) = \bigwedge\nolimits^p V \otimes H^{n+1-p-q}(Y, \Omega^{n+1-b+p}_Y(\log X_t)).$$

Step 6: Green's generalized Jacobi ring Let $P^1(L)$ be the first jet bundle of L. It fits into an exact sequence

$$0 \to \Omega^1_Y \otimes L \to P^1(L) \to L \to 0. \tag{2}$$

There exists a map $j^1 : L \to P^1(L)$ that associates to a section s of L its 1-jet $j^1(s)$. If we dualize the exact sequence (2) and tensor it by L we obtain an exact sequence

$$0 \to \mathcal{O}_Y \to \Sigma_{Y, L} \to T_Y \to 0 \tag{3}$$

with extension class $2\pi i c_1(L) \in H^1(Y, \Omega^1_Y)$. The bundle $\Sigma_{Y, L}$ is called the first prolongation bundle of L (bundle of first order differential operators on sections of L). Let $\mathcal{F}$ be a coherent sheaf of $\mathcal{O}_Y$-modules. For every $s \in H^0(Y, L)$ we have a map $\mathcal{F} \otimes \Sigma_{Y,L} \otimes L^{-1} \to \mathcal{F}$ that is given by contraction with $j^1(s) \in P^1(L)$. Let

$$g_s : H^0(Y, \mathcal{F} \otimes \Sigma_{Y,L} \otimes L^{-1}) \to H^0(Y, \mathcal{F})$$

be the induced map on global sections. Define

$$J_{Y,s}(\mathcal{F}) = \operatorname{im} g_s, \qquad R_{Y,s}(\mathcal{F}) = \operatorname{coker} g_s.$$

Let $X \in |L|$ be a smooth hypersurface section of Y of dimension n. The Poincaré residue sequence

$$0 \to \Omega_Y^{n-p+1} \to \Omega_Y^{n-p+1}(\log X) \overset{\text{Res}}{\to} i_* \Omega_X^{n-p} \to 0$$

induces an exact sequence

$$0 \to H_{\text{pr}}^{n-p+1,p}(Y) \to H^p(Y, \Omega_Y^{n+1-p}(\log X)) \to H_{\text{var}}^{n-p,p}(X) \to 0.$$

Recall that a property (P) is said to hold for a *sufficiently ample* line bundle L if there exists a line bundle L_0 such that (P) holds for L if $L \otimes L_0^{-1}$ is ample.

Proposition 7.5.4 (Green) *If L is sufficiently ample then*

$$H^p(Y, \Omega_Y^{n-p+1}(\log X)) \cong R_{Y,s}(K_Y \otimes L^{p+1}).$$

Proof Contraction with the 1-jet $j^1(s)$ defines a map $\Sigma_L \to L$ whose kernel is isomorphic to $T_Y(-\log X)$. If we dualize and take exterior powers in the resulting short exact sequence

$$0 \to T_Y(-\log X) \to \Sigma_L \to L \to 0,$$

we obtain a long exact sequence

$$0 \to \Omega_Y^{n+1-p}(\log X) \to \bigwedge^{n-p+2} \Sigma_L^\vee \otimes L \to \cdots$$
$$\to \bigwedge^{n+1} \Sigma_L^\vee \otimes L^p \to \bigwedge^{n+2} \Sigma_L^\vee \otimes L^{p+1} \to 0.$$

Using the identifications $\bigwedge^{n+1} \Sigma_L^\vee \cong K_Y \otimes \Sigma_L$ and $\bigwedge^{n+2} \Sigma_L^\vee \cong K_Y$, we obtain the isomorphism of the proposition by chasing through the spectral sequence of hypercohomology associated to this long exact sequence. $\qquad\square$

The bigraded ring

$$R_{Y,s} = \oplus_{p,q \geq 0} R_{Y,s}(Y, K_Y^p \otimes L^{q+1})$$

is called the *Jacobi ring* associated to s.

Example 7.5.5 Take $Y = \mathbb{P}^{n+1}$, $L = \mathcal{O}_{\mathbb{P}}(d)$. The exact sequence (3) is the familiar Euler sequence

$$0 \to \mathcal{O}_{\mathbb{P}} \to \oplus^{n+2}\mathcal{O}_{\mathbb{P}}(1) \to T_{\mathbb{P}} \to 0.$$

Using this sequence, one checks that $R_{Y,s}(K_Y \otimes L^{p+1}) \cong R_{(p+1)d-n-2}$. Hence the ring $\oplus_p R_{Y,s}(K_Y \otimes L^{p+1})$ coincides with Griffiths's Jacobi ring

Step 7: Koszul cohomology Let $\mathcal{F}$ be a coherent sheaf of $\mathcal{O}_Y$-modules. The Koszul cohomology group $\mathcal{K}_{p,q}(\mathcal{F}, L)$ is the cohomology group at the middle term of the complex

$$\bigwedge^{p+1} V \otimes H^0(\mathcal{F} \otimes L^{q-1}) \to \bigwedge^p V \otimes H^0(\mathcal{F} \otimes L^q)$$
$$\to \bigwedge^{p-1} V \otimes H^0(\mathcal{F} \otimes L^{q+1}).$$

These groups were introduced and studied by Green. A standard technique to obtain vanishing theorems for Koszul cohomology is due to Green and Lazarsfeld. Let M_L be the kernel of the surjective evaluation map $e_L : V \otimes \mathcal{O}_Y \to L$. It fits into an exact sequence

$$0 \to M_L \to V \otimes_{\mathbb{C}} \mathcal{O}_Y \to L \to 0.$$

If we take exterior powers in this short exact sequence and twist by $\mathcal{F} \otimes L^{q-1}$ we obtain a complex

$$0 \to \bigwedge^{p+1} M_L \otimes \mathcal{F} \otimes L^{q-1} \to \bigwedge^{p+1} V \otimes \mathcal{F} \otimes L^{q-1}$$
$$\to \bigwedge^p V \otimes \mathcal{F} \otimes L^q \to \bigwedge^{p-1} V \otimes \mathcal{F} \otimes L^{q+1}$$
$$\to \cdots \to \mathcal{F} \otimes L^{p+q} \to 0.$$

The Koszul complex is obtained from this complex by taking global sections, and one obtains the following result.

Proposition 7.5.6

$$\mathcal{K}_{p,q}(\mathcal{F}, L) = 0 \qquad if \qquad H^1(Y, \bigwedge^{p+1} M_L \otimes \mathcal{F} \otimes L^{q-1}) = 0.$$

Step 8: the double complex By Proposition 7.5.4 we can identify

$$E_1^{p,q}(b)^\vee = E_1^{-p,n+1-q}(b) = \bigwedge^p V \otimes H^{n+1-p-q}(Y, \Omega_Y^{n+1-b+p}(\log X_t))$$

with $\bigwedge^p V \otimes R_{Y,t}(K_Y \otimes L^{b-p+1})$. We can identify this E_1 term with the E_1 term of another spectral sequence. Consider the double complex $\mathcal{B}^{\bullet,\bullet}(b)$ defined by

$$\mathcal{B}^{-i,j}(b) = \bigwedge^i V \otimes K_Y \otimes \bigwedge^{b-j} \Sigma_{Y,L} \otimes L^{j-i+1}, \qquad j - i \geq 0.$$

The complex $\mathcal{B}^{\bullet,\bullet}(b)$ is a second quadrant double complex which consists of the terms $\mathcal{B}^{-i,j}(b)$ with $0 \leq i \leq b, 0 \leq j \leq b$ and $j - i \geq 0$:

$$\bigwedge^b V \otimes K_Y \otimes L \to \cdots \to \qquad K_Y \otimes L^{b+1}$$

$$\uparrow$$

$$\ddots \qquad \qquad \vdots$$

$$\ddots \qquad \uparrow$$

$$\bigwedge^{n+2-b} \Sigma_{Y,L}^{\vee} \otimes L.$$

The horizontal differential in this complex is the differential of the Koszul complex; the vertical differential is given by contraction with the 1-jet $j^1(s)$. Let $\mathcal{B}^{\bullet}(b) = s(\mathcal{B}^{\bullet,\bullet}(b))$ be the associated total complex. Set

$$B^{-i,j}(b) = H^0(Y, \mathcal{B}^{-i,j}(b)), \qquad B^k(b) = H^0(Y, \mathcal{B}^k(b)).$$

We have two spectral sequences associated to $B^{\bullet,\bullet}(b)$, given by filtering along the rows or columns:

$$'E_1^{p,q}(b) = H^q(B^{p,\bullet}(b)) \Rightarrow H^{p+q}(B^{\bullet}(b))$$
$$''E_1^{p,q}(b) = H^q(B^{\bullet,p}(b)) \Rightarrow H^{p+q}(B^{\bullet}(b)).$$

By definition

$$'E_1^{-p,n+1-q}(b) = \bigwedge^p V \otimes R_{Y,t}(K_Y \otimes L^{b-p+1}).$$

It is not difficult to show that the isomorphism

$$E_1^{-p,n+1-q}(b) \cong {}'E_1^{-p,n+1-q}(b)$$

is compatible with the d_1 maps in both spectral sequences; hence it induces an isomorphism on the E_2 terms. But this does not suffice to obtain an isomorphism on the E_∞ terms. In [23] we constructed a morphism of filtered complexes $\mathcal{B}^{\bullet}(b) \to \mathcal{C}^{\bullet}(b)$ that induces an isomorphism of spectral sequences

$$E_r^{-p,n+1-q}(b) \cong {}'E_r^{-p,n+1-q}(b). \tag{4}$$

To obtain the vanishing of the E_∞ terms we look at the second spectral sequence. We have

$$''E_1^{i,-j}(b) = \mathcal{K}_{j,i-j+1}(K_Y \otimes \bigwedge^{b-i} \Sigma_{Y,L}, L).$$

Lemma 7.5.7 *Suppose that* $''E_1^{i,-j}(b) = 0$ *for all* (i, j) *such that* $b - k + 1 \leq i - j \leq b$. *Then* $E_\infty^{p,q}(b) = 0$ *for all* (p, q, b) *such that* $p + q + b \leq n + k$.

Proof It follows from the hypotheses of the lemma that $''E^a_\infty(b) = 0$ for all $b - k + 1 \leq a \leq b$. As the spectral sequences $''E^{p,q}_r(b)$ and $'E^{p,q}_r(b)$ converge to the same limit, we find that $'E^a_\infty(b) = 0$. Hence $'E^{-p,n+1-q}_\infty(b) = 0$ for all (p, q, b) such that $p + q + b \leq n + k$ and the proof is finished using the isomorphism (4). $\qquad\qquad\Box$

We have finally reduced the proof to a statement about the vanishing of certain Koszul cohomology groups. By Proposition 7.5.6 it suffices to show that

$$H^1(Y, \textstyle\bigwedge^{j+1} M_L \otimes K_Y \otimes \bigwedge^{b-i} \Sigma_{Y,L} \otimes L^{i-j}) = 0$$

for all (i, j) such that $b - k + 1 \leq i - j \leq b$. Take exterior powers in the exact sequence (3) and twist by K_Y to obtain an exact sequence

$$0 \to \Omega^{n+2-b+i}_Y \to K_Y \otimes \textstyle\bigwedge^{b-i} \Sigma_{Y,L} \to \Omega^{n+1-b+i}_Y \to 0.$$

Using this exact sequence we reduce to a vanishing statement with exterior powers of the cotangent bundle.

Recall that a coherent sheaf $\mathcal{F}$ on a polarized variety $(Y, \mathcal{O}_Y(1))$ is said to be m-regular if

$$H^i(Y, \mathcal{F}(m - i)) = 0 \qquad \text{for all } i > 0.$$

Note that this definition depends on the choice of a polarization on Y. The *Castelnuovo–Mumford regularity* of $\mathcal{F}$ is the number

$$m(\mathcal{F}) = \min\{m \in \mathbb{Z} | \mathcal{F} \text{ is } m\text{-regular}\}.$$

Let $m_i = m(\Omega^i_Y)$ be the regularity of Ω^i_Y. Following Paranjape we introduce the number

$$m_Y = \max\{m_i - i - 1 | 0 \leq i \leq \dim Y\}.$$

Note that this number is always non-negative if Y is a projective variety, as $H^i(Y, \Omega^i_Y) \neq 0$.

Exercise 7.5.8 Show that $m_Y = 0$ if Y is a projective space and that $m_Y = 1$ if Y is a smooth quadric.

If $Y = \mathbb{P}^{n+1}$ and $L = \mathcal{O}_{\mathbb{P}}(1)$ then $M_L \cong \Omega^1_{\mathbb{P}}(1)$ is 1-regular. Using a lemma of Green [10] one can show (with some work) that $\bigwedge^k M_L$ is k-regular on Y.

Lemma 7.5.9

$$H^i(Y, \Omega^j_Y \otimes \textstyle\bigwedge^a M_L \otimes \mathcal{O}_Y(k)) = 0$$

if $i \geq 1$ and $k + i \geq m_j + a$.

Remark 7.5.10 The symmetrizer lemma from Section 7.3 follows from Proposition 7.5.6 and Lemma 7.5.9 (applied with $Y = \mathbb{P}^{n+1}$).

To finish the proof we have to sort out all the vanishing conditions of this type and to translate these into conditions on the degrees $(d_0, \ldots, d_r)$ of the hypersurfaces. In addition there are conditions coming from Proposition 7.5.6. These can be treated in a similar way at the cost of introducing stronger degree conditions. The final result is:

Theorem 7.5.11 *Let* $(Y, \mathcal{O}_Y(1))$ *be a smooth polarized variety of dimension* $n + r + 1$. *Let* $d_0, \ldots, d_r$ *be natural numbers ordered in such a way that* $d_0 \geq \cdots \geq d_r$. *Define* $E = \mathcal{O}_Y(d_0) \oplus \cdots \mathcal{O}_Y(d_r)$ *and let* $U \subset \mathbb{P}H^0(Y, E)$ *be the complement of the discriminant locus. Let* m_j *be the regularity of* Ω_Y^j *and define*

$$m_Y = \max\{m_j - j - 1 | 0 \leq j \leq \dim Y\}.$$

Let $c \leq n$ *be a non-negative integer, and set* $\mu = [\frac{n+c}{2}]$. *Consider the conditions*

(C) $\sum_{v=\min(c,r)}^{r} d_v \geq m_Y + \dim Y - 1$;
(C_i) $\sum_{v=i}^{r} d_v + (\mu - c + i)d_r \geq m_Y + \dim Y + c - i$.

If condition (C) is satisfied and if the conditions (C_i) are satisfied for all i with $0 \leq i \leq \min(c - 1, r)$, *then for every smooth morphism* $g : T \to U$ *we have* $H^{n+k}(Y_T, X_T) = 0$ *for all* $k \leq c$.

In some cases we can avoid the extra conditions coming from Proposition 7.5.4. Suppose that

(∗) $\qquad H^i(Y, \Omega_Y^j(k)) = 0 \qquad$ for all $i > 0, j \geq 0, k > 0$.

If Y satisfies this condition, we can omit the condition (C). Examples of varieties Y that satisfy condition $(∗)$ are projective spaces and, more generally, smooth toric varieties. Condition $(∗)$ also holds if Y is an abelian variety.

Bibliographical hints A good introduction to Koszul cohomology is [9]. A detailed proof of Theorem 7.5.11 can be found in [23, Theorem 3.13] (where the condition $c \leq n$ should be added in the statement of the theorem). In the case $Y = \mathbb{P}^N$ effective versions of Nori's theorem have been obtained by Voisin [29] and by Asakura and Saito [2].

7.6 Applications of Nori's theorem

We keep the notation of the previous section: let $(Y, \mathcal{O}_Y(1))$ be a smooth polarized variety and set $E = \mathcal{O}_Y(d_0) \oplus \cdots \oplus \mathcal{O}_Y(d_r)$, $S = \mathbb{P}H^0(Y, E)$. As a first application of Nori's theorem, we mention the following result.

Proposition 7.6.1 *Let* $X \subset Y$ *be a complete intersection of multi-degree* $(d_0, \ldots, d_r)$ *and dimension n. Suppose that Z is a cycle of codimension* $p \leq n$ *on Y such that* $Z \cap X_s$ *is rationally equivalent to zero for general* $s \in S$. *If* $\min(d_0, \ldots, d_r) \gg 0$ *then* $\mathrm{cl}(Z) = 0$ *in* $H^{2p}(Y, \mathbb{Q})$.

Proof Let $X_S \subset Y_S$ be the universal family of complete intersections with inclusion map $r : X_S \to Y_S$. Set $Z_S = r^* p_Y^* Z \in \mathrm{CH}^p(X_S)$, and let η be the generic point of S. As we have (cf. the lectures of Lewis, Chapter 6)

$$\mathrm{CH}^p(X_\eta) = \varinjlim_{U \subset S} \mathrm{CH}^p(X_U)$$

there exists a Zariski open subset $U \subset S$ such that $[Z_U] = 0$ in $\mathrm{CH}^p(X_U)$, hence $\mathrm{cl}(Z_U) = 0$. By Nori's theorem the restriction map

$$H^{2p}(Y_U, \mathbb{Q}) \to H^{2p}(X_U, \mathbb{Q})$$

is injective if $\min(d_0, \ldots, d_r) \gg 0$. Hence $\mathrm{cl}(p_Y^* Z) = 0$ and $\mathrm{cl}(Z) = 0$ because $p_Y^* : H^{2p}(Y, \mathbb{Q}) \to H^{2p}(Y_U, \mathbb{Q})$ is injective. $\square$

For cycles of codimension $p < n$, Nori has shown that the result of Proposition 7.6.1 remains true under the weaker hypothesis that $Z \cap X_s$ is algebraically equivalent to zero. This result is a generalization of Griffiths's theorem.

Theorem 7.6.2 (Nori) *(Notation as in Proposition 7.6.1.) Suppose that* $p < n$ *and that* $Z \cap X_s$ *is algebraically equivalent to zero for very general* $s \in S$. *If* $\min(d_0, \ldots, d_r) \gg 0$ *then* $\mathrm{cl}_Y(Z) = 0$.

Proof Set $Z_S = r^* p_Y^* Z \in \mathrm{CH}^p(X_S)$. By the definition of algebraic equivalence, there exist a smooth morphism $T \to S$, a family of smooth curves $C_T \to T$, a relative divisor $D_T \in \mathrm{CH}_{\mathrm{hom}}^1(C_T/T)$ and a cycle $\Gamma \in \mathrm{CH}^p(C_T \times_T X_T)$ such that $Z_T = \Gamma_*(D_T)$. The map $\Gamma_* : \mathrm{CH}^1(C_T) \to \mathrm{CH}^p(X_T)$ is obtained from the diagram

$$X_T \times_T C_T \xrightarrow{p_2} C_T$$
$$\downarrow{p_1}$$
$$X_T$$

by setting $\Gamma_*(D) = (p_1)_*(p_2^* D \cdot \Gamma)$. Set $T' = C_T$. As $T' \to T$ is a smooth morphism, it follows from Nori's theorem that $H^{2p+1}(Y_{T'}, X_{T'}) = 0$ if $p < n$ and

$\min(d_0, \ldots, d_r) \gg 0$. Hence there exists $\gamma \in H^{2p}(Y_{T'}, \mathbb{Q})$ such that $r^*\gamma = \mathrm{cl}(\Gamma)$. Consider the diagram

$$Y_T \times_T C_T = Y_{T'} \xrightarrow{\pi_2} T' = C_T$$
$$\downarrow{\scriptstyle \pi_1}$$
$$Y_T$$

and put $\alpha = (\pi_1)_*(\pi_2^* D_T . \gamma) \in H^{2p}(Y_T, \mathbb{Q})$. By construction we have

$$r^*\alpha = r^* Z_T = r^* p_Y^* Z,$$

hence $\alpha = p_Y^* Z$ by Nori's theorem. If we restrict to the fiber over a point $t \in T$ we obtain

$$\mathrm{cl}(Z) = \pi_1(t)_*(\pi_2(t)^* D_t . \gamma|_{Y \times \{t\}}).$$

As $\mathrm{cl}(D_t) = 0$, it follows that $\mathrm{cl}(Z) = 0$. $\qquad\square$

Remark 7.6.3 We can replace Betti cohomology by Deligne cohomology in the proof of Theorem 7.6.2 to obtain a stronger conclusion: if $Z \cap X_s$ is algebraically equivalent to zero for very general s, then the Deligne class $\mathrm{cl}_{\mathcal{D}}(Z) \in H_{\mathcal{D}}^{2p}(Y, \mathbb{Q}(p))$ is zero. Using this result, Albano and Collino [1] have shown that if $Y \subset \mathbb{P}^8$ is a general cubic sevenfold and if $X = Y \cap D_1 \cap D_2$ is a very general complete intersection of Y with two hypersurfaces of sufficiently large degree, then $\mathrm{Griff}^4(X) \otimes \mathbb{Q} \neq 0$ (they even show that this vector space is not finite dimensional). Note that the non-zero elements in the Griffiths group could not have been detected by the Abel–Jacobi map as $J^4(X) = 0$.

We have seen that Nori's theorem implies the theorems of Noether–Lefschetz and Green–Voisin. One obtains effective versions of these theorems using Theorem 7.5.11 (take $c = 0$ and $n = 2m$ respectively $c = 1$ and $n = 2m - 1$). The degree bounds for these theorems are optimal if $Y = \mathbb{P}^{n+r+1}$, see [21].

Nori's theorem can also be used to study the regulator maps defined on Bloch's higher Chow groups.

Theorem 7.6.4 *Let X be a very general complete intersection of multi-degree $(d_0, \ldots, d_r)$ in Y. If $\min(d_0, \ldots, d_r) \gg 0$ and $2p - k \leq 2n - 1$, the image of the (rational) regulator map*

$$c_{p,k} : \mathrm{CH}^p(X, k)_{\mathbb{Q}} \to H_{\mathcal{D}}^{2p-k}(X, \mathbb{Q}(p))$$

is contained in the image of $i^* : H_{\mathcal{D}}^{2p-k}(Y, \mathbb{Q}(p)) \to H_{\mathcal{D}}^{2p-k}(X, \mathbb{Q}(p))$.

The proof of this theorem is analogous to the proof of Theorem 7.4.3; it can be found in [20]. We consider the applications of this result to the higher Chow groups $CH^p(X, 1)$ and $CH^p(X, 2)$. Recall that the group $CH^p_{dec}(X, 1)$ of decomposable higher Chow cycles is defined as the image of the natural map

$$CH^p(X) \otimes \mathbb{C}^* \to CH^p(X, 1).$$

The cokernel of this map is denoted by $CH^p_{ind}(X, 1)$ (indecomposable higher Chow cycles). Similarly we define $H^{2p-1}_{\mathcal{D},dec}(X, \mathbb{Z}(p))$ as the image of the natural map

$$H^{2p-2}_{\mathcal{D}}(X, \mathbb{Z}(p-1)) \otimes H^1_{\mathcal{D}}(X, \mathbb{Z}(1)) \to H^{2p-1}_{\mathcal{D}}(X, \mathbb{Z}(p)).$$

A higher Chow cycle $z \in CH^p(X, 1)$ is said to be R-decomposable (regulator decomposable) if $c_{p,1}(z) \in H^{2p-1}_{\mathcal{D},dec}(X, \mathbb{Z}(p))$.

From Theorems 7.6.4 and 7.5.11 we obtain the following result.

Theorem 7.6.5 *Let* $X = V(d_0, \dots, d_r) \subset \mathbb{P}^{2m+r+1}$ *be a smooth complete intersection of dimension* $2m$ ($m \geq 1$, $d_0 \geq \dots \geq d_r$), $i : X \to \mathbb{P}^{2m+r+1}$ *the inclusion map. If* X *is very general and if*

(C_0) $\sum_{i=0}^r d_i + (m-1)d_r \geq 2m + r + 3$
(C_1) $\sum_{i=1}^r d_i + m\, d_r \geq 2m + r + 2$

the image of the (rational) regulator map

$$c_{m+1,1} : CH^{m+1}(X, 1)_{\mathbb{Q}} \to H^{2m+1}_{\mathcal{D}}(X, \mathbb{Q}(m+1))$$

coincides with the image of the composed map ($N = 2m + r + 1$)

$$CH^{m+1}(\mathbb{P}^N, 1)_{\mathbb{Q}} \xrightarrow{\sim} H^{2m+1}_{\mathcal{D}}(\mathbb{P}^N, \mathbb{Q}(m+1)) \to H^{2m+1}_{\mathcal{D}}(X, \mathbb{Q}(m+1)).$$

As $CH^{m+1}(\mathbb{P}^N, 1) \cong \mathbb{C}^*$ it follows that every element $z \in CH^{m+1}(X, 1)$ is R-decomposable modulo torsion. The exceptional cases include quartic surfaces and cubic fourfolds. For these cases the regulator map has been studied in [20] and [6]; in both cases the group of regulator indecomposable higher Chow cycles is non-zero, and it is not even finitely generated.

One can think of $CH^p_{dec}(X, 1)$ as a kind of analogue of $CH^p_{alg}(X)$ for higher Chow cycles, and of $CH^p_{ind}(X, 1)$ as an analogue of the Griffiths group. Collino [6] has proved an analogue of Griffiths's theorem for higher Chow cycles.

Theorem 7.6.6 *Let* $(Y, \mathcal{O}_Y(1))$ *be a smooth polarized variety and let* X *be a very general complete intersection of multi-degree* $(d_0, \dots, d_r)$ *and dimension*

$n \geq 2$. Let $Z \in \mathrm{CH}^p(Y, 1)$ be a higher Chow cycle such that $Z \cap X_s$ is decomposable for very general s. If $p \leq n$ and $\min(d_0, \ldots, d_r) \gg 0$ then Z is R-decomposable.

Proof After passing to a suitable covering $T \to S$ of the base space S of the universal family X_S, we may assume that $Z_T = r^* p_Y^* Z \in \mathrm{CH}^p_{\mathrm{dec}}(X_T, 1)$, hence $c_{p,1}(Z_T) \in H^{2p-1}_{\mathcal{D}, \mathrm{dec}}(X_T, \mathbb{Q}(p))$. Look at the commutative diagram

$$H^{2p-2}_{\mathcal{D}}(Y_T, \mathbb{Q}(p-1)) \otimes H^1_{\mathcal{D}}(Y_T, \mathbb{Q}(1)) \to H^{2p-1}_{\mathcal{D}}(Y_T, \mathbb{Q}(p))$$
$$\downarrow \qquad\qquad\qquad\qquad\qquad\qquad\qquad\qquad \downarrow$$
$$H^{2p-2}_{\mathcal{D}}(X_T, \mathbb{Q}(p-1)) \otimes H^1_{\mathcal{D}}(X_T, \mathbb{Q}(1)) \to H^{2p-1}_{\mathcal{D}}(X_T, \mathbb{Q}(p)).$$

By Nori's theorem the vertical maps in this diagram are isomorphisms if $p \leq n$ and $\min(d_0, \ldots, d_r) \gg 0$. Hence $p_Y^* Z$ is R-decomposable, and by restricting to a fiber $Y \times \{t\}$ we find that Z is R-decomposable. $\qquad\square$

Remark 7.6.7 Collino applied this theorem to a cubic fourfold $Y \subset \mathbb{P}^5$ and obtained that $\mathrm{CH}^3_{\mathrm{ind}}(X_s, 1) \otimes \mathbb{Q}$ is non-zero for a very general hypersurface section $X_s \subset Y$ of sufficiently large degree. (He even showed that this vector space is infinite dimensional.)

For the higher Chow group $\mathrm{CH}^p(X, 2)$ we obtain the following result.

Theorem 7.6.8 Let $X = V(d_0, \ldots, d_r) \subset \mathbb{P}^{2m+r}$ be a smooth complete intersection of dimension $2m - 1$ ($m \geq 1$, $d_0 \geq \ldots \geq d_r$), $i : X \to \mathbb{P}^{2m+r}$ the inclusion map. If X is very general and if

(C_0) $\sum_{i=0}^r d_i + (m-1)d_r \geq 2m + r + 2$
(C_1) $\sum_{i=1}^r d_i + m\,d_r \geq 2m + r + 1$

the image of the (rational) regulator map

$$c_{m+1,2} : \mathrm{CH}^{m+1}(X, 2)_{\mathbb{Q}} \to H^{2m}_{\mathcal{D}}(X, \mathbb{Q}(m+1))$$

is zero.

Remark 7.6.9 The proof of Theorem 7.6.8 contains a small subtlety. To prove it we do not need the full strength of Nori's theorem, but the following result: if $F^{m+1} H^{2m+1}(Y_T, X_T) = 0$ and $F^m H^{2m}(Y_T, X_T) = 0$, then the restriction map

$$H^{2m}_{\mathcal{D}}(Y_T, \mathbb{Q}(m+1)) \to H^{2m}_{\mathcal{D}}(X_T, \mathbb{Q}(m+1))$$

is surjective. A proof of Theorem 7.6.8 for plane curves appeared in [5, (7.14)].

Example 7.6.10 Let $(Y, \mathcal{O}_Y(1))$ be a polarized abelian variety. As the tangent bundle T_Y is trivial, condition $(*)$ is satisfied by the Kodaira vanishing theorem. As Ω_Y^m is trivial for all m, the proof of Theorem 7.5.11 shows that we can subtract $m_Y + \dim Y + 1$ on the right-hand side of the inequality of condition (C_i). Hence condition (C_i) can be replaced by the weaker condition

(C_i') $\sum_{v=i}^r d_v + (\mu - c + i)d_r \geq c - i - 1$.

This condition is empty if $c \leq 2$. For $c = 2$ we obtain a result on the Abel–Jacobi map for complete intersections in abelian varieties without degree conditions, see [23, Theorem 4.8].

Example 7.6.11 We consider an example mentioned in the introduction of Nori's paper. Let $Y \subset \mathbb{P}^7$ be a smooth quadric, and let $X = Y \cap V(d_0, d_1)$, $d_0 \geq d_1$, be a smooth complete intersection in Y. In this case condition $(*)$ is not satisfied. As $m_Y = 1$, Paranjape's results show that $H^{n+k}(Y_T, X_T) = 0$ for $k \leq 3$ if $d_1 \geq 9$. As $c = 3$ and $r = 1$, the conditions of Theorem 7.5.11 read: (C) $d_1 \geq 6$; (C_0) $d_0 + d_1 \geq 10$; (C_1) $2d_1 \geq 9$. Using precise vanishing theorems for the groups $H^i(Y, \Omega_Y^j(k))$, we find that the bound in condition (C) can be improved to $d_1 \geq 5$. By Theorem 7.6.2 we obtain that $\mathrm{Griff}^3(X) \otimes \mathbb{Q} \neq 0$ if X is very general and $d_1 \geq 5$. Note that $\mathrm{Griff}^3(X) \otimes \mathbb{Q} = 0$ if $d_0 + d_1 < 6$ by a result of Bloch and Srinivas [3].

As we have seen, Nori's theorem cannot be applied to zero-cycles. Nevertheless it is possible to obtain results on zero-cycles using Nori's techniques. As an example we mention a theorem of Voisin [28]. For a smooth variety X we write $A_0(X)$ for the Chow group of zero-cycles of degree zero on X.

Theorem 7.6.12 (Voisin) *Let $S \subset \mathbb{P}^3$ be a very general surface of degree $d \geq 5$, and let $C \subset S$ be a smooth plane section with inclusion map $i : C \to S$. Then the kernel of the map*

$$i_* : A_0(C) \to A_0(S)$$

coincides with the subgroup $\mathrm{Tors}(A_0(C))$ of torsion points of $A_0(C)$.

Proof Over the moduli space B of pairs (S, C) as above we have universal families C_B and S_B. By passing to a covering T of B we can spread out an element $z_0 \in \ker i_*$ to a relative cycle $Z_T \in A_0(C_T/T)$ such that $Z_T(t) \in \ker i_*$ for all $t \in T$ and such that $Z_T(t_0) = z_0$. As $Z(t)$ is rationally equivalent to zero for all $t \in T(\mathbb{C})$ we may assume that $r_* \mathrm{cl}_\mathcal{D}(Z_T) \in H^4_\mathcal{D}(S_T, \mathbb{Q})$ is zero, by replacing T by a suitable Zariski open subset of T. Set

$$U_T = \mathbb{P}_T^3 \setminus \mathbb{P}_T^2, \qquad V_T = S_T \setminus C_T.$$

Consider the commutative diagram

$$
\begin{array}{ccccc}
H_{\mathcal{D}}^3(U_T, \mathbb{Q}(2)) & \to & H_{\mathcal{D}}^3(V_T, \mathbb{Q}(2)) & \to & H_{\mathcal{D}}^4(U_T, V_T, \mathbb{Q}(2)) \\
\downarrow {\scriptstyle r} & & \downarrow & & \downarrow \\
H_{\mathcal{D}}^2(\mathbb{P}_T^2, \mathbb{Q}(1)) & \xrightarrow{k^*} & H_{\mathcal{D}}^2(C_T, \mathbb{Q}(1)) & \to & H_{\mathcal{D}}^3(\mathbb{P}_T^2, C_T, \mathbb{Q}(1)) \\
\downarrow & & \downarrow {\scriptstyle r_*} & & \downarrow \\
H_{\mathcal{D}}^4(\mathbb{P}_T^3, \mathbb{Q}(2)) & \to & H_{\mathcal{D}}^4(S_T, \mathbb{Q}(2)) & \to & H_{\mathcal{D}}^5(\mathbb{P}_T^3, S_T, \mathbb{Q}(2)).
\end{array}
$$

Nori's theorem cannot be applied to the groups $H^3(\mathbb{P}_T^2, C_T)$ and $H^5(\mathbb{P}_T^3, S_T)$, but using a variant of Nori's techniques we can show that

$$
H^k(U_T, V_T) = 0 \qquad \text{for all } k \leq 4 \text{ if } d \geq 5.
$$

Combining this result with the vanishing of $H^k(U_T)$, we find that $H^k(V_T) = 0$ for all $k \leq 3$. Hence $H_{\mathcal{D}}^3(V_T, \mathbb{Q}(1)) = 0$ and

$$
r_* : H_{\mathcal{D}}^2(C_T, \mathbb{Q}(1)) \to H_{\mathcal{D}}^4(S_T, \mathbb{Q}(2))
$$

is injective. As $\mathrm{cl}_{\mathcal{D}}(Z_T) \in \ker r_*$ we get $\mathrm{cl}_{\mathcal{D}}(Z_T) = 0$. By restriction to the fiber over t_0 we see that $\mathrm{cl}_{\mathcal{D}}(z_0) \in H_{\mathcal{D}}^2(C_0, \mathbb{Q}(1))$ is zero. As $H_{\mathcal{D}}^2(C_0, \mathbb{Q}(1)) \cong \mathrm{Pic}(C_0) \otimes \mathbb{Q}$ it follows that $z_0 \in \mathrm{Tors}(A_0(C_0))$. The inclusion $\mathrm{Tors}(A_0(C)) \subseteq \ker i_*$ follows from Roitman's theorem. $\qquad \square$

Remark 7.6.13 It is possible to prove a similar result for curves that are obtained by intersecting an ample divisor Y in a threefold W with a surface S that varies in a sufficiently ample linear system on W, see [22].

7.7 Appendix: Deligne cohomology

We recall the definition of the Deligne and Deligne–Beilinson cohomology groups.

Definition *Let X be a smooth projective variety. Set $\mathbb{Z}(p) = (2\pi i)^p \mathbb{Z}$. The pth Deligne complex on X is the complex*

$$
\mathbb{Z}_{\mathcal{D}}(p) = (\mathbb{Z}(p) \to \mathcal{O}_X \to \Omega_X^1 \to \cdots \to \Omega_X^{p-1})
$$

concentrated in degrees $0, \ldots, p$. The hypercohomology group $\mathbb{H}^k(\mathbb{Z}_{\mathcal{D}}(p))$ is called the kth Deligne cohomology group of X with coefficients in $\mathbb{Z}(p)$ and is denoted by $H_{\mathcal{D}}^k(X, \mathbb{Z}(p))$.

Example From the exponential sequence we obtain a quasi-isomorphism $\mathbb{Z}_{\mathcal{D}}(1) \simeq \mathcal{O}_X^*[-1]$. Hence $H_{\mathcal{D}}^k(X, \mathbb{Z}(1)) \cong H^{k-1}(X, \mathcal{O}_X^*)$.

Definition *Let* $f : A^\bullet \to B^\bullet$ *be a morphism of complexes. The cone complex associated to* f *is the complex* $C^\bullet(f) = B^\bullet[-1] \oplus A^\bullet$ *with differential* $d_C(b, a) = (d_B(b) + f(a), -d_A(a))$. *This complex fits into an exact sequence*

$$0 \to B^\bullet[-1] \to C^\bullet(f) \to A^\bullet \to 0.$$

We say that two complexes $A^\bullet$ and $B^\bullet$ are *quasi-isomorphic* if they have the same cohomology groups; notation $A^\bullet \simeq B^\bullet$.

Exercise Let $f : A^\bullet \to B^\bullet$ be a morphism of complexes. Write $K^\bullet = \ker f$, $Q^\bullet = \operatorname{coker} f$. Prove the following:

 (i) If f is surjective, then $C^\bullet(f) \simeq K^\bullet$;
 (ii) if f is injective, then $C^\bullet(f) \simeq Q^\bullet[-1]$;
 (iii) let $\bar{B}^\bullet$ be the quotient of $B^\bullet$ by a subcomplex $C^\bullet$ with inclusion map $i :$ $C^\bullet \to B^\bullet$, and let $p : B^\bullet \to \bar{B}^\bullet$ be the projection map. Then $C^\bullet(p \circ f) \simeq$ $C^\bullet(f - i)$.

The de Rham complex is filtered by subcomplexes

$$\sigma_{\geq p}\Omega_X^\bullet = (\Omega_X^p \to \Omega_X^{p+1} \to \cdots)$$

that start in degree p (this filtration is called 'filtration bête' or stupid filtration). The quotient of $\Omega_X^\bullet$ by $\sigma_{\geq p}\Omega_X^\bullet$ is denoted by $\sigma_{<p}\Omega_X^\bullet$. As the Deligne complex fits into an exact sequence

$$0 \to \sigma_{<p}\Omega_X^\bullet[-1] \to \mathbb{Z}_{\mathcal{D}}(p) \to \mathbb{Z}(p) \to 0$$

we deduce from the previous exercise that

$$\mathbb{Z}_{\mathcal{D}}(p) \simeq \operatorname{Cone}(\mathbb{Z}(p) \oplus \sigma_{\geq p}\Omega_X^\bullet \to \Omega_X^\bullet).$$

Hence the Deligne cohomology groups fit into a long exact sequence

$$H^{k-1}(X, \mathbb{C}) \to H_{\mathcal{D}}^k(X, \mathbb{Z}(p)) \to H^k(X, \mathbb{Z}(p)) \oplus F^p H^k(X, \mathbb{C})$$

$$\to H^k(X, \mathbb{C}).$$

If $k = 2p$ we obtain a short exact sequence

$$0 \to J^p(X) \to H_{\mathcal{D}}^{2p}(X, \mathbb{Z}(p)) \to \operatorname{Hdg}^p(X) \to 0.$$

There exists a *Deligne cycle class map*

$$\operatorname{cl}_{\mathcal{D}}^p : \operatorname{CH}^p(X) \to H_{\mathcal{D}}^{2p}(X, \mathbb{Z}(p))$$

whose restriction to $CH^p_{hom}(X)$ coincides with the Abel–Jacobi map [8]. More generally there exist *regulator maps*

$$c_{p,k} : CH^p(X, k) \to H^{2p-k}_{\mathcal{D}}(X, \mathbb{Z}(p))$$

on Bloch's higher Chow groups $CH^p(X, k)$ that coincide with the Deligne cycle class map if $k = 0$.

For quasi-projective varieties there exists a variant of Deligne cohomology, called Deligne–Beilinson cohomology. Given a quasi-projective variety X, choose a good compactification $j : X \to \bar{X}$ with boundary $D = X \setminus \bar{X}$, and choose injective resolutions $\mathcal{I}^\bullet$ of $\mathbb{Z}_X(p)$ and $\mathcal{J}^\bullet$ of $\Omega^\bullet_X$. Put

$$Rj_*\mathbb{Z}_X(p) = j_*\mathcal{I}^\bullet, \qquad Rj_*\Omega^\bullet_X = j_*\mathcal{J}^\bullet.$$

There is a natural map of complexes

$$\alpha : \sigma_{\geq p}\Omega^\bullet_{\bar{X}}(\log D) \oplus Rj_*\mathbb{Z}_X(p) \to Rj_*\Omega^\bullet_X.$$

The Deligne–Beilinson cohomology groups are defined as the hypercohomology groups of the complex $C^\bullet(\alpha)$. As before they fit into an exact sequence

$$H^{k-1}(X, \mathbb{C}) \to H^k_{\mathcal{D}}(X, \mathbb{Z}(p)) \to H^k(X, \mathbb{Z}(p)) \oplus F^p H^k(X, \mathbb{C})$$
$$\to H^k(X, \mathbb{C})$$

where $F^\bullet$ denotes the Hodge filtration of the mixed Hodge structure on $H^k(X, \mathbb{C})$.

Acknowledgement

I would like to thank the organizers of the summer school, Chris Peters and Stefan Müller-Stach, for giving me the opportunity to present these lectures and for creating a very pleasant atmosphere for discussions with my fellow lecturers and the other participants. I am grateful to the referee for careful reading of the original manuscript.

References

[1] Albano, A. and Collino, A. On the Griffiths group of the cubic sevenfold, *Math. Ann.* **299** (4) (1994) 715–726.

[2] Asakura, M. and Saito, S. Filtration on Chow groups and generalized normal functions, preprint (1996).

[3] Bloch, S. and Srinivas, V. Remarks on correspondences and algebraic cycles, *Am. J. Math.* **105** (1983) 1235–1253.

[4]　Collino, A. The Abel–Jacobi isomorphism for the cubic fivefold, *Pac. J. Math.* **122** (1) (1986) 43–55.

[5]　Griffiths' infinitesimal invariant and higher *K*-theory on hyperelliptic jacobians, *J. Algebraic Geom.* **6** (1997) 393–415.

[6]　Indecomposable motivic cohomology classes on quartic surfaces and on cubic fourfolds. In: *Algebraic K-theory and its Applications (Trieste, 1997)*, pp. 370–402, World Scientific Publishing, River Edge, NJ, 1999.

[7]　Collino, A. and Pirola, G. P. The Griffiths infinitesimal invariant for a curve in its Jacobian, *Duke Math. J.* **78** (1995) 59–88.

[8]　El Zein, F. and Zucker, S. Extendability of normal functions associated to algebraic cycles. In: *Topics in Transcendental Algebraic Geometry (Princeton, NJ, 1981/1982)*, *Am. Math. Stud.* **106**, pp. 269–288, Princeton University Press, Princeton, NJ, 1984.

[9]　Green, M. Koszul cohomology and geometry. In: *Lectures on Riemann Surfaces, Trieste, Italy*, pp. 177–200, World Scientific Press, 1987.

[10]　A new proof of the explicit Noether–Lefschetz theorem, *J. Diff. Geom.* **27** (1988) 155–159.

[11]　Griffiths' infinitesimal invariant and the Abel–Jacobi map, *J. Diff. Geom.* **29** (1989) 545–555.

[12]　Infinitesimal methods in Hodge theory. In: *Algebraic Cycles and Hodge Theory, Lecture Notes in Mathematics* **1594**, Springer-Verlag, 1994.

[13]　Green, M. and Müller-Stach, S. Algebraic cycles on a general complete intersection of high multi-degree, *Composito Math.* **100** (1996) 305–309.

[14]　Griffiths, P. On the periods of certain rational integrals I, II, *Ann. Math.* **90** (2) (1969) 460–495; 496–541.

[15]　A theorem concerning the differential equations satisfied by normal functions associated to algebraic cycles, *Am. J. Math.* **101** (1979) 94–131.

[16]　Infinitesimal variations of Hodge structure III. Determinantal varieties and the infinitesimal invariant of normal functions, *Compositio Math.* **50** (2–3) (1983) 267–324.

[17]　Griffiths, P. and Harris, J. On the Noether–Lefschetz theorem and some remarks on codimension-two cycles, *Math. Ann.* **271** (1985) 31–51.

[18]　Katz, N. Etude cohomologique des pinceaux de Lefschetz, Exposé XVIII. In: (P. Deligne et N. Katz eds.), *Groupes de Monodromie en Géométrie Algébrique, SGA 7II, Lecture Notes in Mathematics*, **340**, pp. 254–327, Springer-Verlag, 1973.

[19]　Lazarsfeld, R. Some applications of the theory of positive vector bundles. In: *Complete Intersections (Acireale, 1983)*, *Lecture Notes in Mathematics* **1092**, pp. 29–61, Springer, Berlin, 1984.

[20]　Müller-Stach, S. Constructing indecomposable motivic cohomology classes on algebraic surfaces, *J. Algebraic Geom.* **6** (1997) 513–543.

[21]　Nagel, J. The Abel–Jacobi map for complete intersections, *Indag. Math.* **8** (1997) 95–113.

[22]　A variant of a theorem of C. Voisin, *Indag. Math.* **12** (2001) 231–241.

[23]　Effective bounds for Hodge-theoretic connectivity, *J. Algebraic Geom.* **11** (2002) 1–32.

[24]　Nori, M. V. Algebraic cycles and Hodge-theoretic connectivity, *Invent. Math.* **111** (1993) 349–373.

[25] Paranjape, K. Cohomological and cycle-theoretic connectivity, *Ann. Math.* **140** (1994) 641–660.

[26] Voisin, C. Lectures at Sophia-Antipolis, 1991.

[27] Transcendental methods in the study of algebraic cycles. In: *Algebraic Cycles and Hodge Theory, Lecture Notes in Mathematics* **1594**, Springer–Verlag, 1994.

[28] Variations de structures de Hodge et zéro-cycles sur les surfaces générales, *Math. Annal.* **299** (1994) 77–103.

[29] Nori's connectivity theorem and higher Chow groups, *J. Inst. Math. Jussieu* 1 (2002) 307–329.

[30] *Hodge Theory and Complex Algebraic Geometry I, II*, Cambridge University Press, to appear.

[31] Zucker, S. Generalized intermediate Jacobians and the theorem on normal functions, *Invent. Math.* **33** (1976) 185–222.

[32] Hodge theory with degenerating coefficients. L_2 cohomology in the Poincaré metric, *Ann. Math.* (2) **109** (3) (1979) 415–476.

[33] Intermediate Jacobians and normal functions. In: *Topics in Transcendental Algebraic Geometry (Princeton, NJ, 1981/1982)*, *Ann. Math. Stud.* **106**, pp. 259–267, Princeton University Press, Princeton, NJ, 1984.

8

Beilinson's Hodge and Tate conjectures

Shuji Saito

University of Nagano

Introduction

The purpose of this note is to give a brief account of the recent works [AS1], [AS2] and [AS3] on Beilinson's Hodge and Tate conjectures. It is an expanded version of the note of the author's lectures at the Grenoble Summer School 2001.

Beilinson's conjectures concern the surjectivity of certain regulator maps and are analogous to the classical Hodge and Tate conjectures for algebraic cycles. Roughly speaking, the main results claim that Beilinson's conjectures hold for 'general open complete intersections'. Here, by 'open complete intersection' we mean a pair $(X, Z = \bigcup_{1 \le j \le s} Z_j)$ where X is a smooth complete intersection in $\mathbb{P}^n$ and $Z_j \subset X$ is a smooth hypersurface section such that Z is a simple normal crossing divisor in X. The precise statements of the results are given in Section 8.2. Following the works of Green, Voisin and Nori, we use as the main technical ingredient the infinitesimal method in Hodge theory, which has played a powerful role in various aspects of algebraic geometry (cf. [G1] for a lucid account for this). We explain the contents of the following sections.

In Section 8.1 we introduce the Koszul complex associated to a family of open complete intersections which is a key technical ingredient. The main theorems describe the cohomology of the complex. The basic technique we use has been developed in [G2]. As a byproduct of the results, we present a theorem that gives an explicit bound for Nori's connectivity [No] in the case of a smooth complete intersection in projective space. We note that the results in this section have also been applied to the study of algebraic cycles in the works [A], [MSS] and [SaS].

Transcendental Aspects of Algebraic Cycles ed. S. Müller-Stach and C. Peters.
© Cambridge University Press 2004.

In Section 8.2 we state the main results on Beilinson's Hodge and Tate conjectures. We note that the results in Section 8.1 play a key role in the proof. We also explain that it has implications for the injectivity of regulator maps for higher Chow groups of algebraic surfaces.

In Section 8.3 we introduce the Noether–Lefschetz problem for K_2 of open complete intersections. This gives a different point of view on the results in Section 8.2 and a series of new problems arise by analogy with the classical Noether–Lefschetz theorem.

8.1 Koszul cohomology for open complete intersections

In what follows we fix integers $r, s \geq 0$ with $r + s \geq 1$, $n \geq 2$ and $d_1, \dots, d_r$, $e_1, \dots, e_s \geq 1$. We fix a field k of characteristic zero and a non-singular affine algebraic variety S over k. Let $\mathcal{C}_S$ be the category of smooth projective geometrically connected schemes over S. We consider

$$(1.1) \qquad \mathbb{P}^n_S \hookleftarrow \mathcal{X} \stackrel{i}{\hookleftarrow} \mathcal{Z} = \bigcup_{1 \leq j \leq s} \mathcal{Z}_j$$

where $\mathcal{X} \in \mathcal{C}_S$ is a family of smooth complete intersections of multi-degree $(d_1, \dots, d_r)$ and $\mathcal{Z}_j \subset \mathcal{X}$ is a smooth hypersurface section of degree e_j such that $\mathcal{Z} \subset \mathcal{X}$ is a relative simple normal crossing divisor on $\mathcal{X}$. We assume the relative dimension $n - r$ of $\mathcal{X}/S$ is greater than or equal to 1. Let $f : \mathcal{X} \to S$ be the natural morphism and write $\mathcal{U} = \mathcal{X} \setminus \mathcal{Z}$. For $x \in S$ let $U_x \subset X_x \supset Z_x$ denote the fibers of $\mathcal{U} \subset \mathcal{X} \supset \mathcal{Z}$.

Define the sheaf on S_{Zar}

$$H^q_{\mathcal{O}}(\mathcal{U}/S) = R^q f_* \Omega^{\cdot}_{\mathcal{X}/S}(\log \mathcal{Z}),$$

where $\Omega^{\cdot}_{\mathcal{X}/S}(\log \mathcal{Z})$ is the complex of sheaves given by $\Omega^p_{\mathcal{X}/S}(\log \mathcal{Z}) = \wedge^p \Omega^1_{\mathcal{X}/S}(\log \mathcal{Z})$ with $\Omega^1_{\mathcal{X}/S}(\log \mathcal{Z})$, the sheaf of relative differentials on $\mathcal{X}$ over S with logarithmic poles along $\mathcal{Z}$. We have the spectral sequence

$$E^{p,q}_1 = R^q f_* \Omega^p_{\mathcal{X}/S}(\log \mathcal{Z}) \Rightarrow H^{p+q}_{\mathcal{O}}(\mathcal{U}/S)$$

which by [D1] degenerates at E_1 and gives rise to the Hodge filtration

$$F^p H^q_{\mathcal{O}}(\mathcal{U}/S) := R^q f_* \Omega^{\geq p}_{\mathcal{X}/S}(\log \mathcal{Z}) \subset H^q_{\mathcal{O}}(\mathcal{U}/S).$$

We have the algebraic Gauss–Manin connection

$$\nabla : H^q_{\mathcal{O}}(\mathcal{U}/S) \to \Omega^1_S \otimes H^q_{\mathcal{O}}(\mathcal{U}/S)$$

that arises from the exact sequence

$$0 \to \Omega^1_S \otimes \Omega^{-1}_{\mathcal{X}/S}(\log \mathcal{Z}) \to \Omega^{\cdot}_{\mathcal{X}/k}(\log \mathcal{Z}) \to \Omega^{\cdot}_{\mathcal{X}/S}(\log \mathcal{Z}) \to 0.$$

It satisfies the transversality (cf. [KO])

$$\nabla(F^p H^q_{\mathcal{O}}(\mathcal{U}/S)) \subset \Omega^1_S \otimes F^{p-1} H^q_{\mathcal{O}}(\mathcal{U}/S).$$

For integers a, b we write

$$H^{a,b}_{\mathcal{O}}(\mathcal{U}/S) = F^a H^{a+b}_{\mathcal{O}}(\mathcal{U}/S)/F^{a+1} = R^b f_* \Omega^a_{\mathcal{X}/S}(\log \mathcal{Z}).$$

We note that $H^{a,b}_{\mathcal{O}}(\mathcal{U}/S) = 0$ if $a + b \neq n - r$ and $s \geq 1$ by the Lefschetz theory. By transversality, the Gauss–Manin connection induces

$$(1.2) \qquad \overline{\nabla}^{a,b} : H^{a,b}_{\mathcal{O}}(\mathcal{U}/S) \overset{\overline{\nabla}}{\to} \Omega^1_S \otimes_{\mathcal{O}} H^{a-1,b+1}_{\mathcal{O}}(\mathcal{U}/S).$$

Another way to understand the above map is the following. We have the Kodaira–Spencer map

$$(1.3) \qquad\qquad \kappa_{(\mathcal{X},\mathcal{Z})} : \Theta_S \to R^1 f_* T_{\mathcal{X}/S}(-\log \mathcal{Z}),$$

where $\Theta_S = \mathcal{H}om_{\mathcal{O}_S}(\Omega^1_S, \mathcal{O}_S)$ and $T_{\mathcal{X}/S}(-\log \mathcal{Z}) = \mathcal{H}om_{\mathcal{O}_{\mathcal{X}}}(\Omega^1_{\mathcal{X}/S}(\log \mathcal{Z}), \mathcal{O}_{\mathcal{X}})$. Then $\overline{\nabla}^{a,b}$ is induced by

$$R^1 f_* T_{\mathcal{X}/S}(-\log \mathcal{Z}) \otimes R^b f_* \Omega^a_{\mathcal{X}/S}(\log \mathcal{Z}) \to R^{b+1} f_* \Omega^{a-1}_{\mathcal{X}/S}(\log \mathcal{Z})$$

induced by the cup product and $T_{\mathcal{X}/S}(-\log \mathcal{Z}) \otimes \Omega^a_{\mathcal{X}/S}(\log \mathcal{Z}) \to \Omega^{a-1}_{\mathcal{X}/S}(\log \mathcal{Z})$, the contraction. The above map gives rise to the following Koszul complex

$$\Omega^{q-1}_S \otimes_{\mathcal{O}} H^{a+2,b-2}_{\mathcal{O}}(\mathcal{U}/S) \overset{\overline{\nabla}}{\to} \Omega^q_S \otimes_{\mathcal{O}} H^{a+1,b-1}_{\mathcal{O}}(\mathcal{U}/S)$$

$$(1.4) \qquad\qquad\qquad \overset{\overline{\nabla}}{\to} \Omega^{q+1}_S \otimes_{\mathcal{O}} H^{a,b}_{\mathcal{O}}(\mathcal{U}/S).$$

In order to state the main results of this section we need to introduce an invariant to measure the 'generality' of the family (1.1), or how many independent parameters S contains.

Definition 1.1 *We define*

$$c_S(\mathcal{X}, \mathcal{Z}) = \max_{x \in S}\{\dim_{k(x)}(\text{Coker}(\kappa_{(\mathcal{X},\mathcal{Z})}) \otimes_{\mathcal{O}_S} k(x))\}.$$

Remark 1.1 We may use the following more intuitive invariant than $c_S(\mathcal{X}, \mathcal{Z})$. Let $P = k[X_0, \ldots, X_n]$ be the polynomial ring over k in $n + 1$ variables.

Denote by $P^d \subset P$ the subspace of the homogeneous polynomials of degree d. Then the dual projective space

$$\overset{\vee}{\mathbb{P}}(P^d) = \mathbb{P}_k^{N(n,d)} \qquad \left(N(n,d) = \binom{n+d}{d} - 1 \right)$$

parametrizes hypersurfaces $Y \subset \mathbb{P}^n$ of degree d defined over k. Let

$$B \subset \prod_{1 \le i \le r} \mathbb{P}_k^{N(n,d_i)} \times \prod_{1 \le j \le s} \mathbb{P}_k^{N(n,e_j)}$$

be the Zariski open subset parametrizing such $(Y_\nu)_{1 \le \nu \le r+s}$ that $Y_1 + \cdots + Y_{r+s}$ is a simple normal crossing divisor on $\mathbb{P}_k^n$. We consider the family

$$\mathfrak{X}_B \hookleftarrow \mathcal{Z}_B = \bigcup_{1 \le j \le s} \mathcal{Z}_{B,j} \qquad \text{over } B$$

whose fibers are $X \hookleftarrow Z = \bigcup_{1 \le j \le s} Z_j$ with $X = Y_1 \cap \cdots \cap Y_r$ and $Z_j = X \cap Y_{r+j}$. Let $T \subset B$ be a non-singular locally closed subvariety of codimension $c \ge 0$ and let $S \to T$ be a dominant map. Assume that the family (1.1) is the pull-back of $(\mathfrak{X}_B, \mathcal{Z}_B)/B$ via $S \to B$. Then we have $c_S(\mathfrak{X}, \mathcal{Z}) \le c$ and we may use c instead of $c_S(\mathfrak{X}, \mathcal{Z})$ in the following theorems.

The first main result in this section is a generalization of Donagi's symmetrizer lemma [Do] (see also [DG], [Na1] and [No]) to the case of open complete intersections at higher degrees.

Theorem 1.1 [AS1] *Let $c = c_S(\mathfrak{X}, \mathcal{Z})$ be as above and assume $n - r \ge 2$. Assume also that either $a < n - r - 1$ or $r + s \le n$. Then the complex (1.4) with $a + b = n - r$ is exact under one of the following conditions, where $\delta_{\min} = \min\{d_i, e_j\}$, $d_{\max} = \max_{1 \le i \le r} \{d_i\}$, $\mathbf{d} = \sum_{1 \le i \le r} d_i$.*

$\underset{\substack{1 \le i \le r \\ 1 \le j \le s}}{}$

(i) $a \ge 0$, $q = 0$ *and* $\delta_{\min} a + \mathbf{d} \ge c + n + 1$.
(ii) $a \ge 0$, $q = 1$ *and* $\delta_{\min} a + \mathbf{d} \ge c + n + 2$ *and* $\delta_{\min}(a + 1) + \mathbf{d} \ge c + n + 1 + d_{\max}$.
(iii) $a \ge 0$, $\delta_{\min}(r + a) \ge q + c + n + 1$ *and* $r + s \le n + 2$.
(iv) $a \ge 0$, $\delta_{\min}(r + a) \ge q + c + n + 1$ *and* $a < n - r - \frac{q}{2}$.

A distinguished phenomenon is that the complex (1.4) with $q = 0$ fails to be exact in the case $(a, b) = (n - r - 1, 1)$ and $s > n - r$. The kernel of the map

$$\overline{\nabla}^{m,0} : H_{\mathcal{O}}^{m,0}(\mathcal{U}/S) \overset{\overline{\nabla}}{\to} \Omega_S^1 \otimes_{\mathcal{O}} H_{\mathcal{O}}^{m-1,1}(\mathcal{U}/S) \qquad (m = n - r)$$

may in general be non-trivial while it is determined explicitly using the defining equation of $\mathcal{Z}_i \subset \mathfrak{X}$.

Definition 1.2 *Let* $G_j \in H^0(\mathcal{X}, \mathcal{O}_\mathcal{X}(e_j))$ *be a non-zero element defining* $\mathcal{Z}_i \subset \mathcal{X}$.

(1) *For* $1 \leq j \leq s-1$ *put*

$$g_j = (G_j^{e_s}/G_s^{e_j}) \in \Gamma(\mathcal{U}, \mathcal{O}_\mathcal{U}^*).$$

(2) *Define the index set* $J = \{\sigma = (j_1, \ldots, j_m) | \ 1 \leq j_1 < \cdots < j_m \leq s-1\}$
where $m = n - r$ *is the relative dimension of* $\mathcal{X}/S$. *For* $\sigma = (j_1, \ldots, j_m) \in J$ *let*

$$\omega_{\mathcal{U}/S}(\sigma) = d \log g_{j_1} \wedge \cdots \wedge d \log g_{j_m} \in \Gamma(S, f_* \Omega^m_{\mathcal{X}/S}(\log \mathcal{Z}))$$

$$= \Gamma(S, H_\mathcal{O}^{m,0}(\mathcal{U}/S)).$$

Theorem 1.2 [AS2] *Assume* $\delta_{\min}(n - r - 1) + \mathbf{d} \geq n - 1 + c_S(\mathcal{X}, \mathcal{Z})$. *Then* $\mathrm{Ker}(\overline{\nabla}^{m,0})$ *is generated as an* $\mathcal{O}_S$-*module by* $\omega_{\mathcal{U}/S}(\sigma)$ *with* $\sigma \in J$.

Theorem 1.2 plays a crucial role in the work [AS2] on Beilinson's Hodge and Tate conjectures explained in the next section. The proof of Theorem 1.2 is based on generalized Jacobian rings introduced in [AS1].

We close this section with the following application of Theorem 1.1. We work over the base field $k = \mathbb{C}$.

Theorem 1.3 *Assuming* $n - r \geq 2$, *we have*

$$F^{t-n+r+1} H^t(\mathcal{U}, \mathbb{C}) = 0 \qquad \text{if } s \leq n - r + 2 \text{ and } \delta_{\min} r \geq t + r + 1$$
$$+ c_S(\mathcal{X}, \mathcal{Z}),$$

$$F^{t-n+r+1} H^t(\mathcal{X}, \mathcal{Z}, \mathbb{C}) = 0 \qquad \text{if } s = 1 \text{ and } \delta_{\min} r + e_1 \geq t + r + 1$$
$$+ c_S(\mathcal{X}, \mathcal{Z}).$$

where $\delta_{\min} = \min\limits_{1 \leq i \leq r, 1 \leq j \leq s} \{d_i, e_j\}$ *and* F^* *denotes the Hodge filtration defined in* [D1] *and* [D2].

The second vanishing of Theorem 1.3 gives an explicit bound for Nori's connectivity [No] in the case of complete intersections in the projective space. Nagel [Na2] has obtained similar degree bounds for complete intersections in a general projective smooth variety. By the same argument as [No], a remark after Theorem 4, Theorem 1.3 implies the following result which is a generalization of [V2, Theorem 2].

Theorem 1.4 *Assume* $n - r \geq 2$, $s = 1$ *and* $(\delta_{\min} + 1)r + e_1 > 2n + c_S(X, Z)$. *Then* $H^t(X, Z, \mathbb{Q}) = 0$ *for* $t \leq 2(n - r - 1)$. *In particular, the restriction map* $H^t(X, \mathbb{Q}) \to H^t(Z, \mathbb{Q})$ *is an isomorphism if* $t < 2(n - r - 1)$ *and injective if* $t \leq 2(n - r - 1)$.

8.2 Beilinson's Hodge and Tate conjectures

Let U be a non-singular algebraic variety over $\mathbb{C}$. For integers $i, j \geq 0$ let $CH^i(U, j)$ denote the Bloch higher Chow group (cf. [Bl]). The main objective of study in this section is the so-called regulator map (cf. [Sch])

$$(2.1) \qquad CH^i(U, i) \longrightarrow H^i(U, \mathbb{Q}(i))$$

where $H^i(U, \mathbb{Q}(i))$ is the Betti cohomology of U with $\mathbb{Q}(i) = (2\pi\sqrt{-1})^i \mathbb{Q}$. The definition of the map (2.1) in general is not easy; its value on an element $\{f_1, \ldots, f_i\} \in CH^i(U, i)$, which is the image of $f_1 \otimes \cdots \otimes f_i$ with $f_\nu \in \mathcal{O}_{Zar}(U)^*$ under the product map $\mathcal{O}_{Zar}(U)^* \otimes \cdots \otimes \mathcal{O}_{Zar}(U)^* \to CH^i(U, i)$, is given by the $d \log$ map:

$$(2.2) \qquad \{f_1, \ldots, f_i\} \longmapsto \frac{df_1}{f_1} \wedge \cdots \wedge \frac{df_i}{f_i}.$$

Recall that Deligne [D1] endowed $H^i(U, \mathbb{Q}(i))$ with mixed Hodge structure and it can be shown that the image of the map (2.1) is contained in

$$F^0 H^i(U, \mathbb{Q}(i)) := H^i(U, \mathbb{Q}(i)) \cap F^i H^i(U, \mathbb{C}).$$

Recall that $F^i H^i(U, \mathbb{C}) = \Gamma(X, \Omega_X^i(\log X - U))$ where X is a smooth compactification of U with $X - U$ a normal crossing divisor on X. Thus it is evident that the right-hand side of (2.2) is contained in $F^i H^i(U, \mathbb{C})$. In general one uses the fact that the map (2.1) factors as

$$CH^i(U, i) \to H_D^i(U, \mathbb{Q}(i)) \to F^0 H^i(U, \mathbb{Q}(i))$$

where the first map is the regulator map to the Deligne cohomology group and the second map follows from [EV, Corollary 2.10]. Now our regulator map induces

$$(2.3) \qquad \operatorname{reg}_U^i : CH^i(U, i) \otimes \mathbb{Q} \longrightarrow F^0 H^i(U, \mathbb{Q}(i)).$$

We remark that the fibers of $\omega_{U/S}(\sigma)$ introduced in Definition 1.2(2) are induced by elements in higher Chow groups by the above regulator maps.

Conjecture 2.1 [Be1] *The map* (2.3) *is surjective.*

We call Conjecture 2.1 Beilinson's Hodge conjecture. It is evident that it is an analogue of the Hodge conjecture for algebraic cycles on a projective smooth variety over $\mathbb{C}$. Beilinson's conjecture in the case $i = 1$ is easily shown using GAGA. It is completely open in the case $i \geq 2$. As far as we know, there have been no non-trivial examples (probably because it is not so famous as the classical Hodge conjecture).

Remark 2.1

(1) If U is proper over $\mathbb{C}$, one can show that $F^0 H^i(U, \mathbb{Q}(i)) = 0$ for $i \geq 1$ using the Hodge symmetry. This fact might be another reason why Beilinson's conjecture has not been studied so much.

(2) As a generalization of Beilinson's conjecture, one may consider the regulator map

$$\text{reg}_U^{i,j} : CH^i(U, j) \otimes \mathbb{Q} \longrightarrow F^0 H^{2i-j}(U, \mathbb{Q}(i)).$$

Jannsen [J2, 9.11] has shown that the map in the case $i = 1$, $j = 2k - 1 \geq 3$ is not surjective in general, using a theorem of Mumford [Mu], which implies the Abel–Jacobi map for cycles of codimension ≥ 2 is not injective even modulo torsion.

One may consider the following variant (Tate version). Let k be a finite extension of $\mathbb{Q}$ and $\overline{k}$ be its algebraic closure. Let U be a non-singular algebraic variety over k. We have the regulator map

$$(2.4) \qquad \text{reg}_{et,U}^i : CH^i(U, i) \otimes \mathbb{Q}_p \longrightarrow H_{et}^i(\overline{U}, \mathbb{Q}_p(i))^{Gal(\overline{k}/k)}$$

where $\overline{U} = U \times_k \overline{k}$ and $H_{et}^i(\overline{U}, \mathbb{Q}_p(i))$ is the p-adic étale cohomology. Then we have the following Tate version of Conjecture 2.1.

Conjecture 2.4 [Be1] *The map* (2.4) *is surjective.*

In order to state the main result on Beilinson's conjectures we consider the family (1.1) in Section 8.1. Write $m = n - r$ for the relative dimension of $\mathcal{X}/S$.

Theorem 2.1 [AS2] *Assume* $\sum_{1 \leq i \leq r} d_i \geq c_S(\mathcal{X}, \mathcal{Z}) + n + 1$.

(1) *Assume* $k = \mathbb{C}$. *There exists* $E \subset S(\mathbb{C})$ *which is the union of countably many proper analytic subsets of* $S(\mathbb{C})$ *such that*

$$\text{reg}_{U_x}^m : CH^m(U_x, m) \otimes \mathbb{Q} \longrightarrow F^0 H^m(U_x, \mathbb{Q}(m))$$

is surjective for all $x \in S(\mathbb{C}) \setminus E$.

(2) *Assume that k is a finite extension of $\mathbb{Q}$ and $S(k) \neq \emptyset$. Let $\pi : S \to \mathbb{P}_k^N$ be a dominant quasi-finite morphism. There exists a subset $H \subset \mathbb{P}_k^N(k)$ such that:*

(i) *reg_{et,U_x}^m is surjective for any closed point $x \in S$ such that $\pi(x) \in H$;*

(ii) *let Σ be any finite set of primes of k and let k_v be the completion of k at $v \in \Sigma$, then the image of H in $\prod_{v \in \Sigma} \mathbb{P}_k^N(k_v)$ is dense.*

We note that Theorem 2.1(1) follows from Theorem 1.2 by a standard Hodge theoretical argument (see [AS2, Theorem 3.2]).

We now discuss an application to the injectivity of regulator maps for higher Chow groups of surfaces.

Let X be a projective smooth surface over a field k of characteristic zero. We consider $CH^2(X, 1)$, which is by definition the cohomology of the following complex

$$K_2(k(X)) \stackrel{\partial_{tame}}{\to} \bigoplus_{C \subset X} k(C)^* \stackrel{\partial_{div}}{\to} \bigoplus_{x \in X} \mathbb{Z},$$

where the sum on the middle term ranges over all irreducible curves on X and that on the right-hand side over all closed points of X. The map ∂_{tame} is the so-called tame symbol and ∂_{div} is the sum of divisors of rational functions on curves. Thus an element of $\mathrm{Ker}(\partial_{div})$ is given by a finite sum $\sum_i (C_i, f_i)$, where f_i is a non-zero rational function on an irreducible curve $C_i \subset X$ such that $\sum_i div(f_i) = 0$ on X.

Now we fix $Z = \cup_{1 \leq i \leq s} Z_i$, a simple normal crossing divisor on X, and write $U = X - Z$. We consider higher cycles supported on Z. We write

$$CH^1(Z, 1) = \mathrm{Ker}\left(\bigoplus_{1 \leq i \leq s} k(Z_i)^* \to \bigoplus_{x \in Z} \mathbb{Z} \right).$$

This is the Bloch higher Chow group of Z (cf. [Bl]) and we have the exact sequence

$$CH^2(U, 2) \to CH^1(Z, 1) \to CH^2(X, 1).$$

An important tool in the study of $CH^2(X, 1)$ is the regulator map. In the case $k = \mathbb{C}$ it is given by

$$\rho_X : CH^2(X, 1) \otimes \mathbb{Q} \to H_D^3(X, \mathbb{Q}(2)),$$

where the group on the right-hand side is the Deligne cohomology group (cf. [EV] and [J1]). In the case where k is a finite extension of $\mathbb{Q}$ we have the regulator map

$$\rho_{et,X} : CH^2(X, 1) \otimes \mathbb{Q}_p \to H^3_{cont}(X, \mathbb{Q}_l(2)),$$

where H^i_{cont} denotes the continuous étale cohomology of X (cf. [J3]).

The conjectures of Beilinson [Be2] imply the injectivity modulo torsion of the above regulator maps.

Theorem 2.2 *Assume that the first Betti number of X vanishes.*

(1) *Assume $k = \mathbb{C}$ and that there exists a subspace $\Delta \subset CH^2(U, 2) \otimes \mathbb{Q}$ such that reg^2_U restricted on Δ is surjective. Then $\text{Ker}(CH^1(Z, 1) \to CH^2(X, 1)) \otimes \mathbb{Q}$ coincides with the image of Δ under $CH^2(U, 2) \to CH^1(Z, 1)$ and ρ_X restricted on the image of $CH^1(Z, 1)$ in $CH^2(X, 1) \otimes \mathbb{Q}$ is injective.*

(2) *Assume that k is a finite extension of $\mathbb{Q}$. The analogous fact holds for $reg^2_{et,U}$ and $\rho_{et,X}$.*

Remark 2.2 In the next section (Theorem 3.7) we show that there are instances where reg^2_U is surjective and the image of $CH^1(Z, 1)$ in $CH^2(X, 1)$ is non-torsion. Thus Theorem 2.1 has a non-trivial implication on the injectivity of reg^2_U.

Here we explain the idea of the proof of Theorem 2.2(1), which we have taken from [J2, 9.11]. We have the commutative diagram (cf. [Bl] and [J1, 3.3 and 1.15])

$$
\begin{array}{ccccccc}
CH^2(X, 2) & \to & CH^2(U, 2) & \to & CH^1(Z, 1) & \to & CH^2(X, 1) \\
\downarrow \alpha & & \downarrow \beta & & \downarrow \gamma & & \downarrow \rho_X \\
H^2_D(X, \mathbb{Z}(2)) & \xrightarrow{\iota} & H^2_D(U, \mathbb{Z}(2)) & \to & H^3_{D,Z}(X, \mathbb{Z}(2)) & \to & H^3_D(X, \mathbb{Z}(2)).
\end{array}
$$

Here the horizontal sequences are the localization sequences for higher Chow groups and Deligne cohomology groups and they are exact. The vertical maps are the regulator maps. It is not so difficult to show that γ is injective. By using the assumption of Theorem 2.2(1) and the exact sequence (cf. [EV, Corollary 2.10])

$$0 \to H^1(U, \mathbb{C})/H^1(U, \mathbb{Z}(2)) \to H^2_D(U, \mathbb{Z}(2))$$
$$\to H^2(U, \mathbb{Z}(2)) \cap F^2 H^2(U, \mathbb{C}) \to 0,$$

one shows that the images of α and β span $H^2_D(U, \mathbb{Q}(2))$ modulo torsion (see [AS2, Section 5] for the details). By a simple diagram chasing, these imply the injectivity of ρ_X on the image of $CH^1(Z, 1)$. $\qquad\square$

8.3 Noether–Lefschetz problem for K_2 of open surfaces

We start with recalling the set-up of the classical Noether–Lefschetz theorem. In what follows every variety is an algebraic variety over the complex number field $\mathbb{C}$. We fix an integer $d \geq 1$. In the moduli space M of smooth hypersurfaces of degree d in $\mathbb{P}^3$, the locus of those surfaces that possess curves which are not complete intersections of the given surface with another surface is called the Noether–Lefschetz locus and is denoted by M_{NL}. One can show that M_{NL} is the union of a countable number of closed algebraic subsets. The classical theorem of Noether–Lefschetz states:

Theorem 3.1 *If $d \geq 4$, M_{NL} has positive codimension.*

Note that the theorem is false if $d = 3$ since a smooth cubic surface has Picard number 7. Fascinating results have been obtained concerning irreducible components of M_{NL}. First we have the following (cf. [G3]).

Theorem 3.2 *For any irreducible component T of M_{NL} we have*

$$d - 3 \leq \operatorname{codim}(T) \leq \binom{d-1}{3}.$$

Indeed the inequality on the right-hand side is easy while that on the left-hand side requires a non-trivial argument. One translates the problem into the language of the infinitesimal variation of Hodge structures on a family of hypersurfaces. Then, by the Poincaré residue representation of the cohomology of a hypersurface, the result follows from the duality theorem for the Jacobian ring associated to a hypersurface.

It is shown in [CHM] that M_{NL} contains infinitely many components of codimension $\binom{d-1}{3}$ and their union is Zariski dense in M. We note that the left inequality is the best possible, since the family of hypersurfaces of degree $d \geq 3$ containing a line has codimension exactly $d - 3$. Green [G4] and Voisin [V1] have shown the following striking theorem.

Theorem 3.3 *If $d \geq 5$, the only irreducible component of M_{NL} having codimension $d - 3$ is the family of surfaces of degree d containing a line.*

The basic technique of the proof is the same as that of Theorem 3.2 while the involved algebraic problem on the Jacobian ring is more delicate. Results of Macaulay and Gotzmann [Go] on Hilbert functions of homogeneous ideals in a polynomial ring play a crucial role.

Now we consider the following analogue of the above problem. Fix integers $d \geq 1$ and $e_j \geq 1$ for $1 \leq j \leq s$. Let M be the moduli space of sets of hypersurfaces $(X, Y_1, \ldots, Y_s)$ of degree $(d, e_1, \ldots, e_s)$ which intersect transversally. Write $Z_j = X \cap Y_j$ and $Z = \cup_{1 \le j \le s} Z_j$ and $U = X - Z$. We now consider the space

$$F^0 H^2(U, \mathbb{Q}(2)) := H^2(U, \mathbb{Q}(2)) \cap F^2 H^2(U, \mathbb{C}).$$

Let $\Sigma(U)$ be the image of the natural restriction map

$$H^2(\mathbb{P}^3 - \bigcup_{1 \le j \le s} Y_j, \mathbb{Q}(2)) \to H^2(U, \mathbb{Q}(2)).$$

It is easy to see that the space on the left-hand side is generated by the cohomology classes of the following elements

$$d \log(G_i^{e_s} / G_s^{e_j}) \wedge d \log(G_j^{e_s} / G_s^{e_j}) \qquad (1 \le i < j \le s - 1)$$

where G_j is a defining equation of $Y_j \subset \mathbb{P}^3$. This implies that $\Sigma(U) \subset F^0 H^2(U, \mathbb{Q}(2))$ and we put

$$F^0 H^2(U, \mathbb{Q}(2))_{pr} := F^0 H^2(U, \mathbb{Q}(2)) / \Sigma(U),$$

called the space of primitive cycles.

Definition 3.1 *For $t \in M$ let $U_t \subset X_t \supset Z_t$ be defined as above for the fibers of the universal family over M. We define the Noether–Lefschetz locus for K_2 to be*

$$M_{NL} = \{t \in M \mid F^0 H^2(U_t, \mathbb{Q}(2)) \ne \Sigma(U_t)\}.$$

The analogy with the classical problem is explained as follows. Instead of the map

$$H^2(\mathbb{P}^3 - \bigcup_{1 \le j \le s} Y_j, \mathbb{Q}(2)) \to H^2(U, \mathbb{Q}(2)) \cap F^2 H^2(U, \mathbb{C})$$

we consider

$$H^2(\mathbb{P}^3, \mathbb{Q}(1)) \to H^2(X, \mathbb{Q}(1)) \cap F^1 H^2(X, \mathbb{C}).$$

By noting that the space on the left-hand side is equal to $\mathbb{Q} \cdot [H]$ with $[H]$, the cohomology class of hyperplanes, and that that on the right-hand side is equal to $NS(X) \otimes \mathbb{Q}$, the (rational) Néron–Severi group of X, the space defined in the same way as in Definition 3.1 is nothing but the classical Noether–Lefschetz locus.

One can show as before that M_{NL} is the union of a countable number of closed analytic subsets. The reason why we cannot claim that these sets are algebraic is that Beilinson's Hodge conjecture is still open while the Hodge conjecture for divisors on surfaces is known. By analogy, a series of problems arises for our M_{NL}. The following analogue of Theorem 3.1 is a direct consequence of Theorem 2.1.

Theorem 3.4 *If $d \geq 4$, M_{NL} has positive codimension.*

The basic strategy of the proof of Theorem 3.4 is the same as that of Theorem 3.2. A new input is the theory of generalized Jacobian rings developed in [AS1], which gives an algebraic description of the cohomology of the open surface U. In particular, the duality theorem for such rings plays a crucial role.

We have not yet succeeded in obtaining the theorems analogous to Theorem 3.2 and 3.3 in a general situation. In what follows, we need to restrict ourselves to the following special case. In order to fix a homogeneous coordinate on $\mathbb{P}^3$, we write $\mathbb{P}^3 = \mathrm{Proj}(\mathbb{C}[X_0, X_1, X_2, X_3])$. We take $G_j = X_j$ for $1 \leq j \leq 3 \ (= s)$ and let $Y_j \subset \mathbb{P}^3$ be the corresponding hyperplane. Take M to be the moduli space of hypersurfaces of degree d in $\mathbb{P}^3$ which intersects transversally with $Y = \cup_{1 \leq j \leq 3} Y_j$. For $t \in M$ let $Z_t = X_t \cap Y$ and $U_t = X_t - Z_t$. Define $M_{NL} \subset M$ as in Definition 3.1. In order to determine the irreducible components of M_{NL} of maximal dimension, we need to introduce some notation. For an integer $l > 0$ let $S^l \subset \mathbb{C}[X_0, X_1, X_2, X_3]$ be the subspace of homogeneous polynomials of degree l.

Definition 3.2 *Let $\mathfrak{S}_3$ be the permutation group on $(1, 2, 3)$. For $\sigma \in \mathfrak{S}_3$ and a set $\underline{c} = (c_v)_{1 \leq v \leq d}$ of roots of unity and a pair (p, q) of non-negative coprime integers such that $d = r(p + q)$ with $r \in \mathbb{Z}$, we let $T^\sigma_{(p,q)}(\underline{c}) \subset M$ be the subset of those hypersurfaces defined by an equation of the form*

$$F = wB + \prod_{1 \leq v \leq r} (X^{p+q}_{\sigma(1)} - c_v X^p_{\sigma(2)} X^q_{\sigma(3)})$$

with $w \in S^1 - \sum_{1 \leq j \leq 3} \mathbb{C} \cdot X_j$ and $B \in S^{d-1}$.

The following is easily seen.

Lemma 3.1

(1) $\mathrm{codim}(T^\sigma_{(p,q)}(\underline{c})) = \binom{d+2}{2} - 5$.
(2) $T^\sigma_{(p,q)}(\underline{c}) = T^\sigma_{(p',q')}(\underline{c}')$ *if and only if $(p, q) = (p', q')$ and there is c such that $c_v = cc'_v$ for every v.*
(3) *For a surface X defined by such an equation F as in Definition 3.2 we have*

$$\mathbb{Q} \cdot \omega \oplus \mathbb{Q} \cdot \xi \subset F^0 H^2(U, \mathbb{Q}(2)),$$

where ω (respectively ξ) is the cohomology class of

$$d \log(X_2/X_1) \wedge d \log(X_3/X_1) \quad (\text{respectively } d \log(X_{\sigma(2)}^p X_{\sigma(3)}^q / X_{\sigma(1)}^{p+q})$$

$$\wedge d \log(w/X_1))$$

Note that $\Sigma(U) = \mathbb{Q} \cdot \omega$ *so that*

$$0 \neq \mathbb{Q} \cdot \xi \subset F^0 H^2(U, \mathbb{Q}(2))_{pr}.$$

Our main theorem is the following.

Theorem 3.5 [AS3] *Assume* $d \geq 4$. *Let* $T \subset M_{NL}$ *be an irreducible component. We have* $\operatorname{codim}(T) \geq \binom{d+2}{2} - 5$ *and the equality holds if and only if* $T = T_{(p,q)}^\sigma(\underline{c})$ *for some* σ, (p,q) *and* $\underline{c}$ *as in Definition 3.2.*

One of the keys to the proof of Theorem 3.5 is a result by Otwinowska [Ot, Theorem 2] on the Hilbert function of graded algebras of dimension 0.

Finally we discuss an implication of the above results for the injectivity of regulator maps. By Theorem 2.2 we have the following result.

Theorem 3.6 *Assume* $d \geq 4$. *Then, for* $t \in M \backslash M_{NL}$, ρ_{X_t} *is injective on the image of* $CH^1(Z_t, 1)$ *in* $CH^2(X_t, 1)$, *where*

$$\rho_{X_t} : CH^2(X_t, 1) \otimes \mathbb{Q} \to H_D^3(X_t, \mathbb{Q}(2))$$

is the regulator map to the Deligne cohomology.

The following result implies that there exists $t \in M \backslash M_{NL}$ for which the image of $CH^1(Z_t, 1)$ in $CH^2(X_t, 1)$ is non-torsion and hence that Theorem 3.6 indeed has non-trivial implications for the injectivity of the regulator map. We need to introduce some definitions. Let the assumption be as in Definition 3.2. For $1 \leq i \neq j \leq 3$ let $T_{ij} \subset M$ be the subset of those surfaces that satisfy the following condition. There exist a point $x \in Z_i \cap Z_j$ ($Z_j = X \cap Y_j$) and lines $L_i \subset Y_i$ and $L_j \subset Y_j$ such that $L_i \cap Z_i = \{x\} = L_j \cap Z_j$. We note that $T_{(p,q)}^\sigma(\underline{c}) \subset T_{ij}$ when $\sigma(1) = i, \sigma(2) = j$ and $p = 1, q = 0$. For a surface X in T_{ij} we consider the following element

$$c_{ij}(X) = (Z_i, (X_j/L_i)_{|Z_i}) - (Z_j, (X_i/L_j)_{|Z_j}) \in \mathbb{C}(Z_i)^* \oplus \mathbb{C}(Z_j)^*.$$

It is easy to see that $c_{ij}(X) \in CH^1(Z, 1)$ and let $[c_{ij}(X)] \in CH^2(X, 1)$ be its image.

Theorem 3.7

(1) *Assume $d \geq 4$. For general $t \in T_{ij}$, $\rho_{X_t}([c_{ij}(X_t)]) \neq 0$.*

(2) *For $t \in T_{123} := T_{12} \cap T_{23} \cap T_{31}$, let $V_t \subset CH^2(X_t, 1) \otimes \mathbb{Q}$ be the subspace spanned by $c_{12}(X_t)$, $c_{23}(X_t)$ and $c_{31}(X_t)$. Assume $d \geq 8$. For general $t \in T_{123}$, $\dim(V_t) = \dim(\rho_{X_t}(V_t)) = 3$.*

Proof First we prove (1). One easily computes $\mathrm{codim}(T_{ij}) = 2d - 3$ so that $\binom{d+2}{2} - 5 > \mathrm{codim}(T_{ij})$ if $d \geq 4$. Hence Theorem 3.5 implies that for any irreducible component T of M_{NL} we have $\dim(T) < \dim(T_{ij})$ so that for general $t \in T_{ij}$ $F^0 H^2(U_t, \mathbb{Q}(2))$ is generated by the image under $\mathrm{reg}^2_{U_t}$ of $\{X_2/X_1, X_3/X_1\} \in CH^2(U_t, 2)$. Hence the desired assertion follows from Theorem 2.2 is shown similarly by noting that $\binom{d+2}{2} - 5 > 3\mathrm{codim}(T_{ij}) \geq \mathrm{codim}(T_{123})$ if $d \geq 8$. $\qquad \square$

Acknowledgement

The author is very grateful to the organizers of the Grenoble Summer School 2001 for the opportunity to present these lectures, and to the referee for careful reading and helpful comments by which the first version of this paper has been much improved.

References

[A]   Asakura, M. On the K_1-groups of algebraic curves, preprint.

[AS1]   Asakura, M. and Saito, S. Generalized Jacobian rings for open complete intersections, preprint.

[AS2]   On Beilinson's Hodge and Tate conjectures for open complete intersections, preprint.

[AS3]   Noether–Lefschetz problem for K_2 of open surfaces, in preparation.

[Be1]   Beilinson, A. Notes on absolute Hodge cohomology; applications of algebraic K-theory to algebraic geometry and number theory. *Contemp. Math.* **55** (1986) 35–68.

[Be2]   Height pairings between algebraic cycles, *Lecture Notes in Mathematics.* pp. **1289**, 1–26, Springer, 1987.

[Bl]   Bloch, S. Algebraic cycles and higher K-theory, *Adv. Math.* **61** (1986) 267–304.

[CHM]   Ciliberto, C., Harris, J. and Miranda, R. General components of the Noether–Lefschetz locus and their density in the space of all surfaces, *Math. Ann.* **282** (1988) 667–680.

[D1]   Deligne, P. Théorie de Hodge II, *Publ. Math. Inst. Hautes Etude Sci.* **40** (1971) 5–57.

[D2]   Théorie de Hodge III, *Publ. Math. Inst. Hautes Etude Sci.* **44** (1974) 5–78.

[Do]   Donagi, R. Generic Torelli for projective hypersurfaces, *Compositio. Math.* **50** (1983) 325–353.

[DG]   Donagi, R. and Green, M. A new proof of the symmetrizer lemma and a
            stronger weak Torelli theorem for projective hypersurfaces, *J. Diff. Geom.*
            20 (1984) 459–461.

[EV]   Esnault, H. and Viehweg, E. Deligne–Beilinson cohomology. In: *Beilinson's
            Conjectures on Special Values of L-functions* (M. Rapoport, N. Schappacher
            and P. Schneider, eds.), *Perspectives in Mathematics*, vol. 4, Academic Press,
            1988.

[G1]   Green, M. Infinitesimal methods in Hodge theory. In: *Algebraic Cycles and
            Hodge Theory, Lecture Notes in Mathematics* **1594**, pp 1–92, Springer, 1994.

[G2]   Koszul cohomology and Geometry. In *Lectures on Riemann Surfaces,
            ICTP, Trieste, Italy*, pp. 177–200, World Scientific Press, 1987.

[G3]   A new proof of the explicit Noether–Lefschetz theorem, *J. Diff. Geom.*
            27 (1988) 155–159.

[G4]   Components of maximal dimension in the Noether–Lefschetz locus, *J.
            Diff. Geom.* **29** (1989) 295–302.

[Go]   Gotzmann, G. Eine Bedingung für die Flachheit und das Hilbertpolynom
            eines graduierten Ringes, *Math. Z.* **158** (1978) 61–70.

[J1]   Jannsen, U. Deligne homology, Hodge-*D*-conjecture, and Motives. In:
            Beilinson's Conjectures on Special Values of L-functions (M. Rapoport, N.
            Schappacher and P. Schneider, eds.), *Perspectives in Mathematics*, vol. 4,
            Academic Press, 1988.

[J2]   *Mixed Motives and Algebraic K-Theory, Lecture Notes in Mathematics.*
            1400, Springer-Verlag, Berlin, 1989.

[J3]   Continuous etale cohomology, *Math. Ann.* **280** (1987) 207–245.

[KO]   Katz, N. and Oda, T. On the differentiation of De Rham cohomology classes
            with respect to parameters, *J. Math. Kyoto Univ.* **8** (1968) 199–213.

[MSS]   Müller-Stach, S. and Saito, S. On K_1 and K_2 of algebraic surfaces, *K-theory*
            (2004), in press.

[Mu]   Mumford, D. Rational equivalence of 0-cycles on surfaces, *J. Math. Kyoto
            Univ.* **9** (1969) 195–204.

[Na1]   Nagel, J. The Abel–Jacobi map for complete intersections, *Indag. Math.* **8** (1)
            1997 95–113.

[Na2]   Effective bounds for Hodge-theoretic connectivity, *J. Algebraic Geom.*
            11 (2002) 1–32.

[No]   Nori, M. V. Algebraic cycles and Hodge theoretic connectivity, *Invent. of
            Math.* **11** (1993). 349–373.

[Ot]   Otwinowska, A. Composantes de dimension maximale d'un analogue du lieu
            de Noether–Lefschetz, *Compositio Math.*, in press.

[SaS]   Saito, S. Higher normal functions and Griffiths groups, *J. Algebraic Geom.*
            11 (2002) 161–201.

[Sch]   Schneider, P. Introduction to the Beilinson conjectures. *In: Beilinson's Con-
            jectures on Special Values of L-functions* (M. Rapoport, N. Schappacher and
            P. Schneider, eds.), *Perspectives in Mathematics*, vol. 4, Academic Press,
            1988.

[V1]   Voisin, C. Une précision concernant le théorème de Noether, *Math. Ann.* **280**
            (1988) 605–611.

[V2]   Nori's connectivity theorem and higher Chow groups, *J. Inst. Math.
            Jussieu* **1** (2002) 307–329.

Printed in the United States
by Baker & Taylor Publisher Services